DESKBOOK OF CONSTRUCTION CONTRACT LAW — WITH FORMS

DESKBOOK OF CONSTRUCTION CONTRACT LAW — WITH FORMS

H. Murray Hohns, P.E.
Michael T. Callahan, J.D.; LL.M.
James V. Dolan, LL.B.
with Barry B. Bramble, J.D.
H. James Wulfsberg, J.D.
with Sandra J. McNabb, J.D.

Prentice-Hall, Inc.
Englewood Cliffs, New Jersey

Prentice-Hall International, Inc., *London*
Prentice-Hall of Australia, Pty. Ltd., *Sydney*
Prentice-Hall of Canada, Ltd., *Toronto*
Prentice-Hall of India Private Ltd., *New Delhi*
Prentice-Hall of Japan, Inc., *Tokyo*
Prentice-Hall of Southeast Asia Pte. Ltd., *Singapore*
Whitehall Books, Ltd., *Wellington, New Zealand*

"This publication is designed to provide accurate and authoritative information in regard to the subject matter covered. It is sold with the understanding that the publisher is not engaged in rendering legal, accounting, or other professional service. If legal advice or other expert assistance is required, the service of a competent professional person should be sought."

—From the Declaration of Principles jointly adopted by a Committee of the American Bar Association and a Committee of Publishers and Associations.

Library of Congress Cataloging in Publication Data
Main entry under title:

Deskbook of construction contract law — with forms.

Includes bibliographical references and index.
1. Building — Contracts and specifications — United States. 2. Building — Contracts and specifications — United States — Forms. I. Hohns, H. Murray.
KF902.D47 343.73'07869 80-27862
ISBN 0-13-202069-6

Printed in the United States of America

The Authors

H. MURRAY HOHNS, P.E.

A 30-year veteran of the construction industry, H. Murray Hohns is a founder and principal of Wagner • Hohns • Inglis • Inc., a diverse, national construction consulting firm. Mr. Hohns also functions as the operating head of the firm's La Crescenta, California office. He is a nationally acknowledged expert in all phases of the construction process, including the unique aspects of construction claims and litigation problems, project management and productivity improvement/cost control programs. Mr. Hohns has been called in to help evaluate construction project disputes in more than 1500 cases, valued in the billions of dollars. President of the Project Management Institute, he is in constant demand as a lecturer at construction conventions and seminars; has written numerous articles which have appeared in technical periodicals; and is the author of "Preventing and Solving Construction Contract Disputes."

MICHAEL T. CALLAHAN, J.D; LL.M.

Michael T. Callahan is an attorney and vice president/operating head of the Kansas City, Missouri office of Wagner • Hohns • Inglis • Inc. Mr. Callahan holds a B.A. degree from the University of Kansas and a J.D. degree from the University of Missouri School of Law. He is also the recipient of an LL.M degree in Government Contracts from the University of Missouri.

As a WHI executive, Mr. Callahan has participated in the resolution of major disputes involving millions of dollars worth of construction. An expert in the utilization of Critical Path Method scheduling (CPM) and project documents in proof or refutation of delay claims, Mr. Callahan frequently speaks at construction conventions and seminars. He is a member of the Bar Associations of Kansas, Missouri, and New Jersey, the American Bar Association (Public Contracts Section, Litigation Section, Forum Committee on Construction Litigation) and a Commercial Panelist for the American Arbitration Association.

JAMES V. DOLAN, LL.B.

Mr. Dolan is a partner in Steptoe and Johnson, Washington, D.C., whose clientele includes major contractors, owners, insurance and bonding companies. His contract analysis and litigation experience covers all phases of the construction industry.

Mr. Dolan was born in Washington, D.C., and admitted to the bar of the District of Columbia in 1964. He was graduated from Georgetown University (A.B., cum laude, 1960); and his legal education was received at Georgetown University (LL.B., 1963).

He is a member of Delta Theta Phi; Editor-in-Chief, Georgetown Law Journal, 1962-1963. Law Clerk to Judge Charles Fahy, U.S. Court of Appeals, D.C. Circuit, 1963-1964. He is also a member of The District of Columbia Bar; American Bar Association (Member, Sections on Insurance, Negligence and Compensation Law and Public Contract Law).

H. JAMES WULFSBERG J.D.

Mr. Wulfsberg is a principal in the law firm of Lempres and Wulfsberg Professional Corporation, Oakland, California. He has handled all aspects of construction disputes from pre-construction and post-award bid disputes to construction litigation on behalf of owners, contractors, subcontractors and design professionals. Mr. Wulfsberg principally handles the defense of design professionals in construction disputes.

Mr. Wulfsberg was born in Long Beach, California, and admitted to the California bar in 1970. His preparatory education was at the University of Southern California (A.B., 1966); his legal education, Boalt Hall, University of California (J.D., 1969). He is a member of Phi Beta Kappa.

His legal affiliations include: Alameda County and American (Member Forum Committee on the Convention Industry, Insurance, Negligence and Compensation Law) Bar Associations; State Bar of California.

BARRY B. BRAMBLE, J.D.

Mr. Bramble is General Counsel of Wagner • Hohns • Inglis • Inc. He also functions as a senior consultant specializing in the utilization of critical path method scheduling to reflect responsibility and duration of construction delays. Mr. Bramble has been responsible for the review, analysis and documentation of construction claims, preparation of evidential record, discovery, and trial plans, as well as presentation and advocacy at trial and arbitration hearings.

Mr. Bramble holds a B.A. degree from Rutgers University and a J.D. degree from Rutgers University School of Law. He is a member of the New Jersey and Pennsylvania Bars. He is also a member of the American Bar Association (Public Contracts Section, Litigation Section, Forum Committee on Construction Litigation) and the National Contract Manager's Association.

SANDRA J. McNABB, J.D.

Ms. McNabb is an associate in the Law Firm of Lempres and Wulfsberg Professional Corporation, Oakland, California.

As a litigation specialist in real estate and construction-related matters, she has resolved major disputes for owners, contractors, and design professionals.

Ms. McNabb was born in Tacoma, Washington and admitted to the California Bar in 1977. She was graduated from the University of Washington (B.A. 1974); and received her legal education at Stanford University (J.D. 1977).

She is a member of Phi Beta Kappa; the Alameda County; American (Member, Forum Committee on Construction Litigation) Bar Association; and the State Bar of California.

What This Deskbook Will Do For You

This book gives you a practical analysis and a proven approach to make contract law an everyday tool to help you run your business, write new contracts, minimize future risks and help solve current contract disputes. The book examines, in detail, the current thinking generally reflected across America in solving and preventing contract claims.

The principles and techniques set forth are applicable for all sizes and types of firms in the construction industry, including those who purchase and design that which is to be built. The book will benefit the executive, the field manager, the office manager, analysts, lawyers, everyone concerned with the performance of a contract or subcontract. Why do you need this book? A few answers are given below.

*The book is written by leading authorities who have hands-on experience in every state of the country. They have handled millions of dollars in problems on billions of dollars in construction, on some of the largest law suits ever filed and some of the smallest as well. The authors have been there, and more important, are asked to go back.

*Chapter 1 traces the roots of the contracting industry. Reasons for current polemics are developed from the historic perspective in a smooth, logical and pervasive manner. An understanding of one's past is necessary for profitable visions of the future.

*Many of today's problems at the job site can be traced to changes that have come about in the past 50 years. Some of the major changes are discussed in Chapter 2 as background to your future and to writing or signing your next contract.

*The contract, indeed every contract you ever sign, is the heart of Chapter 3. The most important points in successful contracting are probed in this approach of what to do or not to do when thinking through a contract situation.

*The big dollars in contract adjustments are time related. Losses and profits are so often tied to scheduling and target dates that a consistent, firm understanding of what the law now says about schedules is essential for the modern executive or supervisor. Chapter 4 provides this.

*Next the laws behind today's contract thinking are examined to forearm and forewarn everyone in the project as to potential pitfalls and traps in modern management. Points vital to profit and success are made on page after page of Chapter 5.

*Chapter 6 discusses the best approaches to resolving contract disputes with the federal government. The ideas, however, work at all levels of contract resolution.

*Those arenas other than the federal contract relief provisions are covered in Chapter 7. The thinking here will give assurances as to how to proceed and what to expect in arbitrations and the federal courts.

HOW THIS BOOK IS ORGANIZED

The law behind construction contracts is organized into seven chapters each representing basic segments of understanding required in today's world to effectively write, execute and successfully complete contracts in the construction industry today. The chapters are comprehensive and include basic information as well as the latest thinking evidenced in contract disputes across the country. Real life examples and current case citations are used throughout. When you are finished with the book, you will have new knowledge of how to get more out of your efforts.

For example, the book will give you an understanding of how courts look at today's contract language on schedules. With this knowledge in hand you'll be able to decide the scheduling procedures to be used on your next project.

Chapters 1 and 2 will give you insight into the industry from the eyes of a man who has probably helped solve more construction problems than any other person. His thinking comes from a perspective few can have.

Chapters 3, 4 and 5 set forth the points to keep in mind to avoid potentially costly problems. Application of these techniques will certainly help minimize some of the exposure you now have.

Chapters 6 and 7 will help to solve, resist or collect current claims. You'll learn what to do with those problems you have right now.

HOW TO USE THIS BOOK

The preceding pages told you what this book will do for you. Now consider the following possible uses for this book.

It provides many principles of the law and clear answers you need to know to have confidence or forewarning as to what existing contracts or contracts in development really mean. The book tells you how contracts are being interpreted today.

The appendices contain the leading standard form contract documents, that are in constant use throughout the industry. These standardized developments, which have been developed and modified over the years by the American Institute of Architects, the National Society of Professional Engineers, and the Federal Government, through committees drawn from the industry and its attorneys, have effected and set the relationships, liability concepts and expectations of the various participants in a team producing major work in construction. Construction contracts and design contracts, which use the thinking of these standard forms as their spine, will be well-founded since they contain the latest thinking available.

Think through your present operating procedures. Then re-think them once you have read and digested these proven practices and procedures. The book is written to be read easily and then shared. Everyone who administers contracts should have a copy within reach at his desk.

The book will define your thinking, cause you to be more sensitive in your approaches to the maze of contract law, and help you develop an action-oriented framework which will aid you on a daily basis.

H. Murray Hohns

Contents

What This Deskbook Will Do For You 9

1. The Roots of Today's Contract Law 17

by H. MURRAY HOHNS, P.E.

The Primacy of the Engineer • The States' Involvement in the Contract • Standardization of Forms • Settlement of Disputes • Evolution of the Design Engineer's Function.

2. Some of the Changes in Contracting Over the Last Fifty Years 25

by H. MURRAY HOHNS, P.E.

The Change in Working Conditions or Cold Weather Construction • Handling Concrete • The Social Revolution as an Influence Change on construction • Critical Path Method Scheduling and Management Information Systems • The Ease of Travel • The Great Society Programs of the 1960's • The Advent of Professional Liability Insurance

3. The Contract — How to Avoid Million-Dollar Mistakes 33

by JAMES V. DOLAN, LL.B.

The Contract • Lesson to Be Learned • Key Ingredients • Interpreting the Language of the Contract • Delay Claims (the Fine Points; Pinpointing the Differences; Site Inspection Clause in Federal Contracts) • The Owner's Obligation • Contractor's Rights • Cases That Clarify Rights • Warranties/Guarantees/Latent Defects.

4. The Law Behind Construction Schedules 49

by MICHAEL T. CALLAHAN, J.D.; LL.M.

Project Scheduling Techniques (CPM Schedules; Precedence Diagramming (PDM); Bar Charts; PERT Charts; Updating) • The Law and Construction Schedules (the Law's View of Schedules; the Schedule Must Be Complete; It Must Be Substantial; the CPM Consultant; the Dates in the Schedule; the Benefit of Float; Mistakes in the Schedule; Pre-bid Schedules; Mutual Responsibilities; Updating the Schedule).

5. The Law Behind Today's Construction Contracts 75

by H. JAMES WULFSBERG, J.D., with SANDRA J. McNABB, J.D.

The Design Phase (the Owner's Role; the Designer's Role; Cost Estimating; Designer's Interface Among Plans, Specifications, Results; the Contractor's Role) • Minimizing Construction Problems (the Owner's Role; the Designer's Role; the Contractor's Role; the Subcontractor's Role; the Supplier's Role; Reconciling the Competing Interests).

6. The Proper Legal Approach to Federal Contract Disputes 87

by JAMES V. DOLAN, LL.B., with BARRY B. BRAMBLE, J.D.

Processing a Dispute (Claim Procedure; Resources) • Presentation (Justification of Damage Calculation) • Settlement (Does Litigation Work?; How and When to Seek Settlement; Why Does It Take So Long?)

7. The Presentation of Disputes in Non-Federal Arenas 103

by H. JAMES WULFSBERG, J.D.

Appendix A — Partial Scheduling Clauses from *Dobson* v. *Rutgers* 116

Appendix B — Scheduling Clauses from *Peter Kiewit Sons* v. *Iowa Southern Utilities Co.* 117

Appendix C — *Southern Fireproofing Company* v. *R.F. Bull Construction* 118

Appendix D — Scheduling Clauses from *C.H. Leavell & Company* 119

Appendix E — American Institute of Architects Forms 120

Appendix F — National Society of Engineers Contracts 261

Appendix G — Contracts for Architectural and Engineering Services 326

Appendix H — Contract Disputes Act of 1978 349

Table of Cases 361

Subject Index 365

DESKBOOK OF CONSTRUCTION CONTRACT LAW — WITH FORMS

CHAPTER 1

The Roots of Today's Contract Law

H. MURRAY HOHNS, P.E.

THE DECADE OF THE SEVENTIES saw a major emphasis on construction claims involving contractors, owners, designers, subcontractors, suppliers, their insurors, sureties and others. Since 1975 or thereabout, the typical industry periodical seems to have had at least one story describing a legal dilemma; almost all construction or design trade magazines have a legal column, and advertisements for seminar after seminar discussing the law behind contract claims fill the mails. The contract claim is advanced as the latest solution to the industry's problems. It would seem everyone is claim conscious.

Despite the recent emphasis on the presentation and resolution of disputes, most members of the construction industry have an inadequate concept of construction law. Often the author has heard the question: "Where is construction law written down? Is there a source to which one can go to find out the law?" Unfortunately, the construction industry does not have a printed book of rules like the Professional Golfers Association's. So lacking are the rules that, to many, construction law is an inventory of an empty cupboard. Yet to those who study the industry from a legal point of view there is an identifiable reason for the position of the metamorphic mix of the law and its application that we are currently seeing.

This book crystallizes the concepts basic to the law behind construction contracts today. The approach taken is first to lay open the roots of today's law from a historical perspective. Then, the positions of the participants in the contracting industry are set forth and the evolution of a particular party to its present position is described to give an understanding of assumptions and pressures upon the participant.

The observation that the construction industry is governed to a high degree by people and personal discretion must be held paramount in one's assessment of what is behind construction agreements.

Once the background has been established, the text considers the contract and the legal aspects of the use of contract time, i.e., the progress schedule. From there, consideration is given to the legal requirements from the viewpoint of the various parties to the dispute. The Claims Act of 1979 and how to deal with the Federal Government are discussed next and the book concludes with a review of how a contract dispute should be presented.

With this preview in mind, to understand why the industry seems to be in an increasing state of dispute requires first that one establish the past from which today's concepts have come. Once the past is understood, one can grasp many of the reasons for today's seeming problems and those more wise than others can perhaps see the future with a vision which will allow them the chance for unusual gain.

Contracting in America is not really an old game. Indeed, our country is only two hundred years old, hardly a point of permanence when viewed in the context of history. The early years of our history meant little to the contracting industry for the concept of contract law or contracting did not exist in any broad fashion as it does today.

The thoughts first put forth and the basic rules in the early courts of America were based on the then already well established English Law of Equity and Property. In our country's early years, the court typically would examine and decide a dispute from the point of view of fairness or equity. To take economic advantage of another was not permissible in the court's view. Therefore, if one found himself in a losing situation, relief was readily available in the court. No learned, self-respecting judge would permit the enforcement of a contract which economically disadvantaged one party for the gain of the other. That was ungentlemanly, even un-American.

There were few, if any, ways of enforcing a contract if one party had a severe loss or hardship. The early courts struck down many agreements on the basis that the consideration therein was inadequate. Courts refused to honor onerous contract clauses. A basic concept taught in every elementary course in contract law today is that there must be consideration to make a contract valid, though the concept of consideration has been watered down from our Colonial equity law to today where one can have legally valid consideration with just a dollar and other unstated valuables.

As the nineteenth century came to a close our country underwent the Industrial Revolution. The railroads went West, steel mills and other industry began to flourish. While the engineering and construction for the early advances in this industrialization largely were done on an in-house basis, severe need soon existed for codes and rules on how business was to be conducted, and how those involved were to be governed.

One could not make the capital commitments industry required without some sort of agreement which was enforceable over the long term. Simultaneously, the need and market for the design professional came into being. The bridge builder became the bridge engineer; the railroad builder, the railroad engineer. In a relatively short period of time, the engineer's knowledge was being sought by many for planning and design. Others could be easily found to physically construct but the designer was scarce and so the engineering profession began to come into its own. The American Society of Civil Engineers published a number of articles, as part of the Bicentennial celebration, which trace the development of specific branches of Civil Engineering and the growth of the engineer as a professional as well as the technological advances in methods of construction made during those years.

Indeed, in those times just 75 years ago, the design engineer was the only individual on the construction site who had enough education or knowledge to direct the work. It was this fact that led to what for many years in the industry was a common phrase, "as directed by the engineer."

The concept of standard construction contract forms developed as the Industrial Revolution wore on. Standards were developed so that business could move forward on an acceptable base. However, while the contract forms were of great value to at least get some standard of practice established on a wide basis, a number of problems rapidly developed. The first problem (and it still is very much with us today) was that a contract once executed is ultimately enforceable only through the courts unless both parties are in agreement throughout the performance of the contract. The second problem is that the construction contract is written not by lawyers for each party who mutually work out an acceptable agreement, but by architects or engineers assisted occasionally by attorneys, accountants and/or others. In the "pure form" years ago, the designer saw himself as the one who conceived the design, depicted it to whatever degree of detail he felt necessary to describe the work and then specified the various requirements of the materials and methods needed to put his creation into being. The designer set the rules for behavior and obligation of the various parties and indeed often believed he alone was the only one who really knew the broad theory of what was going on, let alone the subtleties, nuances and fine points. He was the creator of it all, interpreter of the contract, and judge of the performance of each party.

As far as the design professional was concerned, the contractor was not competent to comment upon design methods or theory. The contractor was a doer, not a reasoner. And in most cases back then, the designer was correct in his assessment of himself and the contractor. Many of the contractors of sixty years ago were artisans, immigrants whose inherited skills and clan backgrounds had been the admixture combining the laborer's willingness to do hard work along with the opportunity to overcome the language barrier and turn it into a possibility for profit. Contractors have always seemed to be able to count no matter how short their formal education may be. Thus the engineer/designer became the Director, i.e., the Director of Public Works. Furthermore, since the engineer had designed and specified what was to be constructed, it soon followed the engineer was obviously the one best qualified to judge the acceptability of the contractor's performance.

The engineer's position was further enhanced by the fact that 75 years ago there were few qualified engineers. There were a lot of eager, skilled tradesmen or those who were just plain willing. The concept of public work of any scale was just beginning; the politician and big businessman had little concern for the common good. Then too, there was no television, no radio, no concerned mass media or method to inform the public of political irresponsibility.

The only person involved in the construction process with a continuing responsibility, on a generic basis, for safety, workmanship, durability and delivery of a workable product was the designer. Therefore, he was the only one to interpret and administer the contract. He was the only person who really knew what was needed or, if that was not clear, at least what the designer intended. Most owners did not have the knowledge to criticize or interface on a technical basis with the designer, and being non-technical people, most owners were more than happy to simply let George do it, particularly if George was being paid to be the engineer. In view of this, it's not hard to see how the engineer often believes his role is still the same today as it was back in the "good old days." Most of those now in charge were trained by men who lived and worked in the good old days.

The early engineer took his responsibility seriously and did his work in great detail and with pride. A cursory review of engineering drawings from the first 20 years of this century shows drafting techniques and care that are awesome by today's standards: ink drawings by skilled draftsmen on special cloth, using shading and other illustrative techniques in unbelievable detail; a quality and quantity of effort that no one can afford today were commonplace and expected. Little wonder that the designer was regarded as the professional on the job site. Little wonder that the pride in such a profession still shines forth in today's designer. It is his heritage.

THE PRIMACY OF THE ENGINEER

The engineer alone was the sole interpreter of the contract document. He wrote the contract and he read it and all others listened and acquiesced. The engineer alone measured progress, determined what was acceptable and what was not. He determined how much was to be paid and clerked the works. The engineer had great authority and tried to be objective in his interpretation of a questionable point so that it would not favor the contractor or the owner. The designer was truly a professional and he believed it and so did the contractor and the owner.

The business community reacted well. Standard rules and contracts came into their own. The standard form became the accepted thing and everyone buying or about to buy construction began using similar forms. Government as well as private business began to use the same standard form contracts. It was no longer necessary to start with a blank sheet of paper each time. Rules were available to start the game. The problems began to develop when it came time to enforce the contract; for there was no central rules committee or group of similarly trained, qualified, and salaried umpires to cry "foul" or "you're out." The ultimate interpreter of the contract's standard form was often a judge found by chance in hundreds of state or lower courts. Each judge had the authority to decide solutions to contract questions and arrived at conclusions that were variable and uneven.

This lack of uniformity has now led to the current situation where the construction industry, the largest segment of our economy, is totally splintered and has no unified basis of law to deal with problems as they develop. Instead, one finds a morass of inconsistent legal precedents, many of which are very narrow in application and all of which vary from jurisdiction to jurisdiction and are commonly in contradiction to each other. The disputes that find their way to an adjudicated end are tried, argued, judged and opinioned in front of many courts and boards.

The largest buyer of construction is the United States Government. It alone has any number of contracting agencies all of which have their own standard forms. All the forms are similar but vary significantly. These forms are first interpreted by lower tier contractors and government employees and then by contracting officers, regional administrators, boards of contract appeals, Federal District Courts, the United States Court of Appeals, United States Court of Claims and even the United States Supreme Court.

THE STATES' INVOLVEMENT IN THE CONTRACT

Next in volume of contract purchase are the state governments. Each state, and virtually every separate state agency, has a standard form contract, all similar but each with the prejudices of those who drew the final wording. The contract market at state levels was further complicated by the doctrine of Sovereign Immunity which was standard state law until relatively recently. In many states, Sovereign Immunity still exists in one form or another.

The attempts to remove Sovereign Immunity or the lack of right to sue a state or other government body have resulted in a variety of relief systems such as the establishment of quasi-judicial Boards of Arbitration and Review, appointment of hearing officers and other similar ideas. The Boards, however, do not establish or publish legal precedent of any standing. Such boards are bound in some states to the legislature for making good their awards. In general, such relief methods contribute nothing to establishing a unified concept of construction law and procedures. They actually exacerbate the situation.

In addition to the hearing boards, state courts, county courts, and other low tier courts also hear construction cases. My observation is that the lower level court always tries to over-simplify complicated construction problems in its decision making. Hence the existing background of common law to be used as the base for arguing a construction dispute is largely a hodgepodge of contradictory opinions. Not only are the decisions a hodgepodge, there are no standards for qualifying the trier's background.

STANDARDIZATION OF FORMS

The most widely used construction contract standard form today outside of those of the Federal Government is the American Institute of Architect's Form AIA

201A. But just as with the states, each professional group such as the National Society of Professional Engineers, the Associated General Contractors and at least ten more similar organizations also have standard forms all similar to the AIA but with differing provisions.

SETTLEMENT OF DISPUTES

Uncertainty abounds. Can there be any doubt that it should? Moreover, the uncertainty is compounded by the use of arbitration to settle disputes as part of most industry group standard forms. The use of arbitration to settle disputes also has the drawback of not providing a usable file of legal precedent on which to draw in future arguments. Instead, each dispute is argued on its equity merits in front of arbitrators who generally have little knowledge of due process.

The equity evaluation or arbitration of a dispute is really a property law concept and the decisions reached by panels are, in many cases, almost as bad today as those by the courts at the start of the Industrial Revolution which led to standard form contracts in the first place.

When our country began to use the standard form contract, the courts and everyone involved began to interpret the written word very precisely. Reliance on contract language became more of a formality. The court and the parties were on safe ground to believe that which was written and to enforce what was said. What could go wrong? The concept of implied duties not explicitly stated was in its infancy in the court's mind back then. Privity was essential for liability to exist. Without a contract, one had no duties to another and there was no right of reliance on anyone without a contract.

Our standard form construction contract started with strict interpretation of the written word by both users and courts. The architect or engineer was not at privity with the contractor. He therefore had no duty to the contractor and the contractor had no right or remedy to seek anything but approval from the designer.

If this lack of ability or right to rely on the designer to any extent put the contractor at risk, the courts felt the contractor's duty and solution was to price the risk in seeking the work. No matter how unreasonable a contract clause might be, if the two parties had agreed to it, the courts could do nothing but enforce the contract.

EVOLUTION OF THE DESIGN-ENGINEER'S FUNCTION

Thus in the early years of contracting, the designer had no duty or could suffer no penalty from anything done by him in his performance or in his assertion of his authority as owner's agent even if his acts were wrong, improper, ignorant or negligent. A designer could not be negligent to a contractor if there was no contract which established a duty. The early years had further restrictions; for example, plan defects did not excuse contractors from performing a contract. Indeed, at least one building fell down after being erected per plans and the contractor was found at fault even though the plans were drawn by an independent engineer. The courts found no implied warranty existed to the contractor as to the correctness of the plans. The court believed no sensible contractor would build something unless he first checked the plans.

If that wasn't bad enough, the early handling of below grade soils problems was worse. Soil defects always fell to the contractor's risk. If an engineer showed what he believed were existing conditions, i.e., bed rock or no water in the excavation, and during the work it was found that the engineer was in error, the contractor had no recourse but to absorb the extra costs. The engineer had no liability for the problem if he sincerely believed that what he set forth on the drawings was the case. While misbelief was unfortunate, it could not be misrepresentation and the contractor had no cause for relief.

If the engineer showed a method of construction which did not work, the contractor had no remedy, only the obligation to do it a better way. If a contract period was stated and it was wrong or impossible to meet, such was the contractor's risk. The contractor was expected to know better than the engineer how long it takes to build something. The contractor did not have to bid after all.

If the contractor had a doubt about a portion of work, he was expected to inquire and research the matter. If the contractor did not make a needed inquiry, he was expected to suffer the consequences. There was no penalty to an engineer in supplying a builder with incorrect quantities. Again, the builder had the risk and the responsibility to correct the problem.

This is the historical background memory base from which today's contracts are written and interpreted. Reactions to today's problems and challenges are often founded in the false security that the concepts set forth above still exist. The influences of teacher, parent, first boss, and our early years have set in us keystone concepts which we believe are right. Hence, we spend our lives adjusting, somewhat unbelievingly, to the real world versus what we have been taught it should be.

The construction industry had its beginnings where the contract was strictly interpreted, the design professional was the director of the work and had no duty or privity, and thus no liability, to anyone but the owner. The contractor had to take the risk and did so. There were no unions, no income taxes, no safety laws, little social awareness, an ample supply of unskilled labor, a supply of skilled labor which did not speak the language, no media influence, advanced education was unnecessary, and there was no equipment as we know it today. Was life simpler back then? The author does not really know but it would seem doubtful, for the industry was already beginning to argue about what the contract really meant in specific case after case. Most often, the best lawyers were hired by the contractor, surety or insurer. The body politic, most often the defendant, had a less important personal zest for the outcome of the lawsuit. It was the taxpayers' money at stake.

Not only did the lawyers and their skills in persuasion bring changes, but many other changes happened as well.

Thus, 70 or 80 years later, everything has changed, including the rules. Only people have stayed the same and most of us presently in the upper echelon of the contracting field were brought up by parents who lived during the period when the fundamental rules described above were set. What are some of the changes that have made it necessary for us to rethink our rules? In the chapters that follow, critical points of law are set forth, followed by clear guidelines for strict compliance.

CHAPTER 2

Some of the Changes in Contracting Over the Last Fifty Years

H. MURRAY HOHNS, P.E.

THE PRECEDING CHAPTER SET IN CONTEXT THE BASIC CONCEPTS which form the memory system that most senior members and users of the construction industry have as a core mechanism from which to plan, react to, and evaluate various problems and events on a construction project. This chapter will point out a number of changes in and influences on the so-called traditional relationships that have been established over the past half to three-quarters of a century.

Not only is our industry changing, but also our entire society and many of our basic concepts are changing. The confusion and conflict we are experiencing in the construction industry is a reflection of the metamorphic change being experienced across the whole of society. The construction industry is growing up or older, and as it matures in an increasingly regulated environment, the concept of the family-run construction company is being replaced by the absolute necessity for the sophistication of professional management just to stay in business.

More changes are coming. Those of us who teach the continuing and basic educational programs see broader changes on the horizon of the industry. Many of our colleges and universities are now producing young people who have spent four years learning to become construction managers. This background is totally different from that of most of today's managers. Professors like Vern Hastings at Arizona State, John Borcherding, University of Texas, Boyd Paulson, Stanford, and Don

Hanscher at Purdue are sending hundreds of trained young men with vision into the industry each year. Senior men dream dreams of what might have been and resist change, since they have finally learned how to run their businesses to keep what they have gathered safe, but these trained young people are starting to come to the foreground.

These students, as they grow in years and enter the industry's hierarchy, will begin to put the concepts they learned into practice. It is a lot easier to teach new concepts to a twenty-five-year-old than to a fifty-year-old. The author sees the coming influence of these people trained as managers as the next wave of new ideas and changes the industry will experience. To some extent, one can see a definite parallel between the problem in construction and the medical profession. Not too many years ago, the medical profession was reeling under an onslaught of hospital malpractice suits. The first reaction of the medical profession was to rewrite the fine print and be chary in the offering of services. This reaction did not work and was followed by a thorough cleaning up of the profession's act. The same will happen in the construction industry. Our act needs cleaning up.

A recent experience of this author includes representing an unfamiliar owner about to build a $30,000,000 recreation facility. The bid documents incorporated AIA 201 by reference. When the architect was asked for a copy of the preprinted referenced standard form, the partner in charge of the local office in charge had no copy in his possession and further advised us he had not looked at such boiler plate clauses for 15 years. This sort of situation will help lead the move to design-build contracts which I see as another part of the new wave as our young people seek ways to prove themselves and find ways to gain stature and professional recognition. The author looks to these young people to do much to clean up the industry's act.

Their start, however, is where we are today and the litigation and confrontation that now abound must be viewed by considering the changes in the past 50 years.

The early years saw the following:

OWNER --------- DESIGNER --------- CONTRACTOR

The owner retained a designer and then later contracted separately with a contractor, but the administration of the contract was through the designer. In modern times we find:

OWNER. DESIGNER. . . . . . .
. . . . CONTRACTOR

Here, the owner hires the designer as before, but the designer no longer totally administers the construction contract. In most cases the designer is above the contractor on the administration, but the owner is now in charge. The designer is still the owner's agent and for all practical purposes is held at the same level of authority as the owner.

Other variations abound today:

```
OWNER ------------------ DESIGNER
      ------------------ Separate Prime
                         Contractor
      ------------------ Separate Prime
                         Contractor
      ------------------ Separate Prime
                         Contractor
      ------------------ Separate Prime
                         Contractor
```

In this method, the designer influence in the contractor-contract administration is minimal and the professional is usually relegated to consultation on technical matters and submittal review only.

Another current variation:

```
OWNER ---------- DESIGNER
      ----------------- CM ---------------- CONTRACTOR(S)
```

In this case, the owner administers the construction contract through a construction management firm and there is interplay on the owner's behalf between the designer and the CM.

These diagrams are very simplistic, for in the real world many parties can get involved, and always do.

The owner may need a:

Master Planner	Industrial Engineer
Soils Engineer	Bond Proceed Trustee
Boring Contractor	Financial Advisor
Surveyor	Claims Expert
Bond Counsel	Real Estate Consultant — Broker
Counsel	Insurance Consultant
Project Manager	and more.
Architect or Design Engineer	

The Architect or Design Engineer may and probably will need a:

Structural Engineer	Sound Consultant
Electrical Engineer	Graphic Consultant
HVAC Engineer	Specification Writer
Plumbing Engineer	Energy Specialist
Acoustical Consultant	Value Engineer
Landscape Consultant	Computer Service
Scheduling Consultant	and more.
Cost Estimator	

The Construction Manager will need:

Estimating Consultants	MIS Consultants
Scheduling Consultants	Engineering Consultants

The Contractor may require some 20 to 30 subcontractors who average 11 major suppliers each. The contractor will also have 35-40 regular suppliers of substantial size. The owner must become the orchestra leader and the success of the overture often will rise or fall on the owner's ability to orchestrate effectively.

Most of the firms involved in the foregoing lists have the same traits: lack of trained people, lack of equipment to do the job, inadequate capitalization, cash flow problems, and credit demands beyond their ability. Most banks and the financial community as a whole have no interest in contractor-construction-related firms. The industry is so widely splintered there is little base from which the financier can measure what is going on in an environment where everyone deals all day long with other people's money in telephone number quantities.

The changes and similar influences that have occurred and are causing much of the legal dilemmas that seem significant to the author are: 1) the concept of working all year regardless of weather; 2) technological advances and new ideas in handling concrete and other materials; 3) the social revolution in labor; 4) Critical Path Method scheduling and Management Information Systems; 5) the ease of travel; 6) The Great Society Programs; and 7) the advent of Professional Liability Insurance.

THE CHANGE IN WORKING CONDITIONS OR COLD WEATHER CONSTRUCTION

Until the 1920's there was no such thing as cold weather construction; work simply stopped for the bad weather of the year. Little impetus was found to do cold weather work until the Second World War. No one cared about the workman and his personal economic needs. It was the Second World War's demands which first set the tone for year round effort. Then, in the 1950's, Greenland was developed by the Corps of Engineers; in the 70's, Aleyeska built the pipeline across Alaska. It was in the 60's and 70's that the industry got used to working in the cold. This new policy, plus the insistence by all concerned on creature comfort for the workman, the need to keep temperatures high enough to accomplish the state of the art installation of the materials selected by the designer, and the fickleness of nature itself all combined to make the risk of cold weather construction one that all contractors must cope with every year. The risk often results in losses and recovery of loss is almost always pursued. Cold weather construction is a major cause of construction dispute.

HANDLING CONCRETE

Like cold weather construction, concrete dates its infancy to the 1920's. Just 50 years ago there was no such thing as ready-mixed concrete except on a scattered basis. Instead, one found a 2-bag mixer on job sites along with a set of scales for measuring wheelbarrows of sand and stone. Concrete was moved around the job in chutes and there was no such thing as vibration to make concrete flow. Instead, the concrete was mixed soft and fluid, so it flowed and filled the forms. There were no cranes, no pumps, no hoists, none of the common ways of handling concrete that exist today. Yet many of the basic ideas about concrete are virtually the same as they were in 1925.

In the last 30 years, some efforts to use high-strength concrete have been seen. Early concepts established the ratio of water and cement as the governing factor for concrete strength, but today's technology includes use of certain sands and stone, fiber additives, and alumina cement. However, most people in the industry are afraid of new ideas and there is little way to assure quality or ability in execution of new design concepts.

The advances in concrete control through batching by automated controls, its higher strengths, its ever-present need throughout a structure for floors and fireproofing, plus its moldability have led to wide use of concrete as the final finish of walls and ceiling. When done with some imagination this use of the media is called architectural concrete. The use of architectural concrete in the writer's experience is almost a sure-fire way to start arguments on a construction site. It never seems to be quite good enough to satisfy the preconceived ideas of those in charge as to how it would or should look, or, it is far more costly to do than foreseen by the estimator when the job was priced. Architectural concrete has caused as much grief to contractors as any other innovation in the author's experience.

Changes have occurred in other building materials. Structural steel fabrication and connection technology have changed. Aluminum has become widely used. Glass exteriors became popular as HVAC technology permitted cooling and heating at what were once reasonable costs, before the energy crisis. Controls and automation have put new requirements on the contractor.

Technology has indeed advanced, but as it advanced problems arose. One major problem is the lack of universality in sizes, connections and product interfaces. Too many manufacturers saw the concept of impossible interface with any other product as the way to capture the market and insure higher profit on a repeat basis. The result today is that each construction job starts in many respects from day zero. Shop drawings, submittals and design calculations are needed on every size project across the country for what should have been standard in the industry years ago.

THE SOCIAL REVOLUTION AS AN INFLUENCE CHANGE ON CONSTRUCTION

The Government has chosen to use its construction purchasing power to augment social change. Contract wage levels are set by Davis Bacon laws, minority employment goals are common and certain contracts call for 10 or some other stated percent of the work to be given to minority business firms.

The minority business requirement is one of the most vexing. Established industry members who have struggled in their early years often resent the idea of being forced, in their eyes, to be a helping hand. Too often, the established member sees his helping hand being grabbed by a raised hand or clenched fist. There are not enough minority people or firms to fill what has been pontificated and the bureaucratic shoulder shrug invites industry defiance. MBE has, however, opened what certainly was a closed door.

The labor problem of the industry is one of the changes causing most of the litigation today. The exploitation of the construction worker in the early years, followed by the job seekers' total dismay during the depression, led to the unions in the late 30's, their building of power in the 40's and 50's, their wage demands and industry acquiescence in the 60's, and the retaliatory merit shop and double-breasted moves of the 70's.

Labor costs and the often impossible rules today are a main cause of construction litigation.

CRITICAL PATH METHOD SCHEDULING AND MANAGEMENT INFORMATION SYSTEMS

The ability to lay out and rapidly display one's plans for the future and the advent of readily available monitoring systems for measuring progress in time and dollars have put the contract manager in the forefront.

Now the effect of a job delay can be rapidly translated into a meaningful conclusion. The owner can measure and follow the progress of a contractor as well as the contractor itself. CPM and MIS have opened up contract performance so that information about the work abounds. Granted, some of the information is valuable and some is not, but to those involved in management, its dissemination is wider and the job progress is more thoroughly understood.

These data processing systems lend themselves to expert evaluation and the ability to prove delay loss, and pinpoint the party responsible. An accepted method for loss determination now exists and is readily available to be used in the construction dispute. This change to automated data processing has led to an increase in construction litigation.

THE EASE OF TRAVEL

One can only be amazed at the effect of mobility on the construction industry. The author's own firm of over 100 people has offices in five states and has worked in all of the states. To be a major concern in the industry requires the willingness to travel to distant, and perhaps foreign, places. The highway system, standard contract forms, our telephone and mail services, and the airplane make this mobility in marketing possible. Unfortunately, the ease of mobility often ignores that the new man in town has to learn the town to really succeed and there are often too few learning resources available in the budget that is required to win the job.

This ease of mobility, coupled with the low bid marketing system, leaves little incentive for a contractor to build a local permanent quality staff at any level. Everything and everyone is here today, gone tomorrow. Contracting firms don't put down roots to grow. They travel and pull up roots to grow. Mobility is a major cause of today's construction dispute volume.

THE GREAT SOCIETY PROGRAMS OF THE 1960's

Government made some great steps forward in the 60's and 70's. In construction one can recall hospital building programs, community college programs, federal office buildings, airports, urban renewals and sewage treatment plants. Everywhere one turned, new demand was being placed on the various levels of the construction industry. Labor reacted in the late 60's by successfully demanding high increases in pay rates which set the pattern for today's 1 percent per month upward spiral in construction costs.

Labor's counterpart, those in charge in the construction industry, committed their firms to growth and goals which were geared only to growth. With the decline in the economy in the early 70's there were no resources in many firms to withstand the resulting financial stress.

One of the major causes of litigation today is the growth years the industry enjoyed in the late 60's and early 70's.

THE ADVENT OF PROFESSIONAL LIABILITY INSURANCE

The advent of professional liability insurance is also a major cause of litigation today. PLI as it is now known didn't exist to any degree before the mid 50's. It then took the next ten or 15 years to convince designers to be insured and it has taken the past ten or 15 years for the insurance industry to determine what really was involved, how much was needed to pay the losses, and how much premium was needed to underwrite the costs. The premium level has now climbed back to profitable levels and deductibles are so high that much of the defense's costs are borne by the designers. Many designers have almost become self-insuring by that high deductible, a still newer wrinkle.

To a major extent, the concept of insurance is a very large cause of construction litigation. The idea of a source of real money (someone else's) being available to pay for error, omission, act of negligence is a lawyer's dream. When everyone on a jury thinks that insurors make money on top of money, the idea of going to court to recover losses increases in appeal.

There is certainly no sense suing someone who can't pay the judgment, but what if there is real cash available? The advent of insurance over the past 50 to 75 years is a major cause of litigation today.

A LAST THOUGHT

So many points cross the author's mind as this chapter on causative change comes to a close. The media, television, increasing taxation, higher levels of education, the energy problem, the ease of travel, our instant knowledge of what is happening elsewhere, among many other elements contribute to the very common idea that every man everywhere just by his existence is owed a living (and a nice one) by the collective efforts of all.

With this concept of life, why should one suffer loss? Somehow the construction industry and its users seem to feel no one should lose, particularly if they were sincere while the work was ongoing. Competence, honesty and pride have lost their allure or urgency to many.

This is the major reason for the rash of construction litigation today. Seemingly, no one is willing to lose, even if he is wrong, though both reader and writer are excluded from this indictment. Unfortunately, in the real world of our industry there are times when one must lose, particularly if the job was not done.

CHAPTER 3

The Contract — How to Avoid Million-Dollar Mistakes

JAMES V. DOLAN, LL.B.

THE FIRST TWO CHAPTERS HAVE TRACED THE BEGINNINGS of, and subsequent modifications to, the behavior and understanding of the parties and preparers of construction contracts. This chapter deals with the legal principles behind the Contract and their role in evaluating and presenting disputes.

THE CONTRACT

Simplistic as it sounds, it's critical to keep in mind that the starting point for developing any basis to put forth or resist a contract claim is the contract itself. The contract is a bargain and, except in extraordinary circumstances, its terms will be enforced as written. Rights that a contractor might think he has under trade practices can be overridden and bargained away by a clear contract clause to the contrary. One example of this is the current shoaling provision in the Corps of Engineers' dredging contracts. Back in 1935, the U.S. Court of Claims held that when additional shoaling occurred after a contractor completed his initial dredging in a particular area the contractor was entitled to be paid an extra for the cost of re-dredging the shoaling. *Duluth Superior Dredging Company* v. *United States,* 81 Ct. Cl. 912 (1935). The Corps of Engineers solved this problem by the simple expedient of changing the language of the contract provision so that the contractor was granted relief in that case. Under the new provision, the contractor is responsible for dredging all shoaling up to the point of the post-dredge survey. See, *Great Lakes Dredge & Dock Co.,* ENGBCA No. 2654, 66-1 BCA ¶5442 (1966). Millions of dollars in potential claims were avoided by this stroke of the pen.

A similar example can be found in the area of excuses for delay. In 1967, in *Schweigert, Inc.* v. *United States,* 181 Ct. Cl. 1184, the Court of Claims held that a contractor could not be assessed liquidated damages for delay caused by the failure of a second-tier subcontractor with whom the contractor in question had no direct contract relationship and over whom it had no control. In 1969, the standard contract default clause was amended to change this result and charge the contractor with responsibility for all subcontractor delays, regardless of the tier at which they occur.

LESSON TO BE LEARNED

The obvious lesson is that a contractor should check the contract provisions, not simply before it prepares a claim — but before it submits its bid. The contractor should know what the bargain is before entering into it. Too often the contractor waits until it's too late. Not only the contractor has this obvious challenge, the owner also must review and understand the terms of the contract as well.

THE KEY INGREDIENTS OF A CONTRACT

What Is the Contract? It is not just the document one signs. It will generally incorporate numerous other documents, some of which a contractor might not even have seen when he submitted his bid. It includes the specifications, both those specifically drafted for the contract in question and the general set of specifications for all contracts within a particular category. For example, a road contract with the State of Virginia will automatically include by reference the Virginia Department of Highways Road and Bridge Specifications — all 528 pages. On more than a few occasions, an out-of-state road contractor bid its first job in Virginia without reading that document and, later, came to regret it.

The contract will also include the drawings and the notes found on the drawings. There are many instances where a seemingly valid claim, strongly supported by a reasonable reading of the specifications, has been defeated by a simple drawing note that, in the view of the Board or Court, placed upon the contractor the responsibility for work that it never contemplated as being part of the contract.

The contract can often include by reference certain published standards such as military or testing standards. These standards are as much a part of the contractor's obligation as if they were fully set forth in the contract.

INTERPRETING THE LANGUAGE OF THE CONTRACT

Basic Rules of Interpretation: The most basic rule of interpretation is that a contract will be enforced in accordance with its clear and unequivocal terms. The usual problem, in most disputes is that the terms in question are not clear and unequivocal. This brings into play the other rules of interpretation.

The most common rule of construction contract interpretation, when a clause is ambiguous and susceptible to differing interpretations, is that it will be construed against the party who drafted it. *Peter Kiewit Sons Co.* v. *United States,* 109 Ct. Cl. 390, 418 (1947). In most public contracts, the drafter is the Government, which means that ambiguous provisions will be interpreted in favor of the contractor. If the Government and contractor have conflicting interpretations of a particular provision and both are reasonable, the contractor will generally prevail. This rule was succinctly stated by the Court of Claims in *Brezina Const. Co.* v. *United States,* 196 Ct. Cl. 29, 33 (1971).

> It is in cases such as this one, where the contract is ambiguous and where there are no extraneous aids to interpretation, that the courts are forced to resort to guidelines based on what is thought to be sound policy rather than on the intent of the contracting parties. The rule, often repeated by this court, is that if a Government contract is ambiguous and if the construction placed upon it by the contractor is reasonable, the contractor's construction will be adopted, unless the parties' intention is otherwise affirmatively revealed. *See, e.g., Morrison-Knudsen Co.* v. *United States,* 192 Ct. Cl. 410, 418, 427 F.2d 1181, 1185 (1970); *W.P.C. Enterprises, Inc.* v. *United States,* 163 Ct. Cl. 1, 6, 323 F.2d 874, 876-77 (1963).

The ambiguity rule is most often applied in instances where there is a dispute over whether particular work is required by the drawings and specifications. Its application is not limited to that type of dispute; it has been applied, to disputes over pay issues as well, such as whether the owner is entitled to withhold retention against a mobilization payment under a contract providing that retention would not be held against any separate building or division of work. See 46. Comp. Gen. 730.

One caveat that should be noted with respect to the ambiguity rule is the principle that the contractor will not be allowed to bait a trap or to bring a claim where there is a patent ambiguity noticeable before a bid and the contractor, instead of bringing it to the Government's attention, submits a bid in silence and later seeks an increase in price based upon the ambiguity. This was noted by the Court in *Brezina Const. Co.* v. *United States,* 196 Ct. Cl., at 34.

> A companion rule to the rule of ambiguities is that, where a bidder faced with a 'patent and glaring discrepancy' fails to seek clarification from the contracting officer, the discrepancy should be construed against him.

Where a contract is unclear as to its requirements, one of the aids most commonly invoked to clarify the ambiguity is reference to practice or custom in the trade. On several occasions, such as *John McShain Inc.* v. *United States,* 199 Ct. Cl. 364 (1972), the Court of Claims granted a contractor an equitable adjustment for performing work which was not clearly indicated in the specifications and which is not generally required by normal trade practices. Trade practice can be particularly helpful in disputes over allowable tolerances where the specifications are vague as to the tolerance required, as in the *Railroad Waterproofing* case, ASBCA No. 12384, 69-1 BCA ¶7456 (1969), when the ASBCA granted the contractor relief for being held to a tolerance not expressly required by the specifications and in excess of normal industry standards.

It should be noted that any trade practice can be overridden by an express contract provision to the contrary. *See, Southwest Mechanical Contractors,* GSBCA No. 1310, 1320, (1964) BCA ¶4482 (1964). On occasion, a trade practice can be used to give contract language a meaning different from that which would normally be construed. This situation is illustrated by the Court of Claims decision in *Buffalo Merchandise Warehouses Inc.* v. *United States,* 115 Ct. Cl. 568, 88 F. Supp. 276 (1950) where the court held that the Government was obligated to pay a full month's storage on goods that were in the warehouse for only a part of the month, in accordance with a clearly established trade custom. The Court explained its decision in the following terms:

> The Government agency which made the contract was not aware of the trade usage. Was the Government nevertheless bound to pay in accordance with the usage? We think it was. Even though, to the uninitiated, the words of the contract would seem to have a fairly plain meaning, that meaning will be altered to correspond with the recognized usage of the language of the trade.

A lesson this decision teaches is that it behooves both contractor and owner alike to make sure they are familiar with all applicable trade practices before entering into a contract since those practices are very likely to become part of their bargain even when one of the parties has no knowledge of the practice.

When, prior to the emergence of a dispute over a particular provision of a contract, the parties have had dealings with reference to the same provision and those dealings reflect the interpretation which the parties were giving the provision before the dispute arose, the Boards and Courts will obviously attach great weight to those dealings and the earlier interpretation which they reflect. As the Court of Claims stated in *Union Paving Co.* v. *United States,* 126 Ct. Cl. 478, 489, 115 F. Supp. 179, 185 (1953), "Acts of parties prior to the time subject to controversy, are to be given considerable weight in interpreting the contract." This rule can also extend not only to dealings between the parties under a prior contract, but also to their subsequent dealings under that contract. Thus, if, after the emergence of a dispute under one contract, the Government, in a follow-on contract, adopts the contractor's interpretation, the contractor is in a position to argue successfully that the Government's subsequent action is the best evidence of what was intended in the original contract. This argument was successful in the case of *Roscoe Engineering Corp. and Assocs.,* ASBCA No. 4820, 61-1 BCA ¶2919 (1961), when the Board stated:

> Secondly, in what the Board finds is a sufficiently analagous situation existing in the arresting gear area (the Government's Contentions to the contrary not being persuasive) the Govern-

> ment subsequently adopted appellant's position where a similar 'over excavation' had resulted from sound construction considerations, and the practical interpretation by the parties in an analagous situation under the same contract carries great, if not compelling, weight.

It should be emphasized that this is only a rule of interpretation, albeit one of particular force. As such, it will not override clear and unambiguous language to the contrary. As one state court has pointed out, if both parties make the same mistake and by their conduct apply the contract in an erroneous manner, the courts are not bound to perpetuate the mistake and at the later insistence of either party will enforce the contract in accord with its clearly expressed intent. In re: *Chicago & E. I. Ry. Co.,* 94 F.2d 296 (7th Cir. 1938).

It should also be noted that the Government can acquiesce in a contractor's interpretation of an unclear contract provision simply by doing nothing and remaining silent. In *Shadrick Contracting Co., Inc.* ASBCA No. 14613, 71-1 BCA ¶8647 (1971), shortly after bid opening, the contracting officer had a conversation with the contractor in which the contracting officer learned that the contractor was interpreting the contract differently than the Government. However, the contracting officer *refused* to express his disagreement and gave no opportunity for the contractor to withdraw its bid. The Board held that the Government in that case had, by its silence, acquiesced in the contractor's interpretation and was, thus, stuck with it.

The reach of interpretation which the parties by their conduct can engraft upon a contract is not limited to the contract in dispute. The outcome of a dispute under one contract can be determined by the parties' conduct with respect to a prior contract involving a similar undertaking. Although it applies fully to construction disputes, the best example of this principle is found in one of the military products cases, *Watkins-Johnson Company,* ASBCA No. 15457, 74-1 BCA ¶10,395 (1973). The Board in *Watkins* summarized this rule as follows:

> It is a well settled rule of law that the prior course of dealing between the parties can be introduced into evidence to explain the interpretation of the language used in the contract, and such prior course of dealings may include reference to the conduct of the parties *under a prior* contract between them involving requirements of a similar nature. Under the circumstances, the dependence of the appellant (contractor) upon its interpretations of the requirements of a prior specification derived from similar language applied in the performance of a predecessor government contract, in the absence of any specific language to the contrary, must be considered reasonable and proper.

This rule is one of interpretation that proves there can be an exception even to a rule as firmly established as the principle that clear and unambiguous provisions of the specifications will be enforced in accordance with their clear terms. For this rule has been carried to the point of holding that even clear and unambiguous specification provisions may, on occasion, by waived by a clear course of Government conduct under a prior contract indicating that the provision in question will not be enforced. *See, L. W. Foster Sportswear Co.* v. *United States,* 186 Ct. Cl. 499 (1969); *Gresham & Co., Inc.* v. *United States,* 200 Ct. Cl. 97 (1972).

DELAY CLAIMS — THE FINE POINTS

In order to understand the theory of delay claims, it is necessary, to understand the meaning of the changes clause and changes claims. In a majority of cases, the critical prerequisite to a delay claim is the establishment of a successful claim under the changes clause.

A "Changes Claim" is a term most familiar within the context of a contract with the federal government. The legal basis for a "changes" claim, however, is far more reaching and extends, in its essence, to all public contracts and private contracts as well. That basis is the proposition that a party's obligation under a contract is limited to performance only of that work which the contract, reasonably interpreted, requires. The concept of a change is a simple one: a change occurs when a contractor is required through no fault of its own to perform work that could not have been reasonably anticipated from the contract documents and the cost of which is accordingly, not included in the contractor's bid price or the resulting contract price. Under traditional common law principles, the contractor can refuse to perform this additional work if the parties can't agree upon acceptable terms for its performance. The effect of the changes clause is to take this option of non-performance away from the contractor and to require him to perform the extra work in exchange for the imposition upon the owner of a reciprocal obligation to pay for the full cost of that additional work, plus a fair profit or return on capital. The changes clause can thus be best understood not by looking upon it as a boon to the contractor, but rather as the least consideration a contractor is entitled to receive for giving up its right to refuse to perform work not required under its original contract.

With this background, there are two types of changes to consider: the express change and the constructive change. For our purpose here, we have disregarded the former. Although probably the most common in contracting, express changes are unlikely to be the subject of dispute except as to costing.

The most dispute-prone change is the constructive change. The types of constructive changes that a contractor can encounter are as varied as the number of problems that can be encountered in the construction of a building that involves complex structural, mechanical, electrical, and finishing problems and the interface of a dozen or more subcontractors. Legally, the types of constructive changes are more basic. The genesis of a constructive change will usually reduce itself in legal terms to either (a) an error or mistaken assumption in the plans and specifications or (b) an erroneous interpretation of the plans or specifications by the owner or his representative.

Although it is impossible to categorize every type of constructive change, the following are important examples:

Additional work necessitated by an error in the specifications. When the owner issues a set of specifications, it warrants that contractor compliance with those specifications will produce an acceptable result. If this is not the case and additional work is required, the contractor is entitled to an equitable adjustment under the constructive changes rule. See *Spencer Explosives, Inc.,* ASBCA No. 4800, 60-2 BCA ¶2795 (1960), in which the Board granted relief stating:

> Appellant...(contends) that those specifications were defective in the sense that they made consistent results impossible.... We think that is enough. Those specifications, in effect, caused extra work beyond contract requirements...with the result that appellant is entitled to an equitable adjustment under the 'changes' article.

Mistaken assumption in plans and specifications. A contract description of work is often premised upon certain assumptions. If the facts emerge differently from the assumption and additional work is required, the contractor is entitled to an equitable adjustment. See *Guest t/a C. M. Guest & Sons,* WDBCA No. 593, 2 CCF 1015 (1944) where the specified excavation method assumed the work could be performed under dry conditions and the contractor was granted relief when this proved not to be the case and additional shoring and dewatering were required.

Unless the contract clearly indicates to the contrary, the contractor may assume that he will have prompt and continuing access to the work site. Where access is denied, if he must provide his own access or if he is forced to work in close quarters with other contractors (who should have been off the job by then), thus restricting his work space and efficiency, he is entitled to an equitable adjustment under the changes clause. See, e.g., *C. J. Longenfelder & Sons,* WDBCA 706, 2 (1944); *Lewis-Nicholson, Inc.* v. *United States,* 550 F.2d 26 (Ct. Cl. 1977); *Dale Const. Co.* v. *United States,* 168 Ct. Cl. 692 (1964).

If the Government specifies a particular source that must be used in performance of the contract and that source becomes unavailable, the cost of having to switch to a more expensive substitute is compensable as a constructive change. See *Kaplan Contractors Inc.,* GSBCA No. 2747, 70-2 BCA ¶8511 (1970), where the specified source became insolvent during the course of performance.

The contract generally allows the contractor to select its own method of performance, subject to certain limitations. Where several methods of performance are possible under the contract, the contractor is entitled to select that which is least costly and, if the Government denies that option by directing use of a more expensive method or prohibiting use of the least expensive, there is a constructive change entitling the contractor to an equitable adjustment. See *Goslin-Birmingham Inc.,* ENGBCA No. 2800, 67-2 BCA ¶6402 (1967); *Otto Randolph Inc.,* ASBCA No. 11539, 66-2 BCA ¶5928 (1966). This rule applies not only to methods of performance, but also to sources of supply. Thus, where a contract allows the contractor to select which of several borrow pits to use and that option is denied him, he is entitled to an equitable adjustment (although, in the case of the borrow pit, the adjustment would probably be granted under the differing site condition clause). *Gordon H. Ball Inc.,* ASBCA No. 5905, 61-2 BCA ¶3177 (1961).

Holding the contractor to more rigid performance standards than required by the contract is often found in disputes over tolerances. If the contracting officer requires a contract to satisfy tolerances that are more rigid than those set forth in the contract and required by trade practice, he is imposing upon the contractor an additional requirement, entitling the contractor to be paid for additional costs. See, e.g., *Railroad Waterproofing Corp.,* ASBCA No. 12384, 69-1 BCA ¶7456.

Excessive Inspection Requirements is a close relative to the one above. The contractor cannot be required to perform and to have his work satisfy testing procedures more rigid than those set forth in the contract or trade practice incorporated therein. Thus, if in a repair contract inspection and testing requirements generally applicable only to new construction are imposed, there is a constructive change. See *Pacific Car & Foundry Co.,* ASBCA No. 4849, 60-1 BCA ¶2505 (1960).

Erroneous interpretation of the contract requirements by the Government encompases all of the previously listed categories and serves as a catchall. It includes every instance where a contracting officer interprets the contract as requiring the performance of certain work that is not required under a fair reading of the contract. See, e.g., *P & B Co. of Washington Inc.,* ASBCA No. 9429, 1963 BCA ¶3997 (1963), where a contractor was given an equitable adjustment for the cost of installing certain support decking which was not shown in the specifications and drawings with sufficient clarity to make it a contract requirement.

The law applicable to delay claims is not complicated. The usual problem in delay claims is proving the facts that will bring the case within the scope of the legal principles authorizing relief. The proof-of-fact problem is beyond the allotted scope of this particular presentation; it was covered in part in the preceding discussion of the application of the contract schedule to claims and will be touched upon again in the presentation on damages. The legal rules against which the facts of a delay claim must be analyzed are as follows.

DELAY CLAIMS: PINPOINTING THE DIFFERENCES

There are three types of basic delay: (i) delays which are responsibility of the contractor; (ii) delays which are beyond the control or responsibility of either contractor or owner; and (iii) delays which are attributable to the owner. For present purposes, we can forget about the first of these except insofar as a contractor delay brings into effect the concurrent delay rule which is discussed below.

Where delays are beyond the fault or responsibility of either owner or contractor, there is a right to a time extension, but not to monetary relief, except in two cases. The first exception is, of course, the constructive acceleration rule which provides that an owner's failure or refusal to grant a proper time extension can constitute a constructive order to accelerate, entitling the contractor to an equitable adjustment for acceleration costs, such as wage premiums for overtime, additional costs of winter construction, etc. See, e.g., *Raymond International of Delaware, Inc.,*

ASBCA No. 13121, 70-1 BCA ¶8341 (1970); *M.S.I. Corporation,* GSBCA No. 2429, 68-2 BCA ¶7377 (1968). The second exception is where a no-fault delay is encountered at a time when the contractor should have been off the job but for an earlier owner-caused delay. In that event, the no-fault delay can be considered the proximate result of the earlier owner delay and the contractor can be entitled to money damages as well as time. See, *J. D. Hedin Const. Co.* v. *United States,* 347 F.2d 235, 254-55 (Ct. Cl. 1965).

The final type of delay, one for which the owner is responsible, affords a basis not only for a time extension, but also for monetary relief. See, *S & E Contractors Inc.,* AEC BCA No. 97-12-72, 74-2 BCA ¶10,676 (1974); *Robert McMullan & Sons Inc.,* ASBCA No. 19129, 76-2 BCA ¶12,072 (1976).

Where there is a combination of owner-caused and contractor-caused delay, the foregoing rules are obviously changed. The mere occurrence of an owner-caused delay does not always entitle the contractor to relief. The contractor must also show that this delay actually held up his work. He cannot make that showing if, in fact, he had himself delayed the job and would not have been able to proceed even had the owner delay not occurred. See, e.g., *Sun Shipbuilding and Dry Dock Co.,* ASBCA No. 11300, 68-1 BCA ¶7054 (1968); *Benbow Plumbing & Heating Co.,* ASBCA No. 13233, 69-1 BCA¶7565 (1969).

Watch This Pitfall:

Another problem is created when you have owner- and contractor-caused delays which are not completely concurrent, but which both contribute to an overall delay. There is a split of authority as to whether the Court or Board will attempt to segregate or apportion the delays and award the contractor relief for the portion of the delay attributable to the owner. See *E. H. Marhoefer, Jr., Co.,* DOTCAB Nos. 70-17, 70-1 BCA ¶8177 (1970) and *Sun Shipbuilding & Dry Dock Co., supra.* If it is possible to segregate and apportion the two delays, the trend of the courts is to do so to avoid complete denial of relief to which the contractor is entitled. If there are concurrent delays which cannot be segregated, the contractor has no right to damages and the owner has no right to assess liquidated damages. See, e.g., Anno. 152 A.L.R. 1349, 1359 (1941).

Finally, it should be noted that, in the great majority of cases, the prerequisite for a delay claim is an underlying claim under the changes clause since it usually is a change, express or constructive, that has triggered the delay. Accordingly, what is said above with respect to changes claims ties directly into delay claim analysis.

Exculpatory Clauses in Delays. The principal exculpatory clause in delay claims is the no-delay-damage type clause found in some non-federal public contracts. Such clauses typically provide that, in the event of a delay, the contractor's only right shall be to an extension of time and that it is not entitled to recover monetary damages from the owner. Such clauses, though harsh, are generally enforceable. See *Johnson* v. *Fenestra Inc.,* 305 F. 2d 179 (3d Cir. 1962). Their exculpatory effect has been upheld even in the case of delays caused by the owner (See 10 A.L.R. 2d 801 (1937) and delays found to be unreasonable. *Hosmer Inc.* v. *Commonwealth of Massachusetts,* 302 Mass. 495 19 N.E. 2d 800 (1939).

Because of their harsh result, however, such clauses will be strictly construed and the courts will often go out of their way to hold them inapplicable. See, e.g., *Maryland Cas. Co.* v. *Tacoma,* 90 P.2d 226, 94 P.2d 217 (Wash. 1939). *Mix Inc.* v. *Columbus,* 171 N.E. 2d 1967 (Ohio, App. 1959). Whether a clause of this type will bar a delay damage claim depends upon the particular language of the clause and the facts relating to the delay. The only safe general conclusion that can be drawn is that, if the clause clearly covers the claim in question, it will usually be enforced, regardless of the harshness of the result. If there is any doubt that the parties intended the clause to reach as far as the claim in question, that doubt will usually be resolved in favor of the contractor.

SITE INSPECTION CLAUSE IN FEDERAL CONTRACTS

The Site Inspection Clause found in federal contracts and in most other public contracts provides for the contractor to inspect the site prior to submitting its bid and familiarize itself with all physical site conditions. This clause is commonly used as a defense against Changed or Differing Site Conditions Claims where it can be shown that a diligent site inspection would have alerted a reasonable bidder to the physical conditions which he claims constitute a changed condition. A failure to make a diligent site inspection will bar a claim for relief. *Walsh Bros.* v. *United States,* 107 Ct. Cl. 627, 69 F. Supp. 125 (1947). However, a failure to make a site inspection will not bar recovery where Government action prevented the inspection *Pavement Specialists Inc.,* ASBCA No. 17410, 73-2 BCA ¶10,082 (1973); or meaningful inspection was a practical impossibility *(Benson Tree Co.,* IBCA No. 812-11-69, 70-2 BCA ¶8549 (1970); or where the particular condition would not have been noticed by reasonable inspection *(Maryland Painting Co.,* ENGBCA No. 3337, 73-2 BCA ¶10,223 (1973).

Another exculpatory clause commonly used in an effort to defeat a changed conditions claim is a provision stating that "The Government does not guarantee that materials other than those disclosed by the explorations will not be encountered, or that the proportions of the various materials will not vary from those indicated by the logs of the exploration." On the whole, Government attempts to avoid changed conditions responsibility through clauses such as this have been unsuccessful. The Court of Claims held in *Thompson Ramo Wooldridge, Inc.* v. *United States,* 175 Ct. Cl. 527, 536 (1966).

> But we have held, in comparable circumstances, that broad exculpatory clauses (identical in effect with this one) cannot be given their full literal reach and do not relieve the defendant of liability for changed conditions as the broad language thereof would seem to indicate....) General portions of the specifications should not lightly be read to override the Changed Conditions Clause. It takes clear and unambiguous language to do that, for the provisions sought to be eliminated or subordinated is a standard mandatory clause of broad application.

Rights and Violations. The foregoing discussion of the contract and its application has covered the question of Rights and Violations under the contract. The three subjects (the contract, rights thereunder and violations thereof) go hand-in-hand. This section will capsulize the respective rights and obligations of the various parties involved in the performance of a construction contract. A simple appreciation of the general undertaking of the various parties aids in providing a common-sense framework and makes the specific rules covered above more understandable when viewed in the context of the whole framework.

The most basic owner's right is to get what was bargained for, i.e., to have the proposed construction completed in accordance with the specification and time demands of the contract. The owner is not required to settle for anything less. If the contract calls for wooden window walls, the owner is entitled to have them even if aluminum window walls would be cheaper and just as effective. The contractor cannot substitute his judgment for that of the owner.

THE OWNER'S OBLIGATION

In contracts containing a changes clause, the owner has a right to change its mind, to add or delete work or make such changes as it deems appropriate. However, this has with it the obligation to pay for resultant extra work and to grant appropriate time extensions. Moreover, the owner cannot make changes of such magnitude or number that they change the total nature of the contract undertaking. Changes of that magnitude reach the level of a cardinal change which constitutes a breach of the contract, relieving the contractor of its obligation to proceed. See *Edward R. Marden Corp.* v. *United States,* 194 Ct. Cl. 799, 442 F.2d 364 (1971); *Air-A-Plane Corp.* v. *United States,* 187 Ct. Cl. 269, 408 F.2d 1030 (1969). Although cardinal change cases are very rare, the principle was recognized as recently as 1974 by the Armed Services Board of Contract Appeals in *Axel Electronics Inc.,* ASBCA No. 18990, 74-1 BCA ¶ 10,471 (1974).

The owner also has obligations under the contract as well as rights. The most obvious and important is the duty to pay in accordance with the contract, including its provision for progress payments. The Court of Claims has held that an improper refusal to make progress payments excuses the contractor from proceeding further. "Construction contractors cannot be expected to continue working if the Government improperly holds back on substantial interim payments which enable the project to proceed." *William Green Const. Co.* v. *United States,* 201 Ct. Cl. 616, 477 F.2d 930 (1973).

The owner is responsible for the specification issued in its name. It warrants the adequacy of those specifications.

The owner has a general duty to cooperate with the contractor. This duty includes an obligation not to withhold superior knowledge and thereby lead the contractor into a disastrous bargain. *Helene Curtis Industries, Inc.* v. *United States,* 160 Ct. Cl. 437, (1963). This includes a duty to share superior knowledge with regard to latent site conditions. *Maryland Painting Co.,* ENGBCA 3337, 73-2 BCA ¶10,233 (1973). This duty of cooperation requires that the owner not interfere with the contractor's performance. *William A. Smith Contr. Co. of Mo.* v. *United States,* 155 Ct. Cl. 1,292 F.2d 847 (1961) (Government stripped ballast pit of usable materials which contractor needed). It also subjects the owner to liability for unreasonable delays in granting necessary approvals or making necessary changes. *George A. Fuller Co.* v. *United States,* 108 Ct. Cl. 70, 69 F. Supp. 409 (1947).

Finally, it should be noted that the owner is responsible for the conduct of its representatives, including engineers and architects. If they impose an unreasonable interpretation upon the contractor, it is the owner who must pay for resulting additional costs.

CONTRACTOR'S RIGHTS UNDER THE CONTRACT

The contractor has rights under the contract. The contractor's principal right is to get paid the contract amount for contract work and to be paid additional sums for work beyond the contract. There is no absolute right, however, to a profit. If the contractor bid the job low or has performed it inefficiently, it has no right to look to the owner for recovery of loss.

The contractor, also has the right to rely upon the adequacy of the plans and specifications and the right to have the owner and its architect act with reasonable dispatch with regard to required owner action, such as drawing approval or provision of site access, a condition precedent to the performance of work.

Other contractor rights have been discussed in the section on Changes and Delays above.

CASES THAT CLARIFY RIGHTS

The contractor has obligations under the contract. The contractor is obligated to construct the project in accordance with plans and specifications regardless of cost, subject to one exception. That exception is the doctrine of practical or legal impossibility. "Even though a contract is not actually impossible, a contractor is entitled to relief if it can be shown that the contract requirements are impractical because of extreme and unreasonable difficulty, expense, injury or loss involved in meeting them." *Tecon Corp.* v. *United States,* 188 Ct. Cl. 436 411 F.2d 1271 (1969). The Court of Claims has also described practical impossibility as a case where the costs and difficulties of performance are so great that it is "commercially senseless" to proceed in accordance with plans and specifications. *Natus Corp.* v. *United States,* 178 Ct. Cl. 1,371 F.2d 450 (1967).

The contractor is obligated to comply with the notice requirements contained in the Changes and Differing Site (Changed) Conditions clauses or elsewhere in the contract. Failure to do so presents the obvious risk that a valid claim will be denied, although, as the Armed Services Board of Contract Appeals has noted, the Court of Claims has been reluctant in recent years to bar claims for failure to comply with notice requirements. *C. H. Leavell & Co.,* ASBCA No. 16099, 72-2 BCA ¶9694 (1972). Where, however, the failure to give notice has prejudiced the Government by denying it an opportunity to take alternative action to avoid the additional claim expense, that reluctance will disappear and the claim will be barred. *Eggers & Higgins* v. *United States,* 185 Ct. Cl. 765, 403 F.2d 225 (1968). Even when there is no absolute written notice requirement, a contractor, when it encounters a claim condition, cannot remain silent, thereby denying the Government the opportunity to avoid the claim even if that can be accomplished only by terminating for convenience. See *Ling-Temco-Vought Inc.* v. *United States,* 201 Ct. Cl. 135,475 F.2d 630 (1973).

The contract rights and obligations of architects and engineers have traditionally extended only between them and the owner. The general obligation of architects and engineers is comparable to the obligation imposed upon other professionals, e.g., doctors: they must possess the ordinary skill and expert knowledge possessed generally by architects and they must exercise reasonable care in the performance of their duties. See *Geer* v. *Bennett,* 237 So. 2d 311 (Fla. App. 1970). The liability of the architect and engineer is measured by negligence standards as applied to the duties imposed upon them by contract. Thus, an architect can be liable to the owner where he fails to possess the requisite skill or is negligent with respect to the preparation of plans and specifications *(Chapel* v. *Clark,* 76 N.W. 62 (Mich. 1898) or the selection of materials *(Bayshore Dev. Co.* v. *Bonfoey,* 78 So. 507 (Fla. 1918) or in failure to conduct proper supervision of construction where he has assumed that duty under the contract *(Avent* v. *Proffitt,* 95 S.E. 134 (S. Car. 1918).

Traditionally, the architect's duties have been regarded as directed to the benefit of the owner only, since there is no privity of contract between the architect and contractor. See, e.g., *Brick* v. *Okaloosa Title & Abstract Co.,* 404 F.2d 889 (5th Cir. 1968). With increasing frequency, the courts have been recognizing the right of a contractor to recover in tort from an architect whose negligence has caused the contractor to suffer economic loss. See *A. R. Moyer Inc.* v. *Graham,* 285 So. 2d 397 (Fla. 1973). *United States ex. rel Los Angeles Testing Laboratory* v. *Rogers & Rogers,* 161 F. Supp. 132 (S.D. Cal. 1958). In the latter case, the contractor's right to sue the architect was upheld in a claim arising from a federal contract, with the Court noting that the architect was not protected from liability because his contract was with the Government.

When architects and engineers make a decision under the AIA-type dispute clause, they enjoy relative immunity from suit. When an architect or engineer decides a dispute under a contract provision referring such disputes to him, the courts have held that the architect or engineer is, in effect, acting as an arbitrator. As such he enjoys the arbitrator's privilege and immunity from liability for his decisions, except in the case of fraud or demonstrable bad faith. See *Craviolini* v. *Scholer & Fuller Assoc. Architects,* 357 P.2d 611 (Ariz. 1961).

Just as the owner is responsible to the contractor for the action of its architects and engineers, so is the contractor responsible to the owner for the action or inaction of its subcontractors. Under the present version of Article 5 of federal contracts, subcontractor delay is no excuse for non-performance. Within the limited context of prime contractor-subcontractor relations, the prime contractor is in the unique position of standing for the owner, absent some subcontract provision relieving the prime of that burden. Just as the owner warrants the adequacy of the specifications to the prime contractor, so does the prime contractor warrant the adequacy of those specification to its subcontractors.

The exception to the general rule of prime contractor liability for the subcontractor is where the subcontract contains a provision limiting the prime's liability for acts attributable to the owner to the amount recovered by the prime contractor (on the sub's behalf) from the owner by reason of the particular owner acts involved. There is no question that such provisions are enforceable and that, in such cases, the subcontractor's only rights are that (i) the prime must prosecute on the behalf of the subcontractor any claims which the subcontractor has based upon owner action and (ii) that the prime contractor pay to the subcontractor the amount (less prime contractor overhead) recovered under such claims.

The most common problem presented by the prime-subcontractor relationship relates to delay claims. In this area as well, the prime contractor's liability to the subcontractors for delays is comparable to the owner's liability to the prime. What was said above with respect to "no-damage-for-delay" clauses in the prime contract applies equally to the subcontract.

WARRANTIES/GUARANTEES/LATENT DEFECTS

The basic question is what obligations does the contractor have beyond the acceptance of his work. Up to the point of acceptance, there is no question that the contractor must correct any deficiencies in his work that are discovered and not waived. This obligation is fundamental and affirmed by the express language of the federal inspection clause. Moreover, this obligation is not terminated when the work in question is incorporated in the structure and covered up. Under the inspection clause, the Government has the right to direct removal of work to inspect its adequacy. If the covered work in question is found to comply in all material respects with the contract requirements, the Government pays the cost of removal and restoration. If it doesn't, the contractor pays those costs as well as the cost of correction.

Government acceptance, following final inspection generally marks the end of the contractor's work. It does not mean that the contractor's obligations are terminated. Certain obligations can survive under warranty clauses and under the latent defect, fraud or gross mistake provision of an inspection and acceptance clause. A contractor's warranty or post-acceptance liability can generally arise in one of several ways: the contract can include express warranty clauses that warrant the work or certain parts of it for a period of time after acceptance. Examples are found in ASPR §1-324-7 and §1-324.1-3.

The contractor's liability ıor defects can be extended past acceptance by "correction of deficiency" type clauses such as found in *Keco Indus. Inc.*, ASBCA No. 13271; 71-1 BCA ¶8727 (1971). In that case the contract for the supply of air conditioning units contained a provision which required the contractor to correct all deficiencies of which the Government gave notice within 180 days after acceptance of the last item delivered.

A performance criteria of the specifications can occasionally be used to create a warranty type liability which survives acceptance. The *Keco* case, cited above, illustrates this point as well. In that case, one of the performance criteria was that the air conditioning units must be "capable of operating at 500 hours at design capacity without the failure of any major part." The Board held that, under the terms of Section 2-313 of the Uniform Commercial Code "The foregoing obligation can be deemed an express warranty." Section 2-313 of the UCC states as follows:

> ...Express warranties by the seller are created as follows:
>
> (a) Any affirmation of fact or promise made by the seller to the buyer which relates to the goods and becomes part of the basis of the bargain creates an express warranty that the goods shall conform to the affirmation or promise.
>
> * * * * *
>
> (2) It is not necessary to the creation of an express warranty that the seller use formal words, such as 'warranty' or 'guaranty' or that he have a specific intent to make a warranty.

The warranty-type clause which is most often involved in federal construction contract disputes in General Provision Article 10, Inspection and Acceptance. Its final sentence states: "Acceptance shall be final and conclusive except as regards latent defects, fraud or such gross mistakes as may amount to fraud or as regard the Government's rights under any warranty or guarantee." As the final proviso of this sentence indicates, this provision in no way limits owner rights under other warranty provisions of the contract, including any warranty provisions which are not labeled as such, but yet create a warranty-type obligation as did the 500 hour performance criterion of the *Keco* case.

The fact that a defect is not discovered until after acceptance does not mean that the defect is latent. See *G. M. Co. Mfg. Inc.,* ASBCA No. 5345, 60-1 BCA ¶2576 (1960), rev'd. on other grounds. The burden of proving latency is on the Government. See *Hercules Engineering & Mfg. Co.,* ASBCA No. 4979, 59-2 BCA ¶2426 (1959). A latent defect has been defined as one "not discoverable by a reasonable inspection." "Defects which can be discovered by such an inspection are patent and cannot be used to overcome the bar of final inspection (and acceptance)." *Keco Indus. Inc., supra.*

The fraud exception, of course, relates not to the nature of the defect but rather to the conduct of the contractor which the Government contends prevented discovery of the defect. In *Bar Ray Prods. Inc.* v. *United States,* 167 Ct. Cl. 839, 340 F.2d 343 (1964) the Court of Claims stated that "to establish fraud, it is necessary to show a misrepresentation of a material fact, an intent to deceive and reliance on the misrepresentation by the other party to his detriment." Under this definition, the key (and most difficult to prove) element is the intent to deceive. The Armed Services Board of Contract Appeals, however, has suggested that the fraud exception might apply in cases of constructive fraud which it defined as "a breach of a legal or equitable duty which, irrespective of the moral guilt of the fraud feasor, the law declares fraudulent, and neither actual dishonesty of purpose nor intent to deceive is an essential element of constructive fraud." *Catalytic Engineering & Mfg. Corp.,* ASBCA No. 15257, 72-1 BCA ¶9342 (1972).

The final exception to the conclusiveness of acceptance also relates to the conduct of the contractor, rather than the nature of the defect. Few cases have addressed the question of what is a gross mistake amounting to fraud. One of those is the *Catalytic Eng'r* case, supra, which defined the term as follows:

> Gross mistake connotes a mistake so serious or uncalled for as not to be reasonably expected, or justifiable, in the case of a responsible contractor for the items concerned.... (a) gross mistake would...amount to fraud when the misleading statement or action is made by mistake and without an intent to deceive but induces the acceptance of supplies not conforming to the contract requirements to the buyer's detriment.... because the resulting damage to the buyer is the same as it would have been if the statement had been made, or the action taken, with an intent to deceive.

Point to Remember: the law behind the contract is not one-sided, but involves give and take on the part of both parties, each party with its rights and obligations under the contract.

CHAPTER 4

The Law Behind Construction Schedules

MICHAEL T. CALLAHAN, J.D.; LL.M.

I. PROJECT SCHEDULING TECHNIQUES

The "project schedule" serves different functions to the architects, engineers, contractors, subcontractors, materialmen, owners and most especially, courts of law involved in the construction process. The project schedule may mean the contractually stated final completion date or interim completion dates for phases of the work. It may refer to the process of sequencing and scheduling individual activities or tasks which are required to complete the project.[1] It may also refer to a contractual commitment to complete each element of the work at the date designated in the schedule. Before examining the way the law looks at construction scheduling, a review of scheduling principles will be made.

There are several types of construction schedules; the most frequently encountered are gantt (bar) charts and critical path method schedules. PERT charts, a third type, are usually encountered in research and development rather than in construction projects. These types of schedules share many characteristics.

Regardless of type, construction schedules are designed to establish the sequential order in which construction is completed. Establishing sequential order requires an intimate knowledge of construction methods combined with an ability to visualize discrete work elements and to establish their mutual interdependency. Once

---

[1]P. Walstead, "Using CPM in Contract Disputes," in *Construction Claims and Disputes* 173, 181-182 (1978).

sequential order is established, manloading, equipment, material, or other resources may be easily added for greater degrees of control.

Construction scheduling is a time-consuming and difficult task. Most construction projects are extensive, involving thousands of activities with intricate and complex interrelationships. The use of any scheduling technique will produce a schedule only as good as the time invested. If a schedule is based on faulty logic among activities or faulty estimates of activity duration, its utility will be limited.

Construction scheduling is also an attempt from the present to look into the future. As to men, "time and chance happeneth" to construction schedules. A schedule must be revised continuously to reflect the actual conditions of the future as it changes to the present. Without the ability to change, the schedule becomes useless. A construction schedule must be based on the best present information. When assumptions are changed because of superior information, the schedule must also change.

A. CPM Schedules

The critical path method of scheduling contributes significantly to construction contracting. It is an effective tool for day-to-day direction and control of work. It permits advance work schedules for material procurement, equipment availability and manloading. Preparing the network diagram forces one to think through the job from start to finish, permitting early identification of potential problem areas on the project. The CPM schedule is easy to pass to others. The diagram permits efficient communication between field and office forces. The diagram can assist in smooth transition with project management staff if personnel changes are made during the life of the project. It makes job coordination with material suppliers, contractors, subcontractors, owners, and architects/engineers easier.

Preparing a critical path method schedule requires three steps: (1) determination of the construction activities that must be performed in order to construct the project; (2) ascertainment of the sequential relationships or "logic" among these activities; and (3) the presentation of this information in the form of a network diagram.[2]

The elements into which a construction project is subdivided for CPM planning purposes are called "activities." An activity is a single work step that has a recognizable beginning and end. It is a time-consuming task. No definition exists as to the number of activities any CPM schedule should have. One needs only enough activities to sufficiently control the work.

The order in which the activities are to be accomplished is the CPM "logic." Logic should be developed with safety, space (12 men cannot wire one motor control system), and structure in mind. The start of some activities obviously depends upon the completion of others. For example, a roof membrane cannot be installed until the

---

[2]R. Clough, *Construction Project Management* 58 (1st ed. 1972).

structure is complete. Yet, there are many activities that are independent of one another and can proceed concurrently. Much of job logic follows from well established work sequences that are customary in construction contracting. Nevertheless, there is always more than one way to approach a project; no unique "logic" exists. There are many ways to construct a project and, consequently, many different schedules are possible.

The network diagram is nothing more than a graphical display of job logic, presenting in pictorial form all of the dependencies among activities.[3]

To be realistic, a CPM schedule must reflect the practical restraints that apply to the job activities. The restrictions of the job logic itself are restraints. The completion of the structure restrains the installation of the roofing membrane. These restraints arise from the necessary order of construction operations and are simply a part of job logic.

But there are other types of restraints. Structural steel cannot be erected until the steel is available at the site, and steel availability depends upon approval of shop drawings, steel fabrication, and its delivery to the job site. Thus, the start of steel erection is restrained by the necessary preliminary actions of shop drawing approval, steel fabrication, and delivery. Restraints must be shown on the schedule.

Restraints or "dummies" are represented on the network as a dotted line. The dummy's purpose is only to explain certain relationships among activities. The restraint does not represent a work activity. Although the activities which restrain the logic can be thought of as time-consuming in nature, the restraint for scheduling purposes is considered to require "no time." Thus the dummy does not itself represent performance time.

There are two methods of drawing CPM schedules: "I-J" and precedence (PDM).

In I-J, the tail of the arrow is the starting point of the activity and the head of the arrow represents its completion. The starting and ending points of an activity are known as an event or "node." The I node is the beginning point; the J node is the ending point. In a logic diagram, a J node (activity completion) also represents an I node (activity beginning) of the following construction activity. In the usual network diagram the length and orientation of the I-J arrows have no significance. Their directions simply indicate the flow of work and time.

To illustrate the mechanics of I-J diagramming, consider a simple project such as the construction of a gas station. A partial I-J diagram illustrating one method of sequencing the construction activities in building a gas station is illustrated in Exhibit No. 1. Layout and Excavation and the Procurement of Fuel Tanks are all activities that can begin immediately after the contract is signed. They can proceed independently of one another. After the excavation has been completed, construction of the foundations can start. Installation of underground pipes will follow the building foundations but the interior cannot be backfilled until both the hydraulic lift founda-

---

[3]*Id.* at 60.

tions and the hydraulic lift have been completed. Concrete floors will be poured after the building interior has been backfilled. (See Exhibit 1 on page 53.)

Each arrow or activity is identified by the number located in the I node and J node. For instance, the activity "lay out and excavate" is designated as activity 12-14. The activity "lay out and excavate" has an I node of 12 and a J node of 14.

The I-J method of critical path scheduling is based on the premise that a given activity cannot start until all of those activities immediately preceeding it have been completed. In Exhibit No. 1 "form and pour concrete floor" cannot start until "backfill building interior" has been finished. This is not necessarily true in the field but is a limitation of the scheduling process.[4]

Every activity in the I-J network must have a definite event to mark its beginning. It is not possible to have the finish of one activity overlap the start of the following activity. If the start of one activity can overlap the finish of another, the activities should be further subdivided.

The I-J network must be continuous with no gaps, discontinuities, or dangling activity. All activities must have at least one activity following, except those that terminate the project.

In Exhibit No. 1, the dashed arrow, 14-16, is not a time-consuming activity, but is an example of a restraint. The use of restraints is the most difficult aspect of I-J diagramming. Restraint 14-16 is necessary if job logic is to be portrayed correctly. A correct sequential relationship stipulates that the start of "install fuel tanks" must await completion of "order and deliver fuel tanks." But "install fuel tanks" must also await completion of "layout and excavate." The restraint 14-16 shows that the start of "install fuel tanks" depends upon the completion of not only "order and deliver fuel tanks" but also of "layout and excavate." The direction of the restraint designates the flow of activity dependencies.[5]

It is customary that the I node, at the head of the arrow, be numbered lower than the activity at its tail, the J node. J is normally greater than I for all activities. Originally, this was done because of computer program limitations. Although it is no longer necessary, the practice persists because it makes it easier to locate events and activities on a diagram.

In Exhibit No. 1, activity identification is *below* the line. Although some Government specifications require the activity identification to be *above* the line, identification below the line eliminates physical restrictions on the length of description which may be encountered from the arrow. Activity time or duration is indicated next to the I node. Contractor responsibility is indicated next to the J node. The activity arrow may also carry additional information. Many Government CPM specifications require a dollar and manloading value to be assigned to each activity. The more information which the activity is required to carry, however, the less useful the CPM schedule becomes to measure time since the computer will process the greater amount of information at reduced speeds, arrival in the field is delayed further and the

---

[4]H. Hohns, *Preventing and Solving Construction Contracts Disputes* 34 (1st ed. 1979).
[5]R. Clough, *supra* note 2, at 64.

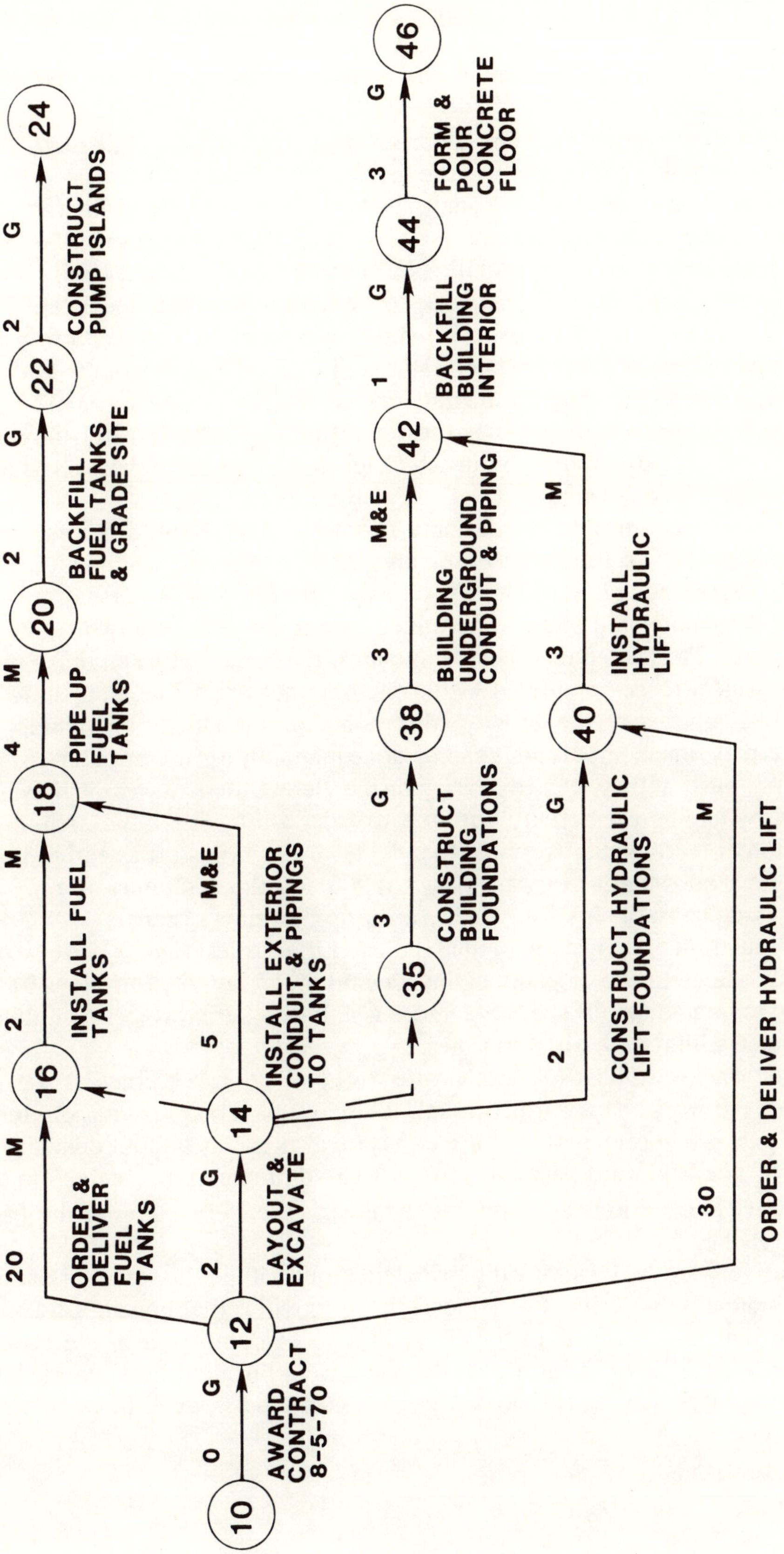
10
0 G
AWARD CONTRACT 8-5-70
12
20 M
ORDER & DELIVER FUEL TANKS
16
2 M
INSTALL FUEL TANKS
18
4 M
PIPE UP FUEL TANKS
20
2 G
BACKFILL FUEL TANKS & GRADE SITE
22
2 G
CONSTRUCT PUMP ISLANDS
24
2 G
LAYOUT & EXCAVATE
14
5 M&E
INSTALL EXTERIOR CONDUIT & PIPINGS TO TANKS
35
3 G
CONSTRUCT BUILDING FOUNDATIONS
38
3 M&E
BUILDING UNDERGROUND CONDUIT & PIPING
42
1 G
BACKFILL BUILDING INTERIOR
44
3 G
FORM & POUR CONCRETE FLOOR
46
2 G
CONSTRUCT HYDRAULIC LIFT FOUNDATIONS
40
3 M
INSTALL HYDRAULIC LIFT
30 M
ORDER & DELIVER HYDRAULIC LIFT

Exhibit 1

volume of material may "overwhelm" the manager, reducing his effective use of the information generated.

Once the network diagram has been developed, time is added to the CPM schedule. During the development of the logic, time should not have been taken into account. The logic diagram should have been developed using unlimited time and resources. Adding activity durations at this stage, however, should be based on equipment, material, and labor limitations. Durations should be reasonally optimistic. As time is introduced into the network, it becomes necessary to redefine certain activities, condense others into fewer activities, or expand them into additional ones.

After a time duration has been estimated for each activity of the network, the "critical" or longest path through the network is calculated. The "critical" or longest path through the project determines the period in which the project may be completed and the period of time within which each activity must be accomplished if completion time is to be met.

This calculation involves the determination of four times for each activity. The "early start" (ES) is the earliest time an activity can possibly start after the completion of the preceeding activities. The "early finish" (EF) is the earliest time an activity can be completed and is determined by adding that activity's duration to its early start time. The "late finish" (LF) is the latest that an activity can finish and allow the entire project to be completed within the contract time. The "late start" (LS) is calculated by subtracting the activity's duration from the latest finish time.

Activity time computations can be done manually or by computer.

First, each activity's early start value is determined. Its early finish time is calculated by adding the activity duration to early start. The basic assumption for computation of "early" activity time is that every activity will start as early as possible. This process is known as a "forward pass." It starts at the beginning of a project and works forward. Thus, the first activity has an early start time of zero.

Calculation of early activity time indicates the earliest time the last activity will finish. *If* a competent job of planning has been done, *if* activity durations have been accurately estimated, and *if* everything goes well in the field, project completion can be anticipated within the early start times.[6]

The "backward pass" calculates late start and late finish times. It starts at the project end and works backwards through the network. The presumption during the backward pass is that each activity finishes as late as possible without delaying project completion. The backward pass calculation is based on the early completion time for the last activity. Durations are subtracted through the Network until the beginning activity is reached.

The forward and backward pass calculations will be the same for some activities. This indicates the "critical activities," those activities that must begin and end on their early start and finish times for the project to be completed within the contract time. Critical activities will form a continuous string through the network known as the "critical path." Critical activities are necessarily the most difficult or those that

---

[6]R. Clough, *supra* note 2, at 95.

may seemingly be the most important job elements. The critical activities merely form the longest path.

Activities with early and late start times that do not match are flexible; they do not have to begin and end within the *early* start and finish times to meet project completion dates. This flexibility is a measure of the capability for a given activity to have its performance extended. This flexibility is called "float."

There are many paths through a schedule. Not all will be critical, but there may be more than one critical path. Even those paths which are not critical because they contain float, however, may become critical if the float is eliminated.

Two classifications of float are generally recognized. The "total float" of an activity is the difference between its late start and its early start. Subtracting the late finish from the early finish gives the same result.[7] Total float of an activity is the amount of time by which an activity can be delayed without affecting the project completion date. Total float is a path concept. Activities share total float; using some in one activity will reduce it in all that share it.

The second type of float, "free float," is found by subtracting an activities early finish time from the early start time of the subsequent activity. The free float of an activity is the amount by which that activity can be delayed without delaying the early start of the following activity or affecting any other activity in the network. Elimination of free float also eliminates total float in an equal amount.

To assist project scheduling, "milestone" dates are often established. Milestone events are points in time identified as being important reference points in accomplishing the work. Milestone events can be dates imposed by the owner for the accomplishment of certain tasks or target dates set by the contractor.

A time-scaled network is an I-J network drawn to a horizontal time scale. A time scale is obtained by plotting the early start and early finish times for every activity. The length of the connecting arrow is the duration of that activity. A time-scaled network has several advantages over a logic diagram. Time-scaled diagrams are convenient devices for checking daily project needs of labor, equipment, material, and for the advance detection of conflicting demands among activities for the same resource. A network plotted to a time scale also permits greater ease in making cash flow studies and for the recording of field progress.[8]

Any delay in a critical activity automatically extends the critical path. Any delay in the finish date of a critical activity for whatever reason, automatically extends project completion by the same amount. Identification of the critical activities is an important aspect of job scheduling because it locates activities that must be timely performed if the project is to be completed within the contract time. Alternately, once identified, critical activities may be rescheduled or have their durations reduced to permit quicker project completion.

To properly use the critical path method, the initial data input must be accurate and the assumptions upon which it is based must be reasonable. If the method is

---

[7]R. Clough, *supra* note 2, at 97.
[8]R. Clough, *supra* note 2, at 115.

introduced carelessly, reluctantly, or with unconcern it will become an added burden to the job which hinders rather than helps.[9]

B. Precedence Diagramming (PDM)

In precedence diagramming, the nodes represent the activities and the arrows between them simply indicate the logic dependencies among the activities. Each activity has only one number — the I number. It is drawn by connecting the work activities from either their starting or finishing points, presenting a start-finish sequence without breaking down the activity to eliminate overlap as required in I-J. Other than the difference in the way the diagram is drawn, I-J and precedence diagramming are essentially identical.

Precedence diagramming provides more flexibility than I-J. The length and direction of arrows in precedence diagramming have no significance because they indicate only the dependency of one activity on another. Precedence diagrams eliminate the need for restraints and extensive details to prevent overlap by utilizing "lag factors." Lag factors indicate an activity cannot start until after the passage of time indicated by the lag factor.

It is easier to draw a precedence diagram than it is to draw an I-J diagram. But it is more difficult to update because one must change "overlap" as well as duration.

C. Bar Charts

Bar charts, the more sophisticated forms of which are called "gantt charts," have been the traditional method for scheduling of construction projects. On uncomplicated jobs involving a minimum of tasks, the bar chart does a reasonably good job. However, on larger scale projects, the bar chart has a fundamental weakness; it does not show dependencies among activities. Each activity may receive the same consideration with no indication of where management attention should be focused. It is possible, however, to develop a critical path from a bar chart.[10]

The bar chart contains a list of the major activities on the project usually corresponding to the specification headings. It is also possible to break activities down by floor, system, or area. A bar for each activity reflects start and end dates, although an activity may have its bar interrupted to reflect temporary activity cessation. The duration of each activity, as reflected by the bar, however, does not indicate that work in the activity is being continually performed. A bar indicating the mechanical work does not mean the mechanics are working on the project on each work day represented by the bar. Rather, it represents the first day the mechanic has work to be performed on the project until the mechanical work is completed. The number of days actually worked during this duration is not specified. No matter how detailed the bar chart, the interdependency of the activities is not made clear. The manner in which one activity follows another is not specified.[11]

---

[9]Associated General Contractors of America, *CPM in Construction — A Manual for General Contractors* 7 (1975).
[10]H. Hohns, *supra note* 4, at 59.
[11]P. Walstead, *supra* note 1, at 179.

There is usually no rational basis for the development of the bar chart. It usually does not portray a detailed, integrated plan of operations. Bar charts are ineffective for project shortening and resource or equipment management. Normally, a bar chart is made up by adjusting the individual work items to fit within an overall time frame.

The visual clarity of the bar chart, however, makes it a very valuable medium for displaying job schedule information. It is immediately intelligible to people who have no knowledge of CPM or network diagrams. It affords an easy and convenient way in which to present information developed from a CPM study. For these reasons, bar charts undoubtedly will continue to be widely used in the construction industry.[12]

D. PERT Charts

Program Evaluation Review Technique (PERT) was developed by Booz, Allen & Hamilton shortly after CPM. It was intended to assist in scheduling construction of naval vessels that had not been built before, where no historical or cost data was available as a guide. PERT is more widely accepted and utilized in the area of systems and supply procurement than in construction.

CPM utilizes a single time estimate for each network activity. This is because in construction each activity is "deterministic." Similar construction work has been performed many times before. Such prior experience enables the contractor to estimate with reasonable accuracy the time required to carry out the scheduled activity. PERT, on the other hand, utilizes three time estimates for each activity: (1) an optimistic or minimum time; (2) a most likely or model time; and (3) a pessimistic or maximum time. This is a "probablistic" approach that lends itself to activities for which there is no historical record or experience, activities that in all likelihood are being done for the very first time.

PERT charts are usually scheduled in weeks rather than days as is CPM.

E. Updating

No schedule can be perfect. The scheduler cannot anticipate every future job circumstance and eventuality. Contingencies arise every day that could not have been foreseen. More attention is paid to the scheduling process as construction advances. Indeed, as portions of the project are completed, better ideas arise. Adverse weather, delivery delays, labor disputes, change orders and differing site conditions may disrupt the original plan and schedule. It is not unusual that activities additional to those included in the original plan prove to be necessary. As a result, it is normal that construction projects deviate from their developed schedules.

During the construction period, considerable time and effort are also required to check the progress of the job and to take whatever action is appropriate and necessary to bring the work back on schedule or report the effects of unrecoverable delay. These actions constitute the monitoring and rescheduling phases of the job schedule known as "updating."

---

[12]R. Clough, *supra* note 2, at 118.

A single activity is generally so limited in scope that any loss of time on that activity is irrecoverable. Corrective action, if required and possible, is based on making up the lost time through rescheduling of subsequent activities. Indispensable to the assessment of project delays and the devising of corrective rescheduling is an updated schedule of those activities yet to occur. These updatings often reveal shifts in critical path and substantial changes in the float of activities. To maintain a schedule as a realistic tool updatings *must* occur with regularity.

How often field progress should be measured and reported depends upon the degree of time control that is considered desirable. Depending upon size, complexity and characteristics of the work, progress reporting may occur anywhere from daily to monthly or even longer.

II. THE LAW AND CONSTRUCTION SCHEDULES

A. The Law's View of Schedules

Courts and administrative boards throughout the United States have dealt with schedules and scheduling techniques. Most have recognized the superior ability of the critical path method to coordinate and schedule. The United States District Court for the Western District of Missouri in *Natkin & Company* v. *George A. Fuller Company,*[13] observed that "The Critical Path Method is a valuable tool on a complex job, saving time and money for owners and contractors."

Generally, the courts have accepted the construction industry definition of the critical path method. As will be seen later, however, courts have not consistently applied the technique to construction disputes in the same manner as it is applied as a planning and scheduling tool in the industry.

The Board of Contract Appeals in *Continental Consolidated Corporation,*[14] viewed the critical path method of scheduling as:

> The CPM scheduling technique is one which requires a breakdown of the entire work into individual tasks and an analysis of the number of days required to perform each task. The analysis is then programmed into a computer which produces a chart showing the tasks and a line which controls the completion of the overall work. The line through the nodes, the junction points for completion of these essential tasks, is known as the Critical Path. In addition, there are numerous side paths for subordinate tasks which normally can be performed without affecting the critical paths. However, these subordinate tasks, if improperly scheduled or unduly delayed in performance, can on occasion become critical and thus change the critical path of the entire work.

---

[13] 347 F. Supp 17 (W.D. Mo. 1972).
[14] ENGBCA No. 2743, et al., 67-2 BCA ¶6624 (1967).

> The critical path method of scheduling requires the logical analysis of all the individual tasks entering into the complete job and the periodic review and re-analysis of progress during performance. It is essential that any changes in the work and time extensions due to the Contractor be incorporated into the progress analysis concurrently with the performance of the changes or immediately after the delay and thus integrated into the periodic computer runs to reflect the effect on the critical path. Otherwise the critical path chart produced by the computer will not reflect the current status of work performed or the actual progress being obtained.[15]

The New Jersey Superior Court in *Dobson* v. *Rutgers,*[16] utilized a definition of CPM from "Critical Path Methods in Construction Practice," by Antill Woodhead (2d edition 1970):

> The Critical Path Method is a new and powerful tool for planning and management of all types of projects. Essentially, it is the representation of a project plan by a schematic diagram or network that depicts the sequence and interrelation of all of the component parts of the project, and the logical analysis and manipulation of this network in determining the best overall program of operation. It is a method admirably suited to the construction industry, and it provides a far more useful and precise approach from the conventional bar graphs and progress charts that previously formed the basis of construction planning and control. Furthermore, it permits the ready evaluation and comparison of alternative works programs, construction methods, and types of equipment. When the best plan has been prepared in this way, the critical path diagram clearly indicates the site operations that control the smooth execution of the works. Finally, as construction proceeds, the diagram provides the project manager with precise information on the effects of each variation or delay in the adopted plan, thus enabling him to identify the operations that require remedial action.[17]

In *Peter Kiewit Sons Co.* v. *Iowa Southern Utilities Co.,*[18] the U.S. District Court for the Southern District of Iowa showed a slightly different view of the method:

> ...the Critical Path Method (CPM), a method of planning, which, as the name suggests, is designed to assure that the most critical

---

[15] *Id.* at 30, 715.
[16] 157 N.J. Super. 357, 384 A 2d 1121 (1978).
[17] *Id.* at 1126.
[18] 355 F. Supp 376 (S.D. Iowa 1973).

> construction work is done in the most expeditious manner, while at the same time assuring that all contractors on the site can proceed with their work in a coordinated manner.[19]

Blackhawk Heating and Plumbing Co., Inc.,[20] recognized that "activities not regarded as critical in the planning stage can become so if actual delay is excessive."[21]

Courts, recognizing the inherent limitations of the bar charts, have consistently held them to be less effective as a project schedule:

> ...bar charts...were [not]...designed to afford an overall coordinated schedule of the total work covered by the contract.[22]

The General Services Board of Contract Appeals in *Minmar Builders, Inc.*[23] refused to find a delay to project completion where the contractor's proof was limited to two bar chart schedules and testimony by the contractor's president that the ceiling work was the "pacing" work element affecting project completion. The contractor alleged that Government delay in issuing ceiling change instructions delayed the project. The Board found the bar chart schedules failed to show whether the ceiling work was indeed the "pacing" element or that the delay affected project completion:

> Since no interrelationship was shown as between the tasks, the charts cannot show what project activities were dependent on the prior performance of the plaster and ceiling work, much less whether overall project completion was thereby affected. In short, the schedules were not prepared by the Critical Path Method (CPM) and hence are not probative as to whether any particular activity or group of activities was on the critical path or constituted the pacing element for the project.[24]

The *Minmar* Board found the tasks affecting completion of the project had no relationship to the ceiling work. The Board concluded that project completion was not affected by Government delay in ordering ceiling changes.

Because of the high visibility of bar charts, however, they may be used to demonstrate a critical path analysis. For instance, in *Dawson Construction Co. Inc.,*[25] the CPM consultant illustrated an as-built CPM in bar chart form to reflect the effect of delay on the excavation and pouring of footings, grade beams, piers and slabs.

Regardless of the scheduling techniques utilized, they can be of considerable assistance in supporting claims for price adjustments and time extensions. Similarly,

---

[19] *Id.* at 88.
[20] GSBCA No. 2432 75-1 BCA ¶11, 261 (1975).
[21] *Id.* at 53,685.
[22] *Natkin & Co.* v. *George A. Fuller Co., supra* note 13 at 21.
[23] GSBCA No. 3430 72-2 BCA ¶9599 (1972).
[24] *Id.* at 44, 857.
[25] GSBCA No. 3998 75-2 BCA ¶11, 563 (1975).

they can aid in the defense of claims and in justifying assessments of liquidated damages for late contract performance.[26] The contractor's own schedule can even be used against it. For instance, the *Dobson* court used the schedule submitted by the contractors to prove owner's delay, to show some of the claimed inefficiency for lack of stairways could not have occurred because the schedule showed no stairways were planned to be available until well within the claimed inefficiency period.

Although courts uniformly recognize the value of scheduling, whether bar chart or CPM, they will not find a scheduling requirement if none was provided in the construction agreement. In *Drew Brown Ltd.* v. *Joseph Rugo, Inc.,*[27] Rugo, the general contractor, and Brown, the structural steel subcontractor, executed a contract which required Brown to "carry to completion the work undertaken by the agreement as the progress of the work requires."[28] No other scheduling clauses were present. Brown alleged Rugo breached the contract by failing to coordinate and schedule Brown's work. The court found Rugo's responsibilities to Brown as enumerated in the subcontract did not include the *advance* scheduling of work for Brown's convenience. It is interesting to note that Brown was unable to prove that scheduling was a custom of the trade. In refusing to find a breach of contract, the court opined that if Brown had felt scheduling was so important, it should have added a clause in the subcontract requiring it.

Under the American system of law, past cases or decisions are looked to as providing assistance in solving present disputes. Consequently, attorneys, when promoting one position, will look to the past for similar cases which support its conclusion. Attorneys attempting to dispute the conclusion will look for cases which take an opposite view. The facts of each particular past case and the Court's reasoning thus take on considerable significance. Distinctions of facts, situations, or reasonings between past cases and the present dispute are carefully examined. Thus, the facts of all past scheduling cases must be studied for they provide the basis for decision. Scheduling clauses, the subject of the case decision, are reproduced in addenda where available.

B. The Schedule Must Be Complete

No schedule will be accepted by a court until it is complete. *Dobson* illustrates the importance of complete information in a construction schedule. The Court there found that it was not until the third update that sufficient data was available for a complete plan for the project to exist, although until the third update the schedule lacked only a procurement schedule from the General Contractor. The scheduling clause in *Dobson* is reproduced in Appendix A.

At the time of the first update, the *Dobson* schedule reflected the scheduling

---

[26] J. Wickwire, P. Walstead, and T. Asselin, "Project Scheduling," in 3 The Briefing Papers Collection, 233, 237 (1975). [hereinafter cited as J. Wickwire].

[27] 436 F. 2d 632 (1st Cir. 1971).

[28] *Id.* at 637.

consultant's manifestation of the Mechanical, Electrical, and General Contractor's intent to construct the building. The schedule was complete through the close-in or until the building was "tight." The Mechanical and Electrical Contractors indicated that the schedule as reflected at the time of the first update was the basis for planning of their activities. The schedule at the time of the second update had advanced to include information on Laboratory Equipment, Owner Furnished Equipment, Finishing Details, and Procurement Items for the Electrical Contractor. But it was not until the third update when the schedule reflected the General Contractor's procurement schedule that the Court found sufficient information included within the schedule to be considered a plan for the project.

A court may even accept the most complete schedule even if it wasn't what the contractor used to construct the building. In *Blackhawk Heating and Plumbing Co., Inc.,* the Board of Contract Appeals utilized an outside scheduling consultant's schedule prepared after project completion rather than Blackhawk's because Blackhawk's "did not.. .break certain activities down into components which would necessarily be helpful in analyzing subsequent project delays."[29] The scheduling consultant had broken down Blackhawk's CPM activities into more detail, permitting more meaningful analysis of project delays. In adding activities the Board found the consultant was only making the contractor's schedule more useful while maintaining the basic construction plan. This was so even though the consultant changed logic from the "planned" floor-to-floor sequence to work being performed on several floors simultaneously because the project had actually been completed in that manner.

A schedule may be accepted by a court as indicating manner and method of completion despite lack of formal approval. The *Dobson* decision illustrates the significance a court places on the "formal approval" of a construction schedule. In *Dobson,* there was never a formally approved working plan as required by General Conditions (see G4-F.4(c) in Appendix A). The court accepted the schedule, as manifested by the third update, as the working plan because it represented the most complete schedule and the one most frequently used throughout the job. But more important, *without* utilizing the *most* complete schedule, despite the absence of "formal" approval, there would be nothing upon which to measure any rights or damages and no complete plan for the work as called for under the contract. The absence of approval did not inhibit the court's use of the construction schedule actually used to build the project as *the* project schedule.

C. The Schedule Must Be Substantiated

Critical Path Schedules are not entitled to automatic acceptance. Courts require a CPM to satisfy certain fundamental tests, before presented, to illustrate the method and manner of construction — the authenticity or validity of the data used to prepare the schedule; the intended purpose of the schedule, whether to estimate or construct; and how the schedule was actually used must be established.

---

[29] *Blackhawk Heating & Plumbing Co. Inc., supra* note 20 at 53, 676.

In *Chaney & James Construction Co., Inc.*[30] the Federal Aviation Agency Contract Appeals Panel found a critical path schedule could not be accepted as establishing either the facts portrayed or the reasonableness of the contractor's assertions as to the influence of specific incidents of work progress. Chaney & James had submitted a bar chart, as required by the contract, which was approved by the contracting officer. Later, Chaney & James submitted a critical path schedule but failed to state why the CPM had been submitted. A CPM had not been required by the contract. At the hearing, no satisfactory evidence as to the origin of the data that went into the preparation of the chart, the purpose that the chart was meant to serve, or what use was actually made of it was presented. In fact, the contractor's president and project manager testified that the CPM was not used either in preparing the bid or in managing the project.

During the hearing, Chaney & James submitted an updated version of the CPM and an additional critical path diagram purporting to show the relationship of the work elements and the impact on work progress of the specific instances of claimed Government caused delays. The Board, however, refused to consider the CPM schedules as evidence. It found the work sequence shown was not demonstrated to be the only possible sequence in which the work could have been accomplished; nor was it demonstrated that the sequence presented in the charts was necessarily the best one. Further, the CPM work sequence shown was not used in estimating and bidding the job; and the sequence shown on the CPM chart was not followed in performing the contract work.

Although the *Chaney & James* decision refusing to consider the CPM can be accepted on the basis of the contractor's failure to utilize or rely on it during the construction process, the CPM should not have been rejected because it was not the "only possible" or "best" sequence in which the work could have been accomplished. By industry definition, a CPM schedule is developed to illustrate only one method of construction. Industry standards recognize many ways in which a project can be completed. In insisting upon an "only possible" or "best" sequence as conditions of acceptance, the Board failed to understand the critical path technique.

The Armed Services Board of Contract Appeals in *Lane-Verdugo*[31] in finding the CPM did not prove Government responsibility for claimed job delays noted that although the contractor's CPM consultant testified that the critical path analysis was only as good as the information upon which it is based, the contractor offered no evidence establishing the authenticity or validity of the data that was used in the preparation of the critical path charts and accompanying printouts. Further, no testimony or documentation establishing how the critical path networks indicate the several claimed delays was introduced.

D. The CPM Consultant

What is the responsibility of a CPM consultant? A CPM consultant may assist in the preparation or updating of a schedule. The *Dobson* Court, in evaluating the

[30]FAACAP No. 67-18, 66-2 BCA ¶6066 (1967).
[31]ASBCA No. 16327 73-2 BCA ¶10,271 (1973).

liabilities of a scheduling consultant, held that the CPM scheduling consultant was no more than a testing laboratory. The CPM consultant there had no power toward any contractor to do anything. The Court found that in respect to time, the CPM consultant's role was to perform a function similar to that of an independent testing laboratory, responsible for material testing. It also found that all the parties involved treated the CPM consultant as an independent expert. The general conditions which defined the role of the scheduling consultant in *Dobson* are reproduced in Appendix A.

The Court's decision in *Dobson* can be criticized. The CPM consultant's responsibility should be considered more in the nature of *professional* services. Scheduling requires an understanding of all phases of construction. It requires the ability to understand mathematics, computers, data processing and drafting. It requires sufficient construction background to recognize the limitations of labor and equipment. It requires sufficient design background to understand how the elements of a project should go together. Although the *Dobson* decision recognized the CPM consultant as an "independent expert," it failed to hold it to any standards but that of a testing laboratory. Scheduling consultants should be held to "expert" standards. A scheduling consultant should be held to the standards of care of similar schedulers.

E. The Dates in the Schedule: a Commitment or Guide?

The law does not uniformly apply scheduling techniques to disputes in the same manner as the construction industry applies scheduling techniques to project management. Some courts see project schedules as being a definite commitment, rather than a flexible, planning tool. Some courts adopt this position despite accepted "industry" definitions of the schedule as being as much subject to change as time changes scheduling assumption to fact.

In *Dobson,* electrical, mechanical, and general prime contractors were to "input" into the construction schedule prepared by the scheduling consultant. The Court held that when the Contractors gave their input to the CPM consultant, they essentially agreed with one another that the work would be performed in that manner, sequence and time. The Court adopted the view that the CPM schedule may not vary with time. This is contrary to the accepted understanding in the construction industry that the schedule is a *flexible* device based on the best known information at the time the schedule is drawn.

In *Natkin,* Western Electric Company negotiated, then assigned a subcontract for mechanical construction with Natkin and Company to George A. Fuller, the general contractor. The Fuller-Natkin contract required the critical path method of scheduling be followed throughout the job. Although a CPM was prepared, it was not continually updated. A computer was not used for calculations. All work order changes were not incorporated into the schedule at schedule updatings. Completion dates for Natkin work were established without Natkin's consent. The CPM was subsequently abandoned when Western Electric demanded early occupancy of certain areas and would not accept the later completion dates which the CPM indicated.

The *Natkin* court found that no accurate or reasonable completion date was established in accordance with contract scheduling methods; and that the failure to schedule and control the work in accordance with the CPM sequence caused Natkin manpower, shortages, congestion, material shortages and loss of efficiency the value of which Natkin was entitled to recover from Fuller *and* Western Electric.

The *Dobson* Court adopts the principle that where Contractors among themselves agree that segments of the work shall be done by certain dates, then the Contractor which fails to perform on time is liable in damages for breach of contract, whether the Contractor has overall supervision or is responsible only for a portion of the work. The *Natkin* court adopts a similar position. It states that failure to schedule and control work "in accordance with the sequence set by the Critical Path Plan"[32] permits recovery of additional costs.

The Court, in *Dobson,* states that Contractors have the ability to enforce this contract right by lawsuit:

> Fortunately, most Contractors honor their commitment or the courts would be overwhelmed.[33]

It is submitted that courts escape being overwhelmed not because most contractors maintain schedules, rather because most Contractors recognize the CPM schedule as a *flexible* identification of sequence, start and finish; a general tool to coordinate a complicated construction project, not an agreement that work must start and finish as indicated. One of the courts to recognize the flexibility of the schedule was the U.S. District Court for the Southern District of Iowa in *Peter Kiewit Sons Co.* v. *Iowa Southern Utilities.*

In *Iowa Southern,* the utility's engineer prepared a prebid construction schedule in the form of a modified bar chart, known as G-1, to assist the multiple prime contractors in completing the work within the period required by Iowa Southern Utilities Company. G-1 set out estimated dates for delivery of major components and the schedule and sequence for constructing the plant. The engineers intended the schedule to coordinate the work under the various prime contracts and to establish the basic working periods, construction sequence, and completion dates for certain phases of the work. G-1 was made a part of the general, mechanical, and electrical contracts but not made a part of the other prime contracts because they had been awarded prior to the issuance of G-1[34]

In addition to G-1, the prime contracts required the contractors to submit "a detailed construction schedule showing the time schedule for completion of each element of the work"[35] after construction contracts were let. The detailed schedule was to be based on G-1 but expanded to include a complete breakdown of all elements of work. The contract clauses which defined the scheduling requirements are reproduced in Appendix B.

Work fell behind the G-1 schedule. In an attempt to bring the project back on schedule, the engineer began making modifications in the construction schedule to

---

[32] *Natkin & Co.* v. *George A. Fuller Co., supra* note 13 at 30.

[33] *Dobson* v. *Rutgers, supra* note 16 at 1146.

[34] *Peter Kiewit Sons Co.* v. *Iowa Southern Utilities, supra* note 18 at 80.

[35] *Peter Kiewit Sons Co.* v. *Iowa Southern Utilities, supra* note 18 at 80.

adjust to existing conditions. At monthly intervals the engineer devised a modified overall schedule using the critical path method. The engineer attempted to assure the critical construction work was done while at the same time permitting all contractors on the site to proceed with their work in a coordinated manner. Because new problems continually arose on the site, the CPM's became obsolete almost as soon as they were issued. The engineer attempted to make modifications as often as new problems arose.

The engineer used as criteria for the monthly schedules the project completion date necessary, its previous experience in supervising construction of power plants, and the economic impact of the scheduling decisions on all contractors. The engineer's scheduling decisions did not continuously favor one contractor over another but rather were attempts to get critical construction work done first. However, the general contractor, Peter Kiewit, encountered interference because the delays forced the engineer to schedule work of more contractors to be done simultaneously than was contemplated by G-1.

Kiewit alleged that Iowa Southern had breached the contract by deviating from the construction dates in G-1. The Court found no breach by Iowa Southern for failure to meet the G-1 schedule. The Court stated that the purpose of G-1, as indicated in the Special Conditions under "Construction Schedule" (Appendix B), was to give a *basic* idea of when the various phases of Kiewit's and other contractor's work should start and how rapidly it should be prosecuted in order for all contractors to make the May 1, 1968, completion date.[36] The presence of a Suspension of Work Clause, Delay Clause, and Extensions of Time Clause in the General Conditions further indicated that G-1 was only a basic idea of when the phases must start and how rapidly they should be prosecuted. The Court concluded that there was no express agreement that G-1 was binding on Iowa Southern Utilities Co., but rather many items in the documents suggested the construction schedule could change as circumstances dictated. The Court refused to find breach of contract by Iowa Southern, stating that Kiewit had no right to rely upon the dates contained in G-1 as being absolute time periods for construction. This view corresponds closely with the construction industry acceptance of CPM as a flexible scheduling tool rather than a firm commitment.

Another court which recognized schedules as only a flexible tool to coordinate a complicated construction project is the Eighth Circuit Court of Appeals in *Southern Fireproofing Co.* v. *R.F. Ball Construction Co.*[37] Southern Fireproofing and Ball executed a subcontract in which Southern, as subcontractor, agreed to perform interior and exterior masonry for two high schools near Cedar Rapids, Iowa, for Ball, the general contractor. Significant portions of the subcontract are reproduced in Appendix C.

Southern Fireproofing sued Ball to recover damages allegedly due to Ball's delay. Among other reasons for recovery, Southern alleged that Ball's progress

---

[36] *Peter Kiewit Sons Co.* v. *Iowa Southern Utilities, supra* note 18 at 91.
[37] 334 F. 2d 122 (8th Cir. 1964).

schedules "required enclosure by December 1, 1956, to relieve Southern of the greater expenses which would be occasioned by exterior" masonry work in the winter.[38] Ball replied that the subcontract contained no warranty as to dates for Southern's work; that the subcontract in fact indicated the possibility of delay; that the progress schedules were not warranties; and that Southern had not seen the schedules until after the subcontract was executed.

Ball had prepared schedules both before and during construction and revised them from time to time in accordance to Article 3 of the General Conditions which stated the schedules "were subject to change from time to time in accordance with the progress of the work."[39]

In finding that Ball was not responsible for Southern's additional costs due to delay, the Court found the progress schedules to be "nothing more than serious estimates and calendars, revised from time to time, for work coordination purposes and perhaps as a disciplinary and incentive measure for all concerned."[40] The Court found nothing in the schedules which indicated guaranty, warranty, or positive representation. Like the court in *Drew Brown* v. *Joseph Rugo,* the *Southern Fireproofing* court reasoned that if completion of exterior masonry were vital, language to that effect should have appeared in the subcontract. The *Southern Fireproofing* court further found that there had been no reliance on the schedule by Southern.

Examination of the subcontract clauses in *Southern Fireproofing* reproduced in Appendix C shows the subcontract itself denied specificity in Southern's work dates. Although the printed form possessed blanks for the statement of any desired definiteness as to the completion of Southern's work, Ball's overall work, and Ball's estimate of the time when Southern's work might begin, these three blanks were filled in, respectively, with "as required," "as required by owners," and "when notified." These are terms of indefiniteness. They fall far short of any date assurance to Southern. They fit instead into a concept of reasonable flexibility and into the practical necessities of the overall progress of the work. They give, as seems proper, the prime contractor some right as to the placement of Southern's work into its appropriate niche in the project's development. Scheduling "within the bounds of reason" is the most Southern could expect.

F. The Benefit of Float

A key issue in project scheduling is the question of whether the contractor or owner receives the benefit of float time. Some construction contracts expressly state who has the control or benefit of float. Some state no extension of time will be granted unless the delay directly affects the critical path, thus requiring the delay to consume any float present, transforming a noncritical activity to a critical one before a time extension will be granted.[41] Most construction contracts, however, make no reference to the issue.

---

[38]*Id.* at 125.
[39]*Id.* at 127.
[40]*Id.* at 126.
[41]J. Wickwire, *supra* note 26 at 238.

Ownership of float affects the method and manner of calculation of delay. If the contractor owns the float, the beginning of a delay should start at the early finish date, maintaining the contractor's control over the difference between the early finish and late finish times. If the owner has the benefit of float, delay calculations should begin at late finish dates. If the contractor owns float, noncritical activities which are delayed should receive time extensions. If the owner has the benefit of float, only delayed critical activities should be granted extensions of time.

Courts and Boards that have considered the ownership of float have not reached similar conclusions. Contributing to the inconsistent decisions is the conflict between two fundamental concepts in construction law. One, that the risk of construction lies with the contractor, tends to support owner claims to float. The other, that the contractor is responsible for the means, methods, and techniques of construction, tends to support contractor claims to float.

The Board in *Blackhawk Heating and Plumbing Co., Inc.* would not permit extensions of time to be granted unless the affected activities were on the critical path. The Board stated "the real question is whether the...delay caused delay in project completion."[42] Similarly, the Board in *Chaney & James Construction Co., Inc.* refused to permit contractor recovery for Government-caused delays not affecting the project completion date. These decisions conform to the general principle accepted by administrative boards that the duration of a time extension is calculated as the amount of time a delay increases the period of contract work as a whole.[43]

Dobson accepted the parties proposition that "late" finish dates are the dates from which delay is measured, rather than early finish dates. *Dobson,* however, made no attempt to determine if delayed late finish dates must be on the critical path.

In contrast, *Natkin* stated that total float may be used by all contractors, but that free float belongs to the particular contractor for scheduling that activity. *Natkin* additionally stated that neither total nor free float is to be used for changes requiring additional performance time.[44]

The Armed Services Board of Contract Appeals in *Heat Exchangers, Inc.*[45] held that a contractor's original "cushion" of extra time, not necessary for performance, maintained in an earlier time extension by the Government should be retained in a subsequent time extension for other design defects. The first extension was calculated on the basis of the time between the contractor's request for clarification and the Government's mailing of the requested information. The Board held the second extension should be calculated in the same manner. Thus the second time extension maintained the contractor's right to a "cushion" of extra time.

The editor in a previous work[46] suggested float belongs to the project rather than exclusively to either party. This is the better view. The editor pointed out that some "use of a portion of the float by the owner will not affect the contractor nor will

---

[42] *Blackhawk Heating & Plumbing Co. Inc., supra* note 20 at 53,676.

[43] *Montgomery-Macri Co. and Western Line Construction Co.,* 1963 BCA ¶3819 (1963).

[44] *Natkin & Co.* v. *George A. Fuller Co., supra* note 13 at 31.

[45] ASBCA No. 8705, 1963 BCA ¶3881 (1963).

[46] H. Hohns, *supra* note 4 at 73.

some use of float by the contractor affect its work for the owner."[47] Determining at what point one's use of float begins to cause the other damage is not easy. When "the line" is crossed, an adjustment is due. But who (and how) should decide this? The Contracting Officer, subject to review by the Board of Contract Appeals, and the Architect, in their positions as defined by standard form contracts, have the power, if not the ability, to make such decisions.

Support for this position can be seen in *Continental Consolidated Corporation* where the Corps of Engineers Board of Contract Appeals in determining the responsibility for concurrent delays permitted the contractor to use float to eliminate the effect of delays it had caused

> Although cable installation under this scheduling would be started 18 days later than planned, appellant could probably have met the original completion date... by using float time...[48]

and at the same time refused to permit contractor reimbursement for Government-caused delays unless the delays affected activities critical to project completion. Float in *Continental Consolidated Corporation* was treated as belonging to the project.

G. Mistakes in the Schedule

The critical path method of scheduling requires the logical analysis of all the individual tasks entering into the complete job. To be logical, a CPM schedule must reflect the practical restraints that apply to the job activities. Failing this, the CPM is ineffective as a scheduling tool. *Joseph E. Bennett Co.*[49] illustrates how this principle has been accepted by courts and administrative review boards.

In *Bennett,* the contractor attempted to show the effect of Government delays by utilizing its CPM. The General Service Board of Contract Appeals refused to permit the CPM to be used for several reasons. First, six mathematical errors were found, one of which generated six additional errors. Second, the schedule ignored foreseeable New England winter weather conditions. The schedule showed compacted fill to be installed in December, January, February and March. The contractor conceded the folly of attempting to place and compact fill on top of frozen ground. Third, changes were made to the original schedule when the contractor began its computerized study of the effect of Government delays. Float time was reduced causing a change in the critical path. Because of these deficiencies, the Board rejected the CPM as a "fully reliable standard for performance under the contract."[50]

The contractor in *William Passalacqua Builders, Inc.*[51] similarly attempted to support its claim before the General Services Board of Contract Appeals by utilizing a "Change Study Analysis" which showed a series of small critical path networks

---

[47] *Id.*
[48] *Continental Consolidated Corporation, supra* note 14 at 30, 716.
[49] GSBCA No. 2362, 72-1 BCA ¶9364 (1972).
[50] *Id.* at 43, 457.
[51] GSBCA No. 4205, 77-1 BCA ¶12,406 (1977).

representing the sequence and effect of the different change orders and change requests issued for the project. The Analysis contained the change order completion dates and the late start date of the next sequential activity on the original CPM. If the change order work was completed subsequent to the claimed late start of the next activity, a delay was shown in the critical path of the original schedule. The Change Analysis also contained change request dates, Passalacqua proposal dates, and Government review times. The study did not, however, apportion responsibility for delays.

The Board found the Change Study Analysis of limited value because it was:

> ...replete with inaccuracies and draws some questionable conclusions. As a result the most complimentary word that we could use to describe it is 'suspect.'[52]

Some examples of the mistakes made in the Change Order Analysis are:

1. The Change Analysis showed a 223-day delay as a result of Change Order 1. The Analysis erroneously shows only 10 days taken for submission of Passalacqua's proposal and 179 days taken by Government for review and approval of the proposal. Actually, Passalacqua took 131 days for proposal submission and Government review took 58 days.

2. Change Analysis incorrectly showed a period of two days taken for Passalacqua submission of a price for Change Request 16 and 41 days taken for Government approval of the price.

3. A clerical error exists in the Change Analysis study of Change Request 118. The 77-day delay indicated should have been 42 days.

4. Change Request 116 involved replacing an installed manhole with a larger one. The Board found the 65-day duration for the change excessive, but Passalacqua could offer no explanation why so long was taken.

5. Change Request 158, 163 and 165 involved adding small sidewalk areas and lowering an isolated manhole that took one or two days. Their direct cost totaled less than $500.00. All changes were issued after Government acceptance of the project. The Board could not understand how those small changes could have delayed the project from 129 to 164 days as indicated in the Change Study Analysis merely because they were issued after the originally scheduled completion date.

6. The Change Study Analysis contained a 39-day clerical error in Change Request 165.

7. Although Change Request 149 was rescinded 15 days after issuance, the Analysis showed a 60-day project delay. Passalacqua could not explain the reason for such a significant impact.

8. Change Study Analysis indicated a 276-day delay for Change Request 10

---

[52] *Id.* at 60,090.

although the work was completed February 21 and the late start date for the same activity on the original schedule was June 13. The Board characterized this as "grossly erroneous."[53]

9. Government review of Change Request 145 was shown as taking 139 days when it actually took 80 days.

10. Change Request 152 which was alleged to have delayed the project 102 days merely provided for the furnishing of tile in boxes for storage on the work site for future use.

11. Change Request 157 which was alleged to have caused 167 days' delay provided for the installation of 16 feet of railing and was performed within 19 days after Passalacqua issued the Change Order to its subcontractor.

12. Change Request 159 which was alleged to have caused 157 days' delay relocated partitions in some areas of the third floor and was performed nine days after the Change Order was issued to subcontractors.

13. Change Request 167 which was alleged to have caused 166 days' delay covered installation of pipe railing on a retaining wall. The work was completed 30 days prior to the issuance of a formal change order.

14. Change Request 169 required reworking some doors, installation of aluminum angles at the bottom of some matrix panels, and raising some carpeting. It cost only $4,098.00 to perform, yet a 228-day delay was claimed.

Passalacqua's method of calculating delay via the Change Study Analysis was rejected.

Errors in the schedule which do not concern logic or durations may not, however, require rejection. *Dobson* points out that the CPM activity descriptions are not controlling. Obligations among the several Contractors are controlled by the contract documents. In *Dobson,* the specifications required the General Contractor to embed plumbing inserts in the concrete. CPM listed these activities as being performed by Plumbing and Mechanical Contractors. The *Dobson* Court permitted the Plumbing and Mechanical Contractor's recovery of the value of the labor performed despite the general contractor's defense that the CPM schedule had listed the activities as their responsibilities.

The General Services Board of Contract Appeals in *Blackhawk Heating and Plumbing Co., Inc.* agreed that activity descriptions are not absolute. The Board looked behind incorrect descriptions to determine the intent of the identifications. In *Blackhawk,* a CPM schedule had been developed after project completion by an outside scheduling consultant to illustrate the effect of project delays. The schedule had defined and inserted certain "delays" into the schedule in an attempt to determine the effect on the critical path; the project completion date. The outside scheduling consultant had *in error* identified these several delays as "dummies," although the

[53] *Id.* at 60,085.

delays were intended to carry time durations and define an activity. Dummies, according to standard definitions, are *not* activities. Dummies carry no description or *time.* Dummies are utilized only to maintain correct construction logic in the network.

The Board in *Blackhawk* recognized the labeling error as failing to conform to "standard CPM diagramming practice"[54] but refused to find any effect on the validity of the scheduling analysis. The Board found the intended meaning of the "dummies" in *Blackhawk* to be clear: an identifiable delay lasting a definable amount of time which affected certain following activities.

Schedules may be attacked for errors in activity durations or activity detail. In *C. H. Leavell & Company*[55] the Government alleged that the contractor's schedule, submitted *and approved* according to contract requirements, unrealistically provided excess durations to eliminate float and contained insufficient detail to accurately schedule some activities. The scheduling clause is reproduced at Appendix D. Despite its earlier approval, the Government maintained the errors in the CPM schedule prevented proper analysis of alleged delay. The Contractor argued that it is the one required to develop the schedule and no specifications were present which defined the time activities were to take.

In *Leavell,* the Contractor had allocated 285 days for "mechanical electrical rough-in." Government contended 120 days a more reasonable duration and alleged lack of detail in the broad category "mechanical-electrical rough-in." What the duration of mechanical-electrical rough-in should reasonably be was important because the activity was delayed. At 120 days, however, the activity maintained sufficient float to be *unaffected;* at 285 days the activity delayed project completion.

The Government's attack in *Leavell* had limited success. The Board found neither the Government nor the Contractor had demonstrated conclusively what the duration should have been. The Board did find, however, that analysis of the schedule was made difficult by the lack of breakdown of the broad mechanical-electrical category. The Board concluded the extent of project delay lay somewhere between the two positions and rendered a "jury verdict" decision.

H. Pre-Bid Schedules

Dobson, Iowa Southern and *Natkin* present an interesting contrast in the treatment of pre-bid schedules. *Dobson* held that the pre-bid schedule present there was not enforceable as a working plan, that it constituted merely an illustration for bid purposes and was not based on the logic for the project of the contractors. *Iowa Southern* found the pre-bid schedule, G-1, was only to give a basic idea of when the contractor's work would start and how rapidly it should be prosecuted in order for all contractors to make the required completion date. The *Natkin* court found, however, that the failure to meet the pre-bid schedule "delayed the entire job."[56]

---

[54] *Blackhawk Heating & Plumbing Co. Inc., supra* 20 at 53,677.
[55] GSBCA No. 2901, 70-2 BCA ¶8437 (1970).
[56] *Natkin & Co.* v. *George A. Fuller Co., supra* note 13 at 31.

Dobson and *Iowa Southern* present the superior view. A pre-bid schedule available with other bid documents should be treated as only one method to coordinate the work. The contractor may choose others as the contractor controls the means, methods and manner of construction. A pre-bid schedule should be considered an identification of time for purposes of wage rates, material prices, or interest rates much as soils information is provided as assistance to preparation of contractor bids. Remember, the Critical Path Method acknowledges there are many, many ways to build a building.

The primary distinction among the three cases is the presence of an independent contractor scheduling requirement in *Dobson* and *Iowa Southern* which is absent in *Natkin.* A detailed contractor schedule shows the manner the contractor will complete the project as compared to the manner somebody else suggests it be done in the pre-bid schedule. No independent scheduling requirement was present in *Natkin* where the contractor adopted and relied on the pre-bid schedule. Thus, the project owner may escape responsibility for errors in the pre-bid schedule by requiring contractor schedules once contracts have been awarded. This conforms generally to the distinction between contractor responsibility for design and performance specifications.

I. Mutual Responsibilities Under the Schedule

Most scheduling clauses in construction contracts attempt to make the contractor solely responsible for the planning and scheduling of work. By approving contractor schedules, however, the owner assumes certain obligations. For example, the owner agrees that it will fulfill its own performance responsibilities in a timely manner according to the approved project schedule. These may include approval of shop drawings, delivery of owner-furnished equipment, or rendering prompt decisions.[57]

Thus, when the Government in *Fullerton Construction Company* failed to review critical shop drawings within the time period permitted by the Contractor's schedule, it was held to have unreasonably delayed the Contractor.[58] This finding was made even though the Government had advised the Contractor it would require a greater period of time for shop drawing review than provided in the schedule[59] and the Government never approved the Contractor's schedule.

Failure to approve may not release the owner from an obligation to perform within the time required by the schedule. In *Fullerton,* lack of approval did not prevent the Contractor from holding the Government to the submitted schedule. In *Dobson,* similarly, owner-furnished equipment was delayed. Although the owner had not approved the construction schedule, it was found liable for the cost of the delays it had caused.[60]

---

[57] J. Wickwire, *supra* note 26 at 237.
[58] J. Wickwire, *supra* note 26 at 237 citing *Fullerton Construction Co.*
[59] J. Wickwire, *supra* note 26 at 237.
[60] *Dobson* v. *Rutgers, supra* note 16.

J. Updating the Schedule

Natkin v. *Fuller* restated the general scheduling proposition that the "critical path plan may become obsolete unless it is kept current."[61] *Continental Consolidated Corporation* recognized that failure to incorporate changes in the work and time extensions will not permit the CPM to reflect the current status of work performed. An inaccurate schedule cannot be found to show the effects of delay on project completion or represent the actual manner in which the project was constructed.

The *Lane-Verdugo* Board found the critical path schedule there submitted not accurate because of a failure to update the entire schedule. When updated no change was made in the initial estimates of time required to accomplish the work. Further, logic changes made to reflect field conditions were made without changing the initial estimates of activity durations. The Board felt these inaccuracies prevented the contractor from relying on the schedule as an accurate measurement of construction progress.

In contrast, the General Services Board of Contract Appeals in *Dawson Construction Co. Inc.* permitted use of a CPM schedule to analyze delay that had been updated monthly but did not reflect all actual start and finish dates. Only about 25 percent of the activities, primarily the critical ones, were updated with actual start and finish dates. Progress on the remaining activities was not identified beyond the month in which a given activity was started or finished.

The CPM schedule identifies many paths through the schedule. Not all are critical. Those paths that are not critical because they contain float may become critical if the float is eliminated. Likewise, those paths that originally are critical may develop float if a larger, delayed path develops. As the schedule is updated, the critical paths are thus likely to change. Legal commentators have recognized this.[62] At least one Board, the General Services Board of Contract Appeals in *Blackhawk Heating & Plumbing Co. Inc.*[63] has also recognized the proposition.

In *Blackhawk,* the Board considered the contractor's argument that certain ductwork delays affected the critical path of the *original* schedule and should be entitled to extensions of time. The Board refused to grant extensions based on the *original* schedule because subsequent occurrences *changed* the critical path. The Board had before it a schedule which showed how the project was actually built and indicated ductwork delays were not critical. The Board concluded that time extensions should be granted on the best available evidence. The best available evidence is, of course, an updated schedule.

---

[61] *Natkin & Co.* v. *George A. Fuller Co., supra* note 13 at 30.

[62] Wickwire and Smith, "The Use of Critical Path Method Techniques in Contract Claims," 7 Public Contract L.J. 1, 43-45 (1974).

[63] GSBCA No. 2432, 76-1 BCA ¶11,649 (1976).

CHAPTER 5

The Law Behind Today's Construction Contracts

H. JAMES WULFSBERG, J.D.
with
SANDRA J. McNABB, J.D.

THE DESIGN PHASE

The design phase is the critical period in which to prepare a framework which will minimize future project controversies and disputes. Each party has its own interests to advance and protect: for the owner, the goal is a well-designed and constructed project, suited to its needs and within its budget; the designer aspires to create not only a working project, but also an innovative and/or aesthetically pleasing one; and the contractor, who most frequently is unrepresented in the design phase, hopes for a well-built structure at the lowest possible cost. These interests are not necessarily compatible. Awareness and compromise are required in all stages of the planning and construction process to unite these competing interests, thereby minimizing construction disputes. Some problems are inevitable, but careful planning and attention to detail in design and contract drafting should minimize controversy.

The Owner's Role

The first critical factor in the design phase is the selection of the design professional. An owner desires a completed structure which will fulfill certain performance requirements and will also come within applicable budget limitations. An owner's objective may not be met by selection of in-house designers as a cost-saving measure. An outside design firm will often provide the benefit of its experience in working for numerous other owners on similar projects, while at the same time assuming the risks inherent in design work. One should not overlook the option of engaging additional independent consultants to assist in reviewing the past work of a prospective designer or a design proposal for a current project. Prudence, however, should be used in retaining such a review consultant over the life of a project. Continued design review will inevitably fragment responsibility for the project design and at a later stage may result in problems in assigning liability for design deficiencies. The objective in engaging such a review consultant in the early stages of design is to ascertain and eliminate such deficiencies and consequent disputes.

In engaging an independent design professional, an owner not only attempts to obtain a specialist to produce the best design available, but he also attempts to shift the responsibility for design to the designer. Design professionals are generally held to the standard of care prevailing in the community of such professionals. *City of Eveleth* v. *Ruble,* 225 N.W.2d 521 (Minn. 1974), *Aetna Insurance Co.* v. *Hellmuth, Obata & Kassabaum, Inc.,* 392 F.2d 472 (8th Cir. 1968). *Huber, Hunt & Nichols, Inc.* v. *Moore,* 67 C.A. 3d 278 (1977). The standard may be varied, however, by contractual provisions binding the designer to a higher standard of care or by specific rule within a particular jurisdiction.[1]

Few, if any, designers will knowingly agree to express or implied warranties. Some designers might agree to increased levels of responsibility but only with a substantial increase in fees because professional liability insurance generally does not cover such warranties.[2] One should view with caution the designer who fails to concern himself with the standard of care for design imposed on him by the contract since this may indicate an inexperienced designer.

An owner should clearly convey all of the aesthetic and performance standards which are desired for the project and should then avoid participation in the design work. This will ensure that the design responsibility remains with the design professional. Similarly, any redesign which may be necessary should be made the designer's responsibility. This accomplishes several ends: it places the burden for design work on the party with the necessary expertise; it allows for continuity in the design; and it insures that the responsibility for the entire design remains with the designer.

A second critical factor in the design phase is that of reconciling the discrepancies between the project budget and the designer's estimate. This may be painful, but it is, nevertheless, a necessary process. Most owners do not have unlimited access to funding, whether from bond issues or capital. Therefore, it is crucial to insure that the designer's estimate, together with a margin for extras, will be sufficient for the project envisioned. An owner should inform the designer in writing of budget constraints. If later cost increases overrun the project budget, an owner may be able

---

[1]Such a high standard generally may not be premised upon an implied warranty, however. *City of Mounds View* v. *Walijarvi,* 263 N.W.2d 420 (Minn. 1978)

[2]Most architect/engineer professional liability coverage does not include contractually assumed liabilities.

to fall back on this communication with the designer. Depending upon facts and circumstances surrounding the agreement with the designer, the custom and practice in the industry in the project area, and the forum in which the matter may be litigated if necessary, an owner may be able to withhold some or all of the designer's fees or may be able to hold the designer liable for all increased costs which were reasonably foreseeable at the time the contract was entered into. *Hedla* v. *McCool,* 476 F.2d 1223 (9th Cir. 1973).

It should be noted, however, that cost estimating is not a science and an owner should evaluate the designer's estimate in light of economic conditions and the owner's projected needs. Many uncontrollable factors may cause costs to exceed the estimate, such as increased costs for materials, services and financing. Factors within the owner's control are changes in the project itself. A designer bases his construction cost estimate upon the parameters set forth by the owner for the project. Therefore, by not expanding the project beyond these parameters, the owner can help to keep the project within the budget.

An expanding source of project disputes associated with design has been the increasing use of bid protests for federally funded projects. For example, Environmental Protection Agency regulations authorize bid protests by interested parties where plans and specifications are claimed to be restrictive. Current EPA regulations may prevent a project award until bid protests are resolved. Thus, unless careful attention is paid to requirements for nonrestrictive specifications, an owner may be forced to have the project design revised, even though it is not technically deficient, in order to satisfy the applicable funding agency. The entire area of compliance with various standards of state or federal funding agencies is often assumed or, more frequently, neglected entirely, in negotiations for design work. These standards should be ascertained and discussed in order to prevent "design" problems, which ordinarily come to light only after a project has been put out to bid and "interested parties" first ascertain the "restrictive" character of the plans and specifications.

The Designer's Role

A designer has no assurance that he will be called upon to execute any more design work than in the contract at hand. One designer may be required to prepare construction plans, specifications, and cost estimates based in part upon the preliminary design work and preliminary estimating of another designer. The tendency to criticize prior design work to lessen or avoid "buying" another designer's mistakes in such circumstances is almost irresistible. The designer should take care in that situation to minimize owner suspicion that the designer's goal is first to avoid responsibility, and second to prepare appropriate plans and specifications.

Naturally, a designer should carefully review the prior work of another before commencing design work based upon conclusions of that professional. In providing an estimate of design costs to an owner, a designer should take into account the cost of carefully evaluating prior design work, recognizing that the owner's decision to change designers at this stage may be related to dissatisfaction with the efforts of the former designer.

As discussed earlier, a designer should be extremely reluctant to agree to contractual provisions which set a higher standard of care than that imposed by common law. This should not, of course, be taken to mean that one would utilize anything less than the highest degree of skill in performing design work. Proving that errors or omissions are not negligent, but are within the acceptable standard of care, is much more difficult than skillfully avoiding mistakes in preparing plans and specifications. However, guarantees of performance should not be written into design contracts nor should they be made orally at any time in negotiations with the owner unless the exact nature and extent of such guarantees are carefully defined and costed accordingly. As the drafter of the plans and specifications, a designer is the one to whom the courts will generally shift responsibility if any ambiguity or mistake is present in those documents. *Blount Brothers Construction Company* v. *United States,* 171 Ct. Cl. 478, 346 F.2d 962 (1965), *Coleman Engineering Co.* v. *North American Aviation, Inc.,* 65 Cal. 2d 396 (1966). Where a designer does guarantee his work, as for example, where he guarantees that a process plant will produce certain results if constructed in accordances with plans and specifications, he should clearly define the nature and extent of that warranty.

The design work itself is the area of the designer's expertise, something in which the designer has received years of training and experience and for which he can rely upon certain established guidelines and principles. However, the designer should clearly define with the owner and note in the specifications if any part of the design is innovative or unusual, thereby alerting both owner and contractor that additional unexpected risks may be encountered in such design applications.[3] A designer does himself and his client a great disservice by failing to recognize his lack of expertise in a particular area. If such is the case, he should either seek the assistance of a qualified designer with the necessary skill or he should simply acknowledge his lack of expertise and pass up the design work. By accepting the work for a design job, a designer in effect represents that he has sufficient skill and training to carry out a suitable design for the project. *City of Eveleth* v. *Ruble,* 225 N.W.2d 521 (Minn. 1974); *United States* v. *Spearin,* 248 U.S. 132 (1918).

Cost Estimating — the Designer's Responsibility

The designer often does not have adequate training in the area of cost estimating. It is crucial for both designer and owner that the designer's errors and omissions coverage includes cost estimating services. Where appropriate, a contractual clause which provides that estimates are not guaranteed should be considered. However, there is usually no liability, even without such a clause, as long as an estimate is not negligently performed, and as long as a designer does not: expressly warrant; either orally or in writing, the accuracy of the estimate; condition payment of fees upon the accuracy of the estimate; or act as the contractor on the project or otherwise have

[3]Many professional liability insurance policies exclude from coverage experimental or innovative designs that are not tried and true designs.

control of the costs. These guidelines are not absolute since courts have considered all of the facts and circumstances in holding designers to their cost estimates or limitations. The owner's control over details and the designer's knowledge of the owner's budget and specific needs are factors which courts have weighed in assessing responsibility. At least one court has found that knowledge of particular budget constraints may impose liability upon the designer for overruns. The liability to which a designer may expose himself may be that of all damages reasonably foreseeable at the time the contract was entered into. *Hedla* v. *McCool,* 476 F.2d 1223 (9th Cir. 1973).

Designer's Interface Among Plans, Specifications, Results

In designing a project, a designer must select performance specifications, or plans and specifications which detail the components necessary to achieve the desired result. Where performance specifications are used, it is the contractor who bears the risk of selecting materials which will meet the stated objectives. *Gurney Industries, Inc.* v. *St. Paul Fire & Marine Ins. Co.,* 467 F.2d 588 (4th Cir. 1972). Of course, a designer must prepare sufficiently clear and precise performance requirements to test construction effectiveness. Additionally, the designer who then reviews the contractor's submittals must assure the owner that the submittals, if properly constructed, will accurately reflect the contract intent. While risk-shifting by the use of performance specifications may be desirable in itself, it decreases designer control over the effectuation of the design. Specifying equipment, materials, and details of construction such as concrete, piping and the like will aid in insuring the integrity of design. However, the designer must have confidence in the specified suppliers, either on the basis of past experiences or with well-documented records, research and warranties.

Where products are being used in a new application, or where new materials are specified, the designer has a duty to request and carefully review all literature available on the materials and to make such inquiries as may be necessary under the circumstances. The designer should question the manufacturer as to the suitability of its products for the specified use. He should not assume that a new product can be installed in the same fashion as a familiar and tested one. A representative of the manufacturer and/or distributor should be on site to instruct the contractor on handling a new material. At the very least, the designer should require that the manufacturer provide the contractor with the written instructions for the product's use. Finally, the designer should read and reread the warranties accompanying any materials to be sure the coverage provided will meet the application requirements.

The Contractor's Role

Contractor input into the design is often limited and frequently nonexistent. However, contractor inquiry and understanding before bid and award may be crucial to making a profit on the project. *The importance of a careful review of the plans, specifications, and contract cannot be overemphasized.* A contractor should not assume that an error in either plans or specifications can be easily remedied after the project is successfully bid. Correction may require delay of work for which an extension will not be granted if the designer can demonstrate that the contractor knew, or should reasonably have known, of the error before bidding. *Acoustics, Inc.* v. *Trepte Constr. Co.,* 14 Cal. App. 3d 887 (1971). Frequently, the plans and specifications contain contract language highly protective of owner and designer. The contractor may be required to represent that he has read *all* of the plans and specifications and that the contractor *can* and *will* complete the work within the contract period fully in accordance with the plans and specifications. Therefore, the contractor should call any known errors or omissions to the designer's attention immediately.[4]

A competent contractor will also be satisfied that he understands the plans and specifications and that, based upon such understanding, he believes that the design can feasibly be carried out. As a general rule, a contractor is not liable for the failure of the project if he reasonably and competently followed the plans and specifications. *Coleman Engineering Co.* v. *North American Aviation, Inc.,* 65 Cal.2d 396 (1966). However, one view is that a contractor has the responsibility to seek an interpretation of any possible ambiguity in plans or specifications, and many contracts so provide.

Review of any site reports made available by the owner, as well as an actual visit to the site, should be performed by the contractor before bidding on the project *unless* the contract contains a clause shifting risks for site and surface conditions to the owner. If there is such a clause and the contractor has relied upon owner information which is in error, he will be entitled to recover for the excess costs incurred. However, this protection may be lost where the contractor relies, not upon the owner's information, but upon his own sources for information prior to bid. Where the owner does assume the risks of unforeseen conditions, there may be additional language in the contract with respect to contractor obligations to timely notify the owner if unexpected conditions are encountered. As with all other contractual provisions, these should be carefully reviewed and complied with during construction, so that recovery may be obtained for any additional work or delays.

The contract sets forth the contractor's rights and responsibilities on the job, and as such should be reviewed both by the contractor and his attorney to be certain that all of the terms are clear as well as to insure that the bid will include a contingency factor for all elements of risks assumed. The contractor's expertise is in construction, not design, so the contractor should not assume responsibility for any design or redesign. If performance specifications are set out, the contractor must be certain that he can supply and install equipment to meet those standards. The con-

[4]Section 4.2.1 of AIA document A201 imposes this obligation upon the contractor as follows: "The contractor shall carefully study and compare the Contract Documents and shall at once report to the architect any error, inconsistency or omission he may discover."

tractor should insist upon the same information from, and protection by, a manufacturer as was discussed earlier with respect to designers.

Familiarization with the notice requirements of the contract, particularly with regard to change orders, is vital. A court or an arbitrator will look first to the contract in attempting to resolve any disputes between the parties. Careful attention should be paid to any contract scheduling requirements imposed by the owner upon the contractor. Provisions for the sequence of construction, as well as submittal of schedules, "C.P.M." (Critical Path Method) diagrams and computer reports, and updates during construction may be set forth in the contract documents. If properly utilized, a construction schedule which accurately depicts sequences and time of construction in complete detail can be a powerful tool for timely project completion and for documenting unavoidable delay. Conversely, a poorly prepared schedule or one ignored by the contractor will document the contractor's deficiencies in performance.

In bidding complex projects, particularly process plants, the contractor's schedule may well spell the difference between profit and loss on the job. Scheduling the construction project with the process in mind not only demonstrates the contractor's understanding of the work — and may be a significant aid in preparing the final bid — but also demonstrates the contractor's compliance with bidding and contract requirements.

MINIMIZING CONSTRUCTION PROBLEMS

The Owner's Role

As mentioned with regard to the design phase, it is desirable to plan the project so as to minimize changes during construction. However, it is inevitable that changes will become necessary or desirable. Often both the designer and the contractor will request additional compensation for such changes, usually based upon claims of changes in scope or extra work. Therefore, the owner should protect himself by spelling out the roles of the designer and contractor with regard to changes.

The owner should insist that the contractor's work be carefully scheduled to prevent any unnecessary delays and to be certain that any necessary changes will become apparent at the earliest possible moment. If the owner assumes the responsibility for scheduling he may also be faced with responsibility for problems which arise as the result of the schedule. Where the owner-issued schedule is used by the contractor, the owner will normally be liable to the contractor for his damages as the result of errors or revisions in the schedule. Consequently, the owner should generally require the contractor to issue the schedule and adhere to it. In that way, the contractor has the responsibility for preparing *and revising* the schedule as necessary.

Similarly, the owner should not assume responsibility for the progress of the work itself by supervision or undefined inspection. The design professional or construction manager should observe construction and report to the owner, while the contractor should be responsible for the scheduling and progress of the work.[5] Again, this responsibility is in accordance with the expertise of the parties.

It may seem that the owner shifts all authority and therefore responsibility to the contractor and the designer and then sits back to passively observe the project taking shape. This is not the case, nor should the owner be lulled into such a false sense of security. It is the owner to whom the others turn not only for payment, but for authority to make any necessary or desired changes. By signing the contract, the owner accepts the designer's rendition of the project and the contractor's proposed method and cost of construction of the project. Therefore, when changes are to be made, it should be the owner who authorizes them. Paragraph 12.5.1 of AIA document A201 authorizes the designer to approve changes which are not intended to affect the contract price or completion date. This may be desirable to prevent project delays occasioned by processing a change order for minor changes through the owner. However, a change so approved arguably has been "constructively approved" by the owner despite his lack of knowledge of the change at the time. It may then give rise to a later "constructive change order" requesting additional compensation for this work.

The change order mechanism should be clearly spelled out in the contract, and it should require that changes be authorized in writing *before* the work proceeds. Variance from the specified procedure is one of the most fertile areas for dispute in construction work. It is frequently the case that the designer is aware, in his supervisory role, that the project will be delayed if work is held in abeyance until written authorization is received. Instead of following the mechanism, the designer simply tells the contractor to proceed while the order is being processed. If the "change" is thereafter approved and completed as planned for the amount of extra compensation requested and without delay, no problem arises.

What happens, however, where something goes wrong? If the contractor has received payment for past work ordered by the designer at variance from the change order mechanism, a court is likely to find that the contractor is justified in relying upon the designer's authority to act in this manner. Since the owner does not want to be required to pay additional compensation for work which he believed was included in the contract or was unnecessary, it should be made clear to the designer that no variance will be accommodated. Immediate objection by the owner to "change orders" performed before approval is necessary to avoid this problem.

The Designer's Role

The owner's relationship with the designer can be difficult because a design professional may serve in three very different roles: (1) as an independent contractor in preparing the design; (2) as an agent of the owner in performing inspection services: and (3) as an independent arbitrator of disputes. In preparing plans and specifications, the designer is an independent contractor, neither acting on behalf of, nor

[5]An owner should analyze the desirability of utilizing the project designer (with a design to protect) to supervise the technical inspection of the work during construction.

exerting authority over, the owner. When observing the construction work as it progresses and reporting to the owner, a designer acts as the owner's agent, with the authority expressly or implicitly given by the owner. It is in this role that the owner acts through the designer and hence should be certain that the designer is abiding by the standards and mechanisms specified in the contract. Finally, the designer may act as a quasi-judicial officer in accordance with the contractor-owner contract in interpreting the contractual rights and duties of those parties. In this role, the designer has control over the owner to the extent of the contractual provisions.

As an independent contractor, the designer is liable for negligence in performing his work. As an agent for the owner or as a quasi-judicial officer, however, the designer may be liable only if he acts maliciously or in bad faith. *Huber, Hunt & Nichols, Inc.* v. *Moore,* 67 Cal. App. 3d 278 (1977).

Although the lines between these three roles are easily and frequently blurred, the owner should bear in mind that the designer has different degrees of authority with respect to the work. He should also remember that the designer is being compensated, not as an employee, but as a professional charged with responsibility for all three roles. The owner may determine that the burden upon the same professional to act as project designer, inspector, and contract arbitrator is too great. In such event, the owner may be better served by hiring one designer for the project design and additional designers or other professionals to inspect and decide contract questions during construction.

The contractor who shrewdly perceives the potential conflict of designer-inspected and arbitrated projects may seek to capitalize on this potential weakness. An effective technique is to accuse the designer of ordering extra work simply to "cover up" design errors. While he is the exception, rather than the general rule, an unscrupulous contractor using such tactics generally insures disputes at project completion and makes the designer's job difficult, if not impossible, in the interim. Frequently, this technique also drives a wedge between owner and designer, depriving each of the assistance of the other during the post construction claims settlement.

The Contractor's Role

The contractor is charged with translating the project from paper to the structure envisioned by both owner and designer on or before the completion date, with as little additional expense or change as possible. Such a Herculean task requires the contractor to be thorough and farsighted during the bidding process. The most important step is a thorough inspection of the plans, specifications, and contract. Errors or ambiguities should be discussed at this stage to avoid later expense and possible delays. The contractor should understand the contractual assignment of rights and duties as well as the requirements for submitting shop drawings, requests for interpretations, change orders and the like.

In light of recent decisions assigning liability to owners for schedules they issue, the construction contract will generally require the contractor to schedule the work. The initial schedule must be submitted by a certain date and approved by the owner. This requirement should not be lightly regarded, since liability for delay may later be assigned to the contractor and/or change order work may be denied upon the basis of the construction methods described in the schedule. It is difficult to schedule work on the contract without a clear understanding of the project work requirements. The contractor should use his best efforts to realistically schedule all work from commencement through completion. Front-end loading is often viewed as an attractive scheduling concept to obtain cash early in the project life for the financing of that project or other projects. This attraction may be considerably diminished if a subcontractor is also front-end loaded and is terminated during the early stages of the contract. When the subcontractor sues to recover the value of the services performed as reflected by the contractor's schedule, the contractor must either acknowledge the subcontractor's right to such high payments and absorb the costs of the replacement subcontractor or admit the inaccuracy of the schedule. Front-end loading can be a seductive but dangerous path for the contractor who fails to see construction slowdowns as well.

The contractor should keep the schedule current, making revisions when necessary. If the schedule is not kept up-to-date, the contractor may be liable to the subcontractors who have relied upon the schedule for processing their work. Furthermore, where the contractor fails to request extensions for delays in the work as originally scheduled, but instead simply maintains the original schedule, a court or arbitrator will generally be unsympathetic to the contractor's untimely plea for an extension. If, on the other hand, the schedule is systematically and realistically revised, it should document the contractor's reasonable processing of the work and provide a basis upon which additional time or compensation can be granted.

In dealing with changes, the contractor should, of course, follow the change order mechanism established by the contract as closely as possible. Where the designer's authority to order changes is limited by the contract, the contractor must beware of a situation in which the designer purports to approve a change beyond that contractual scope. Although in subsequent litigation or arbitration the trier of fact may find that the contractor's reliance on the varied procedure was justified, or even that the contractor may recover the reasonable value of the services, there is always the risk that recovery will be denied for failure to follow the contractually defined procedure. *Crane Const. Co.* v. *Commonwealth,* 195 N.E. 110 (Mass. 1935).

Courts are also becoming increasingly aware of the rights of subcontractors in the construction process, and contractors should formalize their relationship with subcontractors accordingly. To the extent possible, this formalization should extend to any claimed extra work. There is always tension in the contractor-subcontractor relationship. While the prime contractor "controls" the relationship on any particular project, the prime contractor does not want to alienate the subcontractor because he may need the subcontractor's services on future projects. This is particularly evident in respect to change orders. When the subcontractor requests a change order by submitting a claim to the prime contractor, the prime contractor may be tempted to routinely approve the subcontractor's claim. This is not only due to the subcontractor-prime contractor relationship, but because of the markup for overhead and profit the prime contractor may make. However, the prime contractor should always review a change order request, and submit only justified claims to the owner or designer in accordance with the contractual change order mechanism.

The prime contractor should reject any unwarranted claim in a well-documented letter to the subcontractor, with a copy to the designer and/or owner. Otherwise, the prime contractor may need to defend himself against the subcontractor's claim for extra work which the contractor approved, but the owner did not. To a certain extent, this problem may be circumvented by either one of the two decisions: a contractual clause between subcontractor and prime-contractor providing that the subcontractor proceeds at its peril absent the owner's or designer's approval; or, a written acknowledgment of such for each claim submitted, and the contract should generally provide that the general contractor may submit a subcontractor's claim for change order revision, including the prime contractor's markup, without acknowledging the merits of the subcontractor's claim.

The Subcontractor's Role

As mentioned earlier, the judicial system has recognized that subcontractors may frequently be at the mercy of the prime contractor. Therefore, a subcontractor may be allowed to recover directly from the owner, even though there is no contract between the two. *Varco-Pruden, Inc.* v. *Hampshire Constr. Co.* 50 Cal.App.3d 654 (1975). In its *Varco-Pruden* opinion, the court concluded that where the prime contractor/owner contract is incorporated by reference in the subcontractor's contract, the subcontractor may be treated as if he had a contract directly with the owner. The subcontractor may also have recourse to the owner as a third-party beneficiary to the contract of the owner and the prime contractor. Finally, if the owner has dealt directly with the subcontractor, the subcontractor may have a direct cause of action against the owner. Such direct interaction may occur where the owner approves change order requests made by the subcontractor.

The prime contractor generally insists upon incorporating part or all of his contract with the owner by reference in the contract with the subcontractor. The prime contractor seeks to put the subcontractor on notice of all the contractual requirements for the prime contractor in order to assure the subcontractor's performance of the prime contractor's contract obligations. As discussed earlier, this incorporation by reference may be beneficial to the subcontractor. However, the subcontractor also wants a contract which clearly expresses the rights and duties between the subcontractor and the prime contractor. Unless the provisions of both the subcontractor contract and the prime contract incorporated by reference are carefully reviewed and reconciled, discrepancies are bound to arise. Where this occurs, the question of which document is controlling is not easily resolved. It is an established principle of contractual interpretation that the specific prevails over the general. Therefore, if specific terms of the prime contract (or the specifications) are clearly incorporated by reference in the subcontract, those items will generally prevail. However, if the subcontract simply refers to the prime contract, but the subcontract specifically spells out duties which are only generally covered by the prime contract. the subcontract will normally govern.

Generally. the law implies a duty of cooperation between prime contractor and subcontractor to insure that the subcontractor can perform without unreasonable hindrances or delays. However. the prime contractor may desire protection from certain risks. such as payments for extra work of the subcontractor. so a clause may be inserted in the subcontract. requiring the subcontractor to assume such risks. For example. where a contractor provided by contract that it had no obligation to insure that materials were available when scheduled. a lower-tier subcontractor was denied recovery when in fact. the materials were not available as scheduled. *International Erectors* v. *Wilhoit Steel Erectors & R. Serv.,* 400 F.2d 465 (5th Cir. 1968). Hence, it behooves a subcontractor to carefully read both the prime contract and the subcontract, rather than simply assuming that he will be protected.

The Supplier's Role

Often a supplier or manufacturer will bid an equipment item without regard to all of the contract documents. including specifications and drawings. All too frequently. the supplier or manufacturer may submit a bid "in strict accordance with plans and specifications" without ever reading them. The only sure way for a supplier to protect himself when he feels that the amount of his bid does not justify careful scrutinizing of all the contract documents is to limit the bid and purchase order by clearly defined parameters. Unless careful attention is given to the form of purchase order used by the party to whom the bid is given. the manufacturer/supplier may be unexpectedly bound by the terms of delivery. installation and special conditions not included in the original bid.

Reconciling The Competing Interests

While the competing interests of owner, designer and contractor are frequently intensified by competitive bidding upon fixed plans and specifications. with care and consideration. these interests can be reconciled. With the increasing tendency to substitute time-consuming and expensive claims or litigation for compromise and accommodation. such reconciliation becomes exceedingly important.

Many owners, frustrated by the conflicts between designer and contractor, engage a multi-purpose entity to design and construct projects to broad-scope project criteria. Designers constantly faced with the two-pronged attack of owner and contractor complaints seek to further insulate themselves from liability, thereby fueling the fires of doubt of designer integrity. Contractors confronted with spiraling construction costs and increasingly complex projects seek out any available means to insure project profits.

The failure of owner, designer or contractor to recognize and carefully plan for the competing interests of each will insure conflict, dispute and controversy during the construction of the project. Therefore, every effort should be made, at each stage of the construction process, to reconcile those interests.

CHAPTER 6

The Proper Legal Approach to Federal Contract Disputes

JAMES V. DOLAN, LL.B.
with
BARRY B. BRAMBLE, J.D.

INTRODUCTION

The preceding discussion (in Chapter 3) dealing with the contract and the rights and liabilities of the parties thereunder provides the general legal principles governing the contractor's presentation and the Government's evaluation of a construction contract claim. This chapter will deal with the practical aspects of claim presentation. A reading of all of the decisions of the United States Court of Claims, the various Boards of Contract Appeals and the decisions of the myriad of state and federal courts dealing with construction contract disputes will tell us little about how contractors should deal with the Government and its agencies or, for that matter, how the Government and its agencies should deal with the contractor. There are no legal precedents for this. We've now moved from the world of the law and the cold language of the contract to the living world of the practical, sound business judgment, the old art of persuasion and, last but, perhaps most important, common sense, common courtesy and mutual respect.

Reflecting this difference in subject matter, this section of the discussion will be very light on case citations and will draw on the practical experience of the writer and other attorneys. It will describe what has proven successful and what has not. What follows will also reflect more of how things appear from the contractor's side than from the Government's. However, the suggestions and comments that follow will hopefully be constructive to Government representatives as well as to contractors.

Preliminary Points

1. By the time you reach the dispute stage, you, as the contractor, already have a relationship with the Government, with the contracting officer, the resident engineer, and inspectors whose input will have a great deal to do with how your claim is processed. That relationship started with your first day on the job and grows, for better or worse, with every day of performance. Any toes that were unnecessarily stepped on are very likely looking for a chance to get in a kick or two. Any egos that were unnecessarily bruised are going to be looking for an opportunity to show the contractor-claimant where the power lies — that's trouble.

On the other hand, if you dealt with the Government, and particularly its individual representatives, on your job with respect — but firmly — they'll give your claim the same respect.

A Government resident engineer once said, "There are two types of contractors — both have something in common — problems. One, however, cries about them; the other tries to overcome them." It's important that your contracting officer pictures you in the second category. If he considers you a problem-solver, then, when you come in with a claim, he'll give it the respect it deserves. He knows that you don't file a claim when every little problem pops up — that when you file a claim, it means that you've been hurt, that you firmly believe that it was not your fault or responsibility — more important, that you're ready to prove your case. That's the foot you want to be on at the start of the dispute process.

2. *Second Basic Point:* Just as Government's general attitude toward a claim or other dispute starts being formed long before that claim or dispute has crystallized, so also has the claim process itself started before you finally submit it to the Government. *It starts with notice.* The importance of notice can't be overstressed. Many a good claim has died at birth because of lack of notice.

3. Occasionally, you hear a contractor say "I don't file claims — it hurts my relationship with the Government." That's not true. Government contracting officers are in the claim business as much as they're in the construction business. They've handled and granted hundreds of claims before yours. The claim process is an integral part of the procurement process. That's why the changes clause is in federal contracts. It, like the Changed Conditions (Differing Site Conditions) and, Suspension of Work Clauses, is there to be used. The best statement of how claims fit into the procurement scheme was made by the Court of Claims in *Ozark Dam Constructors* v. *United States,* 130 Ct. Cl. 354, 127 F. Supp. 187 (1955). The Court noted there that, with the Changes, Changed Conditions and Disputes clause, every contractor bidding on a contract would have to put in his price a substantial contingency factor to cover unknowns. That would cost the Government more in the long run than it pays out in claims. The claims procedure thus benefits the Government in the long term. Set up your claim with this fact in mind — because, if you can get your contracting officer to look at your claim from that standpoint, the processing of your claim will go much more easily.

4. On any job, contracts between you and the Government are spread over many people. On the contractor's side, the home office administrators of the contract, the project manager, the superintendent, the engineer — on the Government side, the contracting officer or his representative, the resident engineer, the inspectors. Any contact can later assume importance to your claim. An example — the case where the Government's bid package had failed to disclose superior knowledge of facts which were contained in the Government's files and which indicated that the contract could not be performed in the manner indicated. After the award, but before the start of substantial performance, one of the contractor's engineers was talking with one of the Government's engineers who mentioned casually some of the problems which the Government had experienced on work of this type in the past. The contractor's engineer made a note of the conversation but, because the conversation was at a relatively low level in the overall hierarchy and was just a passing, off-the-cuff thing, the home office paid no particular attention to it.

Later, the same problems were experienced by the contractor. When he learned that the Government had experienced similar problems before and had not disclosed that knowledge in its bid package, he made a claim under the Superior Knowledge doctrine of the *Helene Curtis* case. The claim was found valid but for one fact — the Court held that the conversation between the two engineers put the contractor on notice at that point of the Government's failure to disclose the Superior Knowledge and the contractor was, at that point, obligated to put the government on notice of its claim — so that the Government could decide whether to proceed with performance, subject to having to pay a substantial claim, or to terminate the contract for convenience and start all over again. The Court found that, by failing to give that notice, the contractor denied the Government that option and its claim was accordingly barred.

The lesson — (a) Control your informal contacts with Government representatives, (b) make a record of them, (c) analyze every one of them from the standpoint of what it might mean to your long-term rights and position in a later dispute.

PROCESSING A DISPUTE

Claim Procedure

The basic claim procedure is relatively simple. Within the context of the federal contract, the claim must be in writing and is in the first instance submitted to the contracting officer. The contracting officer will promptly refer the claim for technical evaluation to his field representatives and contract administrators who are familiar with the contract and the facts upon which the claim is based. Either at the same time, or following receipt of the technical evaluation, he will probably send a copy of claim to his legal advisors for evaluation of its legal merits.

If those evaluations indicate that the claim has merit, the contracting officer will usually authorize commencement of negotiations with the contractor. Rarely will a difficult claim be allowed without negotiations. Some negotiation is unavoidable unless the claim is of the "black or white" type on its merits and the costs are unquestionable and fully documented.

In substantial claims, there will also be an audit. The audit, depending upon the agency, can either be conducted during the evaluation stage or be deferred until after the evaluation has found merit in the claim.

If the Government's claim evaluation has been favorable and the audit has substantiated the claimed costs, the negotiation process will ultimately lead to a claim settlement which will then be reflected in a contract modification and payment of the amount of the claim.

On claims less than $50,000, the contractor may take advantage of an expedited procedure providing for the contracting officer's decision within 60 days of the contractor's written request for a decision. The larger claims are controlled by a different procedure. The contractor must make certifications that (1) the claims are made in good faith, (2) the supportive data are accurate and complete to the contractor's best knowledge and belief, and (3) the amount of the claim accurately reflects what the contractor believes to be the Government's liability. There are less time constraints upon the contracting officer's consideration of claims greater than $50,000. However, within 60 days of receipt of such a claim, the contracting officer must either issue the decision or notify the contractor of the reasonable time period in which the decision will be made. Should the contracting officer unreasonably delay issuing a decision, the contractor may request the Board to direct the contracting officer to render a decision.

The denial of the contractor's claim may come about in two ways. First, the contracting officer may decide against the claim, based upon his technical and legal evaluations, in a written decision furnished to the contractor. The written decision must state the reasons for the decision, although specific findings of facts are not required. Second, the failure to issue a decision within the time requirements is deemed to be a denial of the claim. Should the claim be denied, the contractor has an option of appealing the contracting officer's decision to the appropriate Board of Contract Appeals or begin an action in the Court of Claims.

If the contractor opts to seek review by the agency's Board, he must appeal the decision within 90 days of receipt of the decision. Under the disputes clause, the appeal can be made in a letter addressed to the head of the agency and sent to the contracting officer, preferably by certified or registered mail to provide proof that the appeal was made. The letter should state clearly that the contractor hereby appeals the decision. Don't say that you intend to appeal. That type of language has caused trouble in the past.

Once the appeal is filed, jurisdiction of the claim vests in the appropriate Board of Contract Appeals and the case will be processed to hearing in accordance with the Board rules. The Board can do one of three things: (a) it can grant relief and make a definite monetary award; (b) it can decide that the contractor is entitled to relief and remand to the contracting officer for a determination as to the amount of the award, or (c) it can deny the claim.

Depending on the amount of the claim, the contractor may be able to take advantage of expedited appeals procedures. On claims less than $10,000, the contractor may elect to utilize the "small" claims procedure providing for simplified rules and resolution within 120 days whenever possible. Such small claims appeals may be heard by only one Board Member. For claims of $50,000 or less, the contractor may seek an accelerated disposition of his appeal. Whenever possible, such appeals are to be resolved within 180 days of the contractor's election to use the expedited procedure. On claims over $50,000, the contractor must patiently await the determination of his appeal.

In the event that the contractor has elected to use the "small" claims appeals procedure on claims less than $10,000, neither the contractor nor the Government may further appeal the Board of Contract Appeals determination except in cases of fraud. However, in all other matters, both the agency and the contractor may seek judicial review of the Board determination by the Court of Claims. Such appeal must be filed within 120 days of receipt of the Board decision. The review need not be limited to questions of law. Fact findings by the Board may be overturned if not supported by substantial evidence, or if the fact findings were the result of fraud, bad faith, arbitrariness, or capriciousness.

The contractor may seek a route bypassing the Board of Contract Appeals by bringing an action in the Court of Claims within one year of the contracting officer's decision. This procedure will be of great value where the contractor anticipates an unsympathetic appeals board or has missed the 90-day deadline for filing an appeal before the board. The contractor is entitled to an entirely new trial of his claim before the court.

Resources

The resources involved in the presentation and review of any claim are twofold: people and documents.

Remember, although your claim is against the Government, you're still dealing with people — individuals who can move your claim along quickly or slowly — who can resolve gray area questions for you or against you. Obviously you need to know who they are and what type of presentation and approach is most likely to appeal to them.

Titles might vary from one Government agency to another — but the positions and roles in claims processing are basically the same.

1. *Contracting Officer:* He is the decision-maker. His decision will be final unless appealed to the Appeal Board or the Court of Claims. Unless you get into a dispute, you'll probably have little, if any, contact with the contracting officer. You'll usually see him only when major problems occur. In some cases, particularly Corps of Engineers work, the Contracting Officer, who is the Division Engineer, will act through a designated representative, such as the Chief of the Construction and Operations Division. Although he's been out of the field for some time, the contracting officer or his designated representative, in most cases, will be an experienced

and practical construction man. He'll understand your real problems and he won't be taken in by a lot of smoke.

The final decision must be that of the contracting officer. Boards of Contract Appeals have, in fact, held invalid decisions made by anyone other than the contracting officer. That situation is not likely to happen very often, and when it does, it's usually when, because of political-type considerations, one of the contracting officer's superiors takes a decision out of his hands. In one such case a contractor on a GSA contract was terminated for default. When the contractor appealed and later inspected the GSA's records, it found that the decision to terminate came from the top and that the contracting officer, in terminating, was merely following orders. When the contracting officer's deposition was taken, he testified that, in his opinion, the contractor was not in default and shouldn't have been terminated. That ended the case and the Government settled.

Although the final decision must be that of the contracting officer, this does not mean that he cannot rely upon others for factual input. Since he hasn't been in the field, he has to. He's going to have to get the facts from one of two sources — either his field representative or from you, the contractor. A cardinal rule is that the contractor should never rely upon the Government to prove the contractor's case. There is a temptation to do this when you assume that the Government knows the job and the problems you experienced as well as you do.

When you submit a claim, don't assume anything. Don't leave any gaps to be filled in by the Government. In the first place, if you leave a gap in your factual presentation, chances are that this gap will be filled in by one of the Government field representatives and that what he puts in that blank will not be as helpful to your case as it could be. Second, no matter how favorably disposed a contracting officer is to your claim, he can't grant you relief unless you demonstrate that you are legally entitled to it. He is responsible for public funds which means he is administering a public trust. He is bound by the contract and regulations — in the case of the United States, usually either the Armed Services Procurement Regulations or Federal Procurement Regulations. His action is subject to review by the legal department, his superiors and to audit by the General Accounting Office. He can't give a contractor money on a claim simply because he thinks the contractor did a good job and got hurt. Sympathy is not enough.

If you don't make out your case for yourself, the contracting officer can't make it out for you. You can ask any contracting officer and he'll tell you of a number of occasions when he thought the contractor was entitled to relief, but he couldn't give it because the claim had not been properly presented.

Don't rely upon catchwords and conclusions. Simply saying that you were constructively accelerated means nothing. You have to show (a) that your work was delayed for causes that entitle you to time extensions, (b) that you requested those time extensions, (c) that they were denied either actually or constructively by a failure to act on them, (d) that, but for those delays that were not your fault, you would have been on schedule, (e) that you, in fact, accelerated, (f) that the acceleration, in fact, caused you to incur additional costs, and (g) you must show with reasonable certainty just what those additional costs were and that they were incurred because of the construction acceleration — not for some other reason. If you fail to cover any of these points — even when the contracting officer knows that you have a valid claim — you can expect your claim to be denied.

2. *Resident Engineers:* To this point, we've been concentrating on the role of the contracting officer in the processing of a claim. However, in a sense, he's only the tip of the iceberg. Essentially, he acts upon the basis of information provided him by others. On the Government side, there are several avenues of input into the processing of a claim. When a claim is submitted, the first thing the contracting officer will do will be to send it into the field to his resident engineer for review and comment as to the accuracy of the facts stated therein. As a general rule, you can expect the resident engineer to argue against your claim. More often than not, he seems to think that when a contractor makes a claim, it is some sort of a veiled attack upon his performance of his job. This problem can be minimized in two ways: First — never attack the work of the Government engineer unnecessarily. Without sacrificing your legal position, present your claim without attacking the ability or motives of those on the other side. For example, when unexpected conditions give cause for an extension of time and the resident engineer recommends denial and keeps pressing for progress, point out that he was trying to get the job done as soon as possible for the Government — a worthy goal. However, under the contract, it is the Government's legal responsibility to pay for it.

Instead of couching your claim in terms of personal competency, couch it in terms of legal obligations. It is better to talk in terms of breach of warranty or legal duty than in personal terms. Let the engineers blame the problems on the law and lawyers.

The second way to minimize resident engineers' opposition is to document your claim fully. A resident engineer can oppose you only on facts. He generally won't argue the law. If you have all facts documented:

1. You'll save a lot of argument and time.

2. In those gray areas where facts can be read either way, he might be a little less likely to challenge you. It's not likely that you'll get the resident engineer on your side, but you can neutralize him somewhat.

What type of documentation should you have? Objective documentation is the best — photographs where appropriate, test reports, progress meeting minutes, pour logs, etc. Your letters to the Government are usually self-serving and the least objective. Nonetheless, they are helpful, particularly if they are factual rather than argumentative and if they were written at the time the facts they refer to occurred.

3. *Inspectors:* In cases where your claim relates to the manner of inspection, e.g., arbitrary or harassing inspectors, there is a temptation to couch a claim as an attack upon the inspectors involved. An over-inspection case can get emotional.

Avoid that; don't talk of harassment. Simply point out that "Government asked for higher quality than required by contract and it must pay for it."

4. *Architects:* A contractor will usually have much more direct contact with architects on state jobs than on federal jobs, although federal work will have architects in the forefront in certain cases, e.g., where GSA is acting as manager for some other federal agency without construction expertise.

The extent of an architect's involvement in a claim will depend upon whether you are attacking design or some other aspect of performance that doesn't directly relate to the architect's work. In the latter case, the architect will serve as an advisor to the owner. In the former, he'll play a more active (and probably antagonistic) role.

In AIA-type contracts,the architect's role in disputes is critical since he is the designated arbitrator of disputes in the first instance. Legally, the architect is supposed to resolve disputes, with complete objectivity and impartiality. This is not usually the case, unfortunately. In most cases, the architect, though independent in name, knows whom he is working for. He's going to side with the owner, at least subconsciously. He also has a tendency to feel that his job, particularly if it involves inspection, is to defeat claims and not to support them.

Finally, many architects simply refuse to recognize that they can make a mistake. I've had an architect tell me that if a spec was wrong, it's not his responsibility — it's the contractor's fault because he didn't pick it up in his pre-bid review of the specs and bring it to the architect's attention.

There is no real secret to success in dealing with the architect. In most cases, unfortunately, he is going to be an adversary of the contractor. All you can do is to state your position clearly and document it fully. Don't inflate your costs. Don't make personal attacks unless absolutely unavoidable. Keep his ego out of the dispute process to the extent that you can.

5. *Lawyers:* When you get to the claim stage, a lawyer is always part of the Government's team. He can be a great help or a great obstacle in the prompt and fair processing of your claim. On the plus side, he's not been personally involved in the dispute. Although the Government lawyer's allegiance is with the Government, he usually recognizes that there is such a thing as a valid claim and that it should be paid. He can be a reasoning force to overcome any personal biases affecting the judgment of the resident engineer or others who were personally involved.

However, if the lawyer finds that an essential legal element of the claim has been omitted, he's going to recommend denial and his recommendation will in almost all cases be followed even when the contracting officer and field team are sympathetic to your claim. No contracting officer is going to grant a claim when his lawyer says that there is no legal basis for payment. Remember that the Government lawyer's job is to look for loopholes, so don't give him any.

6. *Special or Outside Counsel:* On difficult claims, government agencies, particularly state agencies, will occasionally utilize special outside help, such as a private law firm specializing in construction claims, a CPM analyst, or special claims analysts, such as Wagner-Hohns-Inglis. Since this is obviously costly, it is done only when the agency lacks the manpower to review a claim fully and the amount of money at stake will justify incurring the expense.

The involvement of special outside counsel is usually a plus for both sides, if done selectively in the right cases. The special counsel can provide time, expertise and objectivity. If they do their job, claims can be processed faster and reviewed and priced more effectively and there is a greater chance of avoiding unnecessary trials.

Never underestimate the ability of Government personnel. They are competent professionals who spend their lives in construction and, more often than not, have far more claims experience than the contractor. As a general rule, you are better off when the people on the other side, reviewing your claim, are highly competent and experienced. When your claim has merit, they will recognize it, if it is properly presented. It is easier to work a fair settlement with people who know what they are doing than with people who don't.

PRESENTATION

Justification of Damage Calculation

Damages are half of your claim — if that half fails, the claim fails with it.

Even when the contracting officer feels that your claim is valid and you are entitled to something, he's going to assume that you're asking for twice as much as you're entitled to. You have to convince him otherwise.

If you review the Board of Contract Appeals decisions, you'll find that, in at least half the cases where the contractor succeeded on the merits, he got hurt on damages. It doesn't do any good to succeed on the merits if you don't get paid for it.

Proof of damages must come from you. Government personnel may know that you ran into different subsoil conditions from those shown in the drawings, but only you know what that cost you.

No one will take the contractor's word on what a problem cost him. You must prove it. You prove it with records. Payroll records for labor costs, purchase orders and invoices for material costs, invoices for the rental equipment, equipment logs for owner equipment. You can't have too much detail in proving damages.

You can't, as a general rule, submit all the cost back-up with the claim — it's too voluminous. But you should submit enough so that the officials reviewing the claim are convinced that the back-up is there. For example, furnish summaries of weekly payroll where appropriate. Establishing cost credibility is all important.

Prepare your cost presentation with a view to an audit. (Even if audit is not required by law, if you want money from the Government, you'll have one — one way or another.) Establish an audit trail. In your claim write-up, explain how your costs were computed and what records support them. Don't leave room for guesswork or speculation. When you know you're into a change order or claim situation, you should set up a special account on your books to collect all charges against the change order or claim work. The time you spend doing this will save you at least twice that much time later on.

You must establish the causal relationship between the costs you're seeking and the act or omission that is the basis for your claim. This is generally the most difficult aspect of proving damages. Countless factors can increase the cost of work.

You must isolate the factor that is the basis of your claim and show that this alone is what caused your cost overrun. This is particularly a problem in inefficiency claims. An example (outside of construction) of this type of problem was a case tried several years ago under a contract to manufacture smoke cannisters for mortars. The basis for the claim was the fact that the particular smoke mix used caused an extensive outbreak of dermatitis. The Government was liable for this because it knew at bid that this would probably happen and failed to divulge it. The contractor knew the dermatitis had an adverse effect on morale and efficiency. The workmen went somewhere else to work, leaving the contractor the least skilled, least dependable and least efficient. The problem was how to measure that loss of efficiency. In that case, it was done in two ways. First, the loss was shown by reference to experience on comparable work, not involving the dermatitis problem. Those other cases showed what level of efficiency the contractor could achieve on work of that type but for the dermatitis problem. Second, the contractor prepared a chart correlating the man-hours required for a unit of production for each month of the contract and correlated that with the incidence of dermatitis. This confirmed the relationship between the dermatitis and the increase in man-hours per unit, and to that extent quantified the efficiency loss.

If the efficiency loss is for only part of the job, you can use the period before the problem occurred and the period after the problem was solved to show what efficiency you would have achieved but for the problem. That allows you to measure the cost of the inefficiency. This method of proving damages was approved in *Natkin & Co.,* v. *George A. Fuller Co.,* 347 F. Supp. 17, 35 (W.D. Mo. 1972).

In presenting the claim, the contractor should state the damages aggressively without inflating damages. The term "inflated damages" can mean different things to different people. Generally, it means stating a damage figure that's greater than the actual cost incurred as a result of the act or omission that is the basis for your claim, in order to give the claimant more room to bargain.

As previously indicated, on claims over $50,000, the contractor must certify that he believes the damages requested accurately reflect the Government's liability. But on any size claim, the contractor may face a severe penalty for fraudulent misrepresentation of damages. Where the contractor is unable to substantiate any part of the claim due to misrepresentation or fraud, the contractor is liable for the amount of the unsupported part of the claim, as well as Government costs for reviewing that portion of the claim. Misrepresentation in this context requires an "intent to deceive or mislead." Although honest and unintentional errors will not result in penalties, contractors should exercise caution in damage calculations. Antagonistic and zealous Government personnel may level charges of misrepresentation which would greatly complicate the contractor's ability to recover for legitimate damages.

Aside from problems of fraud and monetary penalty, inflation of damages will almost always be discovered and it will destroy the claimant's credibility.

If the contracting officer feels that damages have been inflated, that is, you're trying to get the Government to pay more than the costs you incurred, he's going to think that you're trying to put something over on him. He's not going to trust anything you say. If he doesn't trust you, you can forget about settlement. You're going to have to prove your claim and every penny at hearing, trial or arbitration. Your ultimate recovery is going to be delayed. Your net recovery will be reduced by the costs of trial. *Loss of credibility carries with it a high price.*

All of this is not to say that you should not present your damages aggressively. As I indicated, there's a difference between an inflated damage presentation and an aggressive one.

Where there's room for reasonable doubt, resolve the doubt in your favor, stating damages without exaggeration.

Good example: estimates of inefficiency caused by delay, acceleration, etc. No one can pinpoint that with absolute certainty — e.g., 14-1/2 percent. There's a range in your claim. Use the high side of the range and fight for it. The Government will do the same on the low side and you'll come out somewhere in the middle.

In the pricing out of a claim, always make sure you include all indirect costs associated with the extra work — your field overhead, home office overhead, price escalation due to delays, equipment stand-by. That's proper aggressive damage presentation, not improper inflation of damages.

There is a temptation to use a total cost damage approach in cases involving delay and efficiency loss claims. Under a total cost approach, the contractor doesn't attempt to tie particular cost to a particular Government act that is the basis for a claim, but rather seeks the full difference between his actual cost of performance and his bid, on the theory that the full cost overrun is the Government's fault. To rely completely upon a total cost approach to damages can be dangerous, as it is usually unsuccessful.

The Boards and Courts have allowed it only in exceptional cases. Contracting officers like to use it even less. Use of a total cost approach alone will significantly reduce the possibility of settlement.

When can you use it? *First,* it can be used only when there is no other way of estimating damages, e.g., many Government acts that increase the cost of performance, affect the whole job, and are inextricably intermeshed. See *S & E Contr. Inc.,* AECBCA No. 97-12-72, 74-2 BCA ¶10,676 (1974). *Second,* it must show no underbid. However, one may use a revised pro forma bid to correct a known mistake. *Third,* you must prove that there was no contractor inefficiency.

It is often helpful to make a two-pronged presentation — (a) total cost, (b) backed up by more conventional, individually priced damage calculation. Total cost approach in that manner affords a good way to argue equities of a large claim.

A word should also be mentioned about the jury verdict method of awarding damages. This is a method which Boards and Courts have used to calculate damages where the amount of damages can't be ascertained with certainty and the best that can be done is to make a reasoned approximation. It is used when you know the contractor has been hurt, but can't really determine the actual loss attributable to the Government act. So you make a reasonable guess based upon the information before you.

When a Board or Court uses the jury verdict approach, you know the final number they choose is going to be on the low side — the safe side as far as the Government dollar is concerned.

Accordingly, do everything you can to minimize resorting to jury verdict approaches. If you can't price damages in the conventional way, use common sense and imagination to find some way for measuring the extent of the dollar harm done. For example, on claims for inefficiency due to acceleration, engineering studies have been conducted to show the percentage of efficiency loss associated with acceleration. Get one of those studies. Use it as a basis for estimating efficiency loss. You can use experts to reconstruct what the job should have cost.

Two final points about damage presentation. *Make it understandable.* Make it easy for the person reviewing the claim to know how and why you've arrived at your figures. Use schedules that are easy to follow. Use summaries that tie all your costs together. *How the package is wrapped can help.*

Second, before you submit your damages, step back, take a look at it and put yourself in the shoes of the Government. Look at it the way you would look at a claim of a subcontractor against you. That's the best way of making sure that your presentation is complete and convincing.

SETTLEMENT

Does Litigation Work? How and When to Seek Settlement

A judge once said litigation is, at best, successful only half the time because in every case there's one winner and one loser. The winner probably thinks litigation works fine. The loser has a different view. That's an overly simplistic analysis, but a good starting point, because the first thing to remember about litigation is that there is always some risk. There is almost no such thing as a sure thing in the courtroom.

Second basic point: a million dollars obtained over a settlement table will always have a greater net value than a million dollars obtained at a trial. What you're shooting for in most cases is a good settlement. Litigation is the last resort.

But, as a general rule, you'll never get a good settlement unless you're ready, willing and able to go to trial. The saying about getting the other guy to blink first applies. If you go to the settlement table ready and willing to go to trial and the other side isn't, you're going to win at the settlement table. If it's the other side that's ready for trial and you're not, then you're in trouble. This is the psychological relationship between litigation and settlement — and it's important.

There is a second and more direct relationship. In a sense, every set of settlement negotiations is a mini-trial. Each side is presenting its case to the other. The ideal in settlement discussions is for both sides to review and evaluate the other side's case and proof-evaluate their own at the same time and then sit back and examine the probabilities and see where the case would probably end up after a trial. They are then at a point to agree on that probable result, saving the time and money of actual trial.

The ideal never happens, but there are things you can do to bring your settlement discussions close to that point and maximize the chance of success.

1. Don't start settlement talks until you're fully prepared. All of your ducks should be in a row. Your claim should be finalized, costs finalized and all documents ready, including job records, photographs you're relying on, cost back-up, etc. If you're not ready, it will soon become apparent and you'll have lost an important psychological battle.

2. Make sure you know your case before you go into settlement — both your strong points and your weak points. No case is ironclad. Some part is always weaker than others. Don't be afraid to recognize a weak point. When it's there, not recognizing it won't make it go away. Recognize it and work around it, minimize its importance.

To know your case well, you have to know the other side's case. Again, put yourself in the Government's shoes. What would you do if you were the Government? How would you defend against the claim? If you do this, you can (a) anticipate the Government's position and be in a better position to handle it, and (b) you'll have a better picture of the true value of your own case and be in a better position to evaluate it properly for settlement.

3. Create the proper psychological climate for settlement. Let the Government know that you're not there out of weakness or because you think you can put a hard sell on. You're both there to see if you can resolve your problems quickly and without unnecessary delay and expense.

Before you go to a settlement meeting, get on the telephone with your Government counterpart and tell him what you hope to achieve. Tell him that neither of you will gain by guaranteeing full employment for the lawyers for the next year or so. He'll appreciate that. Although he's not paying for his lawyers, he probably doesn't want them around any more than you do.

4. Be candid — don't hide anything. When you're faced with a point in the other side's favor, admit it. But minimize it as best and as honestly as you can. There is a danger to never admitting a mistake or anything that might help the other side. It can lose credibility.

5. Don't set your sights at an unrealistic level. That will guarantee that settlement won't come.

6. Don't overstate your case. A strong case can be weakened by including unjustified or weak claims. Like a bad apple in a barrel, again this can result in a loss of credibility. Credibility can't be overemphasized. You will never settle a case without it. Remember the penalty for submitting fraudulent, unsubstantiated claims.

7. If you want a meaningful settlement, don't play it too close to the vest. Don't hide part of your case out of fear that the other side will attack it once you're found out about. To hold back part of your case is like playing only four cards in five-card stud.

8. When you submit a claim, always suggest a follow-up meeting, face-to-face.

9. If it's a major claim, don't wait until after settlement to get your lawyers and experts involved. That can be too late.

10. Remember the bargaining aspect. Usually everyone has to give up something in settlement. The initial figure should be aggressive to give bargaining room, but still honest. Again, don't inflate. In evaluating settlement offers, keep in mind the cost and aggravation of litigation.

Keep in mind that litigation is expensive and slow. Both sides pay for one, maybe more, lawyers to prepare for and conduct a trial. They also pay for experts. Your own key personnel are tied up. The fact that they are not able to perform their normal duties while testifying at court can be the highest cost. You never know how long trial will take. Trials that have been predicted to take two weeks have been known, on occasion, to take two months. Most important, remember that you can never be certain of a trial outcome.

Before you go into trial, evaluate in terms of probabilities. A 90-percent probability means that if ten of these come to trial, nine will be winners. You could have the tenth. Keep that in mind when you evaluate settlement offers.

11. Be realistic about risk evaluation. Don't be blindly hard-nosed.

12. Keep personalities out of settlement talks. If your project engineer and owner's resident engineer are like flint and stone, call up the owner and suggest that he keep his resident engineer out of settlement meetings and you'll do the same with your project manager.

When Do You Seek Settlement?

Have in mind that how the claim is presented from the beginning and how it is framed can maximize or minimize settlement. If frontal attacks on basic design or an attack on the competency of the architect or design engineer are made, the claim will be difficult, if not impossible, to settle. If the personalities are provided with an excuse, such as by categorizing the Government action as a response to something unforeseen that required a change, a claim is easier to settle.

Lawyers in most cases are not a roadblock to settlement. They're reasonable. They have settled lots of cases. Avoid the lawyers who say they don't settle any cases.

There are some cases that just can't settle:

(1) Political problems which arise more often in state than federal contracts;

(2) Threshold legal issues which can be resolved only in court;

(3) Large claims, since contracting officers don't want to pay that much unless it is ordered by a court;

(4) Ego problems, such as claims alleging defective design that the architect and owner regard as personal affronts.

In these cases, settlement is probably out of the question. You might as well get ready for trial.

There are things you can do to minimize cost of trial:

(1) Document exchange;

(2) Stipulate facts as much as possible;

(3) Where a case involves several claims, perhaps some of them can be severed and settled.

How Do You Get the Attention of the Government?

1. Start at the beginning of the contract — the image you build will play a large part in type of attention you later get. Don't be a crybaby. Be a problem-solver. Play it straight. Don't automatically blame everything that goes wrong on others. Don't start letter wars. Document your facts, but avoid inflammatory letters. Don't get an image of claimant — someone who starts building a claim file as soon as he gets on the job.

2. Prosecute your claims promptly — if you move with dispatch, the other side will take you seriously.

3. Come in with a thorough, complete presentation. A good presentation will show that a lot of work has gone into the claim. The claim should not appear to be just a wrapping up of old correspondence. Instead, it should appear to be the result of a full investigation.

4. Have a prompt meeting after the Government has had a chance to review the claim. Don't spare the horses at the meeting, the owner won't. Have people there who can answer all questions.

5. Do as much of the owner's work in advance as possible, such as providing documentation.

6. Follow through. After the first meeting, if you don't hear, call — keep the pressure on.

When Should You Present Your Case?

Don't confuse presentation of your case with notice. Give notice right away.

Let the Government know the general basis of your claim right away. Give the owner a chance to avoid the problem if it can be avoided. For example, minimize damages — the smaller the amount in dispute, the easier to settle.

If you can't fully present it for decision until costs are finalized, then wrap it up and submit it as soon as possible.

At all costs, don't let a claim appear to be an afterthought, a 13th hour effort to get money because at the end of the job you discovered that you had lost money.

There are, of course, cases when a claim can't be fully presented until a job is substantially complete. This is particularly true in delay-type claims, when the full and final impact often is not known until completion.

Why Does It Take So Long?

1. A lot of work goes into reviewing a claim just as in preparing one. It takes time.

2. Usually, there are other claims on the contracting officer's desk. Even more questions regarding on-going work are on the same desk and they'll probably get priority.

3. The best way to speed up process is by a good, thorough, well-documented, prompt presentation, by being persistent and by being reasonable in your demands.

CHAPTER 7

The Presentation of Disputes In Non-Federal Government Arenas

H. JAMES WULFSBERG, J.D.

ATTORNEYS ARE FREQUENTLY RETAINED BY CONSTRUCTION CLIENTS only when litigation is imminent or has actually been started and the client's financial security is threatened. Clients may perceive attorneys as unknown beings who routinely charge exorbitant fees, produce reams of unintelligible documents, and ultimately, after years of waiting, present the client with a settlement or judgment that falls far short of the victory the client had anticipated. Consequently, many construction disputants feel frustrated when faced with the need to select an attorney. Although attorneys are now permitted to advertise, a qualified attorney in the field of construction disputes will seldom advertise in the local newspaper. Attorneys in general do not "bid" on a job, submitting a detailed statement of costs and fees upon which a client can rely.

All of this is not to say, however, that an attorney's qualifications and experience cannot be evaluated, or that they should not be. The attorney is an independent consultant for the client and the working relationship between them may be more important than any developed during design or construction. A first place to start in obtaining the names of lawyers who work in the field is to talk with other professionals in the construction industry. Find out not only whether the attorney obtained a satisfactory result, but how an on-going attorney-client relationship was maintained. Was the client kept informed, not only about the events taking place in the litigation, but also about periods when nothing was taking place, with the reasons for such inaction? Was client input solicited and considered in the decision-making process? If the client's suggestions were rejected, was the rationale for that rejection explained? Was the attorney accessible? These questions assume that any such critique of an attorney-client relationship will include an assessment of the attorney's work product, knowledge, confidence and ability to communicate.

In evaluating an attorney-client relationship, accessibility is one of the more important elements to be considered. Obviously, an attorney has more than one case or client and cannot be immediately available to all clients at any hour of the day. As with all consultants or individual independent contractors, it is best to schedule meetings in advance and to apprise the attorney of future requirements in order to allow for proper planning. It is frequently the case in larger law firms that two or more attorneys will be responsible to a client or for a particular problem of a client, so that one will generally be available in any situation. However, with an attorney, it should be possible for a client to relay a message to him and to receive advice or the personal appearance of counsel in an emergency situation. An understanding and appraisal of such needs are desirable from the start.

Other attorneys, personal friends or business acquaintances, may have information about those skilled in construction contract disputes. It should be kept in mind that most attorneys are specialists to a certain extent, and that the attorney who is extremely competent in drafting wills or handling personal injury work will not necessarily be conversant with principles of construction law. It may be that an attorney with whom the client has already established a working relationship can associate with a more experienced construction law-oriented firm to bridge this gap. Both groups of attorneys must be prepared to devote the necessary time and energy to such an arrangement, and the fee arrangement will reflect this pooling of skills and inevitable duplication of effort.

Trade journals or other professional publications in the construction or design fields may contain articles or treatises by attorneys working in this area. *Martindale Hubbell* is a multi-volume legal directory, maintained in county law libraries and law schools, which contains a listing of attorneys by geographical area. The information provided includes a brief description of areas of practice, biographical information on each member of the law firm, including law school and publications, and frequently a listing of representative clients. Finally, the local bar association can provide names of attorneys who have indicated an interest in certain types of legal work.

The option of selecting counsel may be foreclosed where an insurance company covering the litigant selects counsel. While the carrier may select qualified attorneys to represent the party in regard to the claim, this fact should never be assumed. Some companies select counsel on the basis of pre-arranged fee agreements which may preclude the type of representation desired or necessary on a particular case. Frank discussions with the claims manager in charge of assigning counsel is recommended. Dissatisfaction with counsel is the right of the client — particularly when subsequent premiums, coverage or claims above policy limits may be involved.

Attorneys are experts, just as are designers, contractors and owners in their fields. The attorney's expertise lies in planning to avoid disputes, while at the same time protecting the client's interests in the event disputes do arise. Therefore, it is best to involve the attorney in the early stages of the construction process, so that he or she may inform the client of possible problem areas, review or draft documentation as it is necessary and build a record to document the client's position if problems arise later. In this way, the attorney-client relationship is formed and strengthened while the work is being performed on the project. Mutual trust and confidence should then exist if litigation arises following project completion. This also allows the attorney to have a "head start" since he will be familiar with the facts and potential witnesses from the outset.

However, there is a serious problem with such a long-term relationship. In a recent California Supreme Court case, the court held that where an attorney's involvement with pre-litigation negotiations was such that he should be a witness in the litigation, that attorney and his firm may be required to withdraw as the client's attorneys of record in the litigation. *Comden* v. *Superior Court,* 20 Cal.3d 906 (1978). For large designer, owner or construction clients who have in-house counsel, a possible solution is to hire outside counsel for litigation, something which many such firms now do. For those clients who do not have a legal staff, the problem can present extra expense since it seems to require hiring two sets of attorneys — one for pre-litigation, the other for litigation. However, the solution may simply be to limit the attorney's involvement in pre-litigation phases to an advisory one. Advice given in an attorney-client context where no third parties are present is privileged and not subject to the purview of the *Comden* rule. Where it appears that the attorney's involvement may escalate into negotiations or other dealings directly with third parties, then the attorney may be required to recommend that the client obtain independent trial counsel.

In a decision involving the client's interests in a construction dispute, including whether or not to litigate, attorney and client should discuss all aspects of the problem before reaching a conclusion. The attorney is an advisor who can provide input as to issues, problem areas, estimated costs and an assessment of likelihood of success. Prosecuting an action through the courts is not the only solution in resolving construction disputes. Nor is it always advisable to be the instigator of litigation. An attorney can present the pros and cons which the client can then evaluate in light of its financial position, working relationships with the parties, and other factors. There are some decisions which the attorney will be asked to make in regard to the client's interests in the course of litigation, which, because of time pressures, cannot be discussed in advance with the client. These decisions are within the attorney's expertise, but they should be discussed in a general way with the client initially. All major decisions, however, especially whether and when to litigate, must not only be discussed in advance, but must be within the client's control.

Traditionally, little consideration has been given to the forum for claims presentation at the time the parties execute their contracts. However, when disputes subsequently arise, parties may be unpleasantly surprised to discover that they agreed to resolve the problem in a particular forum, and that this uninformed or casually considered choice will be binding upon the parties. In a public construction context, the parties may be required to comply with specific claims procedures which may or may not be specified in private contracts. Once the claims procedure has been satisfied, the parties usually must comply with an arbitration clause, if such clause makes arbitration mandatory. The binding effect of such clauses has been recognized in federal law for contracts involving interstate commerce, 9U.S.C. SS1-14, and under most state laws (See e.g., Cal. Code Civil Procedure SS1280-1294.2.) However, courts have recently recognized that some parties, such as contractors who are ostensibly bound by such clauses, possess little or no bargaining power, and hence may invalidate such clauses as to those parties. *Player* v. *Georgc M. Brewster & Son, Inc.,* 18 Cal. App.3d 526 (1971).

If arbitration is desirable, the contract should provide for the power to join other parties, including subcontractors, (whose contract should reflect that joinder), a description of arbitrable disputes, and a method of appointing the arbitrator. Other provisions may also be included, such as the use of discovery tools, power to produce witnesses and the method of arbitrating.

Abritration is frequently viewed as a relatively speedy, inexpensive means of resolving disputes on a practical basis with the assistance of an impartial individual having some technical expertise in construction work. Whether this is, in fact, the case, is open to question. Many times the contractual arbitration clause limits the amount in controversy, making it advisable to pursue litigation even where it initially appears that the disputed amount is within the contractual limits. Attorneys may be necessary for adequate representation in either case; it is simply not possible to resolve claims of millions of dollars on a one-to-one basis between arbitrator and the parties. Hence, there will be a certain amount of legal skirmishing involved in either situation.

Initial costs are greater for arbitration, although long-range costs may frequently be less. Court personnel and facilities are tax-payer supported, while arbitration and arbitrators are not. Discovery costs will increase the expense of litigation, but they will also occur with arbitration where discovery is permitted by contract or later agreement of the parties. Litigation involves more costs before the trial itself, including both discovery and law and motion matters, while arbitration generally moves into the hearing phase much more quickly, without such legal proceedings and costs.

The location of the hearing or trial may differ in litigation and arbitration. There are well-established principles of law providing for the location of trial that do not take account of the convenience of parties or witnesses. Much pretrial time may be consumed with establishing which of these principles determines the place of trial. In arbitration, the hearing location may be more flexible, depending upon the contract clauses, and the arbitrator may find it much more convenient than the judge, jury and entourage to bisit the site. Location also raises the question of publicity, since court actions are matters of public record while arbitration is not. Adverse local publicity can be a strong deterrent to litigation and should not be overlooked by clients or their attorneys.[1]

The speed with which either arbitration or litigation may be accomplished varies greatly, depending upon the parties and the facts and circumstances of the dispute. Although arbitration will generally be accomplished in less time, it is frequently the case that initial questions, such as the interpretation of the contract clause providing for arbitration, must be determined by a court. In such a situation, much of the time advantage of arbitration is lost.

One of the most important elements of difference between arbitration and litigation is that of discovery. Arbitration clauses frequently do not provide for discovery, with the result that none may be available prior to the hearing. Some states do require that discovery be available, but the methods used may not be as effective as those available in litigation. This is a serious disadvantage for the attorney attempting to present a client's case in arbitration, since there is very little advance notice as to the opposition's information or strategy. Especially where the client has neglected to retain documents received from others, or to properly document its own position, this lack of discovery can be an insurmountable obstacle.

When a contract requires arbitration and no discovery is available, a prudent party *must* prepare his case as he goes along, documenting occurrences as thoroughly as possible. This is necessary because he will not be able to discover from the opposition any facts he has forgotten or overlooked.

Many provisional remedies are unavailable in arbitration unless specifically spelled out in the contract. If a party desires to file an attachment, stop notice or mechanic's lien, and this is not initiated prior to demanding arbitration, the right to such remedies may be lost. On the other hand, since all of these rights require the filing of a court action, in some states the party may lose its rights to demand arbitration by first pursuing a provisional remedy. However, in California, for example, such a party is probably protected as long as his court action is directed merely to preserving his right to one of these remedies. *Homestead Savings & Loan Assoc.* v. *Superior Court,* 195 Cal.App.2d 697 (1961). A temporary restraining order, which is a court order restraining an individual or entity from taking certain actions, is available only in litigation unless specified in the arbitration clause. Whether resorting to such a remedy is a waiver of arbitration is questionable. If the remedy is available according to the arbitration clause of the contract, pursuing it outside of arbitration probably does amount to a waiver.

There are both advantages and disadvantages to the technical expertise which an arbitrator brings to bear on a construction dispute. On the one hand, it may be less time-consuming to present an action where the trier of fact already has some background in the technical aspects of the problem. This may also be reassuring to the parties, who may feel that someone who "knows" construction will vindicate their rights. On the other hand, the role of the trier of fact, whether judge, jury or arbitrator,

---

[1]Note that the final award of an arbitrator may become public knowledge if confirmed in a court proceeding for that purpose.

is to hear all of the evidence, evaluate that evidence and render its decision. In the case of someone who has been involved with the construction process, it may be that the decision is made not as an impartial arbitrator, but simply upon the basis of how that arbitrator would react as a party to the construction contract.

As noted earlier, if some parties are required to arbitrate, it is important that all be required to arbitrate. Otherwise, parties may be faced with expending the time and money for both procedures in order to fully resolve the issues of liability. This concern for inclusion of all parties has become increasingly important in the wake of recent cases which have recognized the concept of comparative indemnity.

California has recently acted to recognize two important considerations involving arbitration. First, under Executive Order Number B50-78, all contracts involving certain designated public agencies must contain a provision that any disputes shall be submitted to arbitration before a panel of certified arbitrators (who must be attorneys) in accordance with specified provisions covering discovery rights and procedures to protect disputants' rights. All subcontractors, equipment purchase orders and contracts with unions concerned with the contract must have a clause incorporating arbitration and permitting joinder in the arbitration.

Second, California code of Civil Procedure Section 1281.2 recognizes that not all circumstances are appropriate for arbitration. In some situations, third parties not bound by arbitration contracts should be brought into the proceeding to avoid inconsistent results or to avoid multiple proceedings. In such circumstances, upon a proper showing, arbitration will be stayed and court litigation permitted to go forward.

Careful attention must be given to the available and appropriate forum before initiating or submitting to any arbitration proceeding. California, as well as other states[2] have recognized the imperfection of both arbitration and litigation and have sought to fairly and practically deal with this problem.

The recent California Superior Court case of *American Motorcycle Assn.* v. *Superior Court* 20 Cal.3d 578 (1978) extends the comparative negligence theory previously adopted in California to permit defendants with potential joint and several liability to seek "equitable indemnity" relief from each other based upon comparative fault. An arbitration clause in a construction contract which fails to compel all potential "indemnitors" and "indemnitees" to join in the arbitration may well create multiple forum litigation and substantially increase the cost of dispute settlement. For example, an attorney who desires speedy arbitration of disputes may be unsympathetic to a contractor or subcontractor seeking equitable indemnity from a supplier or designer who was involved in the dispute. Conversely, parties to an arbitration may waive their right to arbitration if a third party should be brought into the action in order to resolve the entire matter in one proceeding. Comparative fault indemnity cases may bring about the demise of arbitration in most construction contracts.

---

[2]See e.g., *Exber, Inc.* v. *Sletten Constr. Co.*, 558 P.2d 517 (Nev. 1976) requiring consolidation of owner-prime contractor and prime contractor-subcontractor arbitration; *Sullivan County* v. *Nezelek, Inc.*, 366 N.E.2d 72 (N.Y. 1977), encouraging consolidation of all actions; *Mass. Gen. Laws Ann.*, Ch. 251, 52A (West), requiring consolidation in situations of "hardship."

Finally, there is the question of what the judgment or award actually means. An arbitration award is, as a general rule, a final award not appealable by the parties. For the prevailing party, this is obviously a great advantage, but there is unfortunately no way to predict the outcome at the time the forum is selected. A court judgment is appealable either upon the grounds that the findings of fact are unsupported by the evidence, or on the grounds that an error of law has occurred. An appeal increases the time and expense for all concerned, but it does provide protection against an arbitrary decision.

Hearings before government review boards or similar hearing boards at either the local, state or federal level may be remedies mandated by statute or contract which must be exhausted before proceeding with either court litigation or arbitration. Although it may seem advisable to ignore or disregard the claims procedure before such review boards because of the "apparent" lack of impartiality, careful attention must be given to satisfying this first step to insure that a subsequent trial or arbitration proceeding may be pursued at all. Furthermore, the review board may be surprisingly sympathetic to the claimant, as in the case of the Environmental Protection Agency review of bid protests, or the claimant may be able to establish a reputation for credibility with a particular board over the course of several projects.

Many standard specifications in public works projects require the contractor to present his claims to the "engineer of the work" or other similar hearing officers or boards. Usually the period within which a claim must be presented is short and the failure to comply with the claim presentation provisions bars later recovery. Therefore, even when a party is certain that claims will be routinely denied, they must be presented within the required time period or they may well be lost.[3]

Claims which would be denied in more formal settings may be approved by the relatively informal boards. Discovery is generally available, but careful attention to the details of a claim might persuade the hearing board to allow a claim which might not be permitted in another forum. Often decisions are quick and the cost of preparing and presenting the claim relatively minor. Since the claimant may be requested to submit claims as a condition of later asserting them in litigation, a procedure for documenting and forwarding claims, especially calendaring, is essential.

An attorney should have a basic understanding of the design and construction process at the outset of any litigation. However, most attorneys do not have, nor

---

[3]For example, the Standard Specifications for the City of San Diego provides in SS1.04.01 "Claims And Protests":

(a) If the Contractor considers any work demanded of him to be outside the requirements of the Contract, or if he considers any instruction, ruling, or decisions of the Engineer to be unfair, he shall within ten days after any such demand is made, or instruction, ruling, or decision is given, file a written protest with the Engineer, stating clearly and in detail his objections and reasons therefor. Except for such protests and objections as are made of record in the manner and within the time above stated, the Contractor shall be deemed to have waived and does hereby waive all claims for Extra Work, damages and extensions of time on account of demands, instruction, rulings and decision of the Engineer.

(b) Upon receipt of any such protest from the Contractor, the Engineer will review the demand, instruction, ruling or decision objected to and will within 19 days thereafter advise the Contractor in writing of his final decision, which shall be binding on all parties, unless within ten days thereafter the Contractor shall file with the City a formal protest against the said final decision of the Engineer. The City will consider and render a final decision on any such protest not later than 30 days after the date of formal acceptance of the work.

should they be expected to have, a thorough knowledge of the intricacies of design or construction methods. It is better to engage an expert who can provide an in-depth, objective analysis of a particular problem area or possibly of the entire project. Depending upon the attorney's own knowledge, the scope of the project and the type of dispute, the expert may be an architect, engineer, or a specialist in any one of numerous fields, such as manufacturer, computer specialist or a construction specialist. Whatever the expertise of the particular specialist, he or she should have the ability to coordinate, analyze and assimilate the massive amounts of documentation generated during any construction project.

If there is any question in the client's or attorney's mind about the validity of pursuing a claim or the best means of advancing the client's interests, an expert may provide assistance in evaluating a course of action. For example, where the owner knows that the plant is inoperable as built, an expert may be required to pinpoint the difficulty or difficulties and the cost of repair or replacement.

Once litigation or arbitration has commenced, an expert can be beneficial in educating the attorney as to the terminology and the theories underlying design defects, construction methods, fabrication processes and the like. A well-educated attorney who can talk intelligently about such matters may spell the difference between an advantageous early settlement or protracted litigation. Such education also benefits the attorney and the client by enabling the attorney to obtain the best results from discovery. The various discovery techniques and their advantages and disadvantages will be discussed more fully below. At this point, it is sufficient to say that an expert can greatly improve a party's position in a controversy by assisting the attorney in the discovery process.

If an expert is engaged prior to or during the discovery process, the expert's identity and/or reports may be discoverable by the other parties. If the expert is engaged by the party directly and provides reports to the party at its request, the opposing parties may request to see these reports and may well be entitled to do so. Consequently, if at all possible, the expert should be retained by the attorney and should prepare all reports at the attorney's direction. If these procedures are followed, the reports should themselves fall within the work product doctrine and be protected from discovery. A party may request, in writing, the names of individuals assisting in the preparation of responses to interrogatories. If an expert has aided the attorney in responding, the expert's name should be disclosed. Similarly, a party may request the names of experts who will testify at trial. Where the expert is preparing for testimony during the discovery period, his or her identity should be revealed. The disadvantages of such relatively early disclosure include the fact that the other parties will be placed on notice of one expert's participation and may decide to engage experts when they would not otherwise have done so. However, it is generally better to run this risk than to postpone preparation until late in the action or risk the expert's exclusion from trial testimony.

An expert's testimony at trial is the service which is clearly the most important function. Especially when a lay jury is involved, the expert's testimony may be accorded great weight. In line with this, it is important that an expert engaged for testimony at trial, as with any other expert, be an independent and objective source of information. An expert may be questioned on his or her statement by opposing parties as to bias. A trier of fact, whether judge or jury, will discredit the testimony of an expert who appears to be closely linked to the party on whose behalf testimony is given. In addition, while the temptation to engage a "friendly" witness is strong, it should be kept in mind that proper presentation at trial, whether prosecution or defense, requires a thorough advance knowledge of all of the facts, including those adverse to the case. Any expert, whether employed for testimony or general preparation, should be able to point out deficiencies on the client's part as well as on the part of opposing parties.

Experts are expensive and cannot be expected to render services, including preparation for testimony at trial, without compensation. The expert's costs should be discussed in advance and a method of payment arranged to insure that the expert's services will be available when needed. Along this same line, the services to be provided by the expert should be clearly spelled out, including a time frame for completion of reports, appearance dates and the like. By making such arrangements in advance, the course of litigation or arbitration can be planned with the knowledge that all necessary services will be provided when and as needed.

There will frequently be an enormous number of records involved in any construction dispute. The first and most important order of business in any litigation is to review all records of all the parties. The mechanics of arranging for such a review will differ, depending upon whether all of the parties involved in design and construction are also parties to the lawsuit and also upon the amount of cooperation between the named parties. Records of non-parties may be subpoenaed while parties' records may be reviewed pursuant to requests for production or other less formal procedures.

Simply as a function of the number of documents involved, reviewing records is a costly process. An attorney cannot be expected to simply look at documents once, assess their importance and retain the information. Copies of pertinent documents in the possession of other parties must be made to allow proper review and subsequent references. Copying can be done by microfilming or hard copy, depending upon the number of documents, facilities available and the relative costs involved. In considering the question of costs, it is important to bear in mind that all subsequent discovery, in addition to any settlement negotiations, will be premised upon the information obtained from the records of other parties. Costs expended at this point are well spent and it is usually false economy to attempt to cut costs for copying or review of documents.

For the parties producing documents, there is also expense involved in amassing and perhaps transporting them. In order to avoid multiple productions of this nature, it may be advisable for the producing party to offer to cooperate in making available all relevant, unprivileged records to all parties at one time. If a later request is made, the producing party can point to its earlier cooperation in objecting to a subsequent production as unduly oppressive and burdensome. A court will have discretion in compelling a subsequent production, but the request will frequently be denied where there is documentation indicating that the records have already been made available. This is all the more reason to spend the necessary time and money to inspect, and possibly copy, all available records when the opposing parties' records are made available.

There are specialists in the field of records review and it is worthwhile to consider retaining such experts, particularly where voluminous documentation is involved. These experts offer a range of services, including a detailed analysis of the entire project or a specific area of work, based upon the documentation supplied them. As with other experts, their testimony at trial may also be desirable so that they may present their opinions on the evidence analyzed. Because there are so few experts offering these services, one should be certain that no conflicts of interest exist before engaging them. And this analysis is not a substitute for a thorough review of the documents by client and counsel, since the expert's view is only one interpretation of the records and may overlook factors which client or counsel will perceive only if they have reviewed the records themselves.

The deposition is a discovery technique in which one or more attorneys question a witness under oath before a reporter who takes down the testimony verbatim. The deponent's attorney is present and can object to the form of questions asked, such as where the question is unclear or argumentative. The deponent's attorney may also object to the manner in which the deposition is being conducted, as, for example, where the deposing attorney is acting in an overbearing manner. A deposition is taken for a variety of reasons: to obtain information; to evaluate the deponent as a prospective witness; and/or to use as a settlement tool. The first two objectives are self-explanatory, except to note that the deponent should be asked for both helpful *and* potentially damaging information. A deposition is useful as a settlement tool because the questions are unexpected and the responses are recorded under oath. Consequently, the witness may reveal some information of which his attorney is not yet aware. Depending upon the magnitude of such a revelation, settlement negotiations may be substantially enhanced for the deposing attorney.

In preparing for a deposition, the deponent should be briefed as to the nature of the proceeding, possible problem areas, the need to ask for clarifications by the deposing attorney where necessary and the importance of carefully considering each answer before responding. Because of the potential benefit or harm which can flow from a deposition, sufficient time must be spent in preparing witnesses in advance of the deposition.

Another tool utilized in the discovery process is that of interrogatories. Interrogatories are written questions propounded to a party which must be answered by that party under oath. It is a good policy to use interrogatories as a means of discovering pertinent records, so that when a request for production is drafted it will refer to all records, avoiding the necessity of a subsequent production request. Interrogatories are also used to ascertain the factual bases of contentions made by the parties in their pleadings and to discover other relevant information. Since the responses are prepared with the assistance of the responding party's attorney, they are carefully drafted to advance the responding party's case, while protecting its interests, as much as possible. While the time and "coaching" which are allowed for interrogatory responses may seem to decrease their effectiveness as a discovery technique, a court will take those factors into account when the responses are introduced as evidence. It is much more difficult to explain away a written response for which the party had many days to prepare than an unrehearsed answer given in a deposition. Therefore, it is important for both client and counsel to spend the time necessary to carefully prepare both interrogatories and responses.

None of the foregoing litigation aids, either alone or in combination, can substitute for a careful and thoughtful evaluation of one's own records and witnesses. Becoming totally immersed in the opposition's evidence may cause a party to overlook critical facts buried in his own job site records or the recollections of his present or former employees. There is simply no substitute for knowing one's case — and then discovering the opposition's case.

Appendixes

		Page
Appendix A —	Partial Scheduling Clauses from *Dobson* v. *Rutgers*	116
Appendix B —	Scheduling Clauses from *Peter Kiewit Sons.* v. *Iowa Southern Utilities Co.*	117
Appendix C —	*Southern Fireproofing Company* v. *R.F. Ball Construction* .	118
Appendix D —	Scheduling Clauses from *C. H. Leavell & Company*	119
Appendix E —	The American Institute of Architects	120
	Standard Form of Agreement Between Owner and Contractor *(AIA Document A101)*	121
	Standard Form of Agreement Between Owner and Contractor *(AIA Document A101/CM)*	125
	Abbreviated Form of Agreement Between Owner and Contractor *(AIA Document A107)*	131
	Standard Form of Agreement Between Owner and Contractor *(AIA Document A111)*	139
	General Conditions of the Contract for Construction *(AIA Document A201)*	147
	General and Federal Supplementary Conditions of the Contract for Construction *(AIA Documents A201 and A201/SC)*	166
	SUBCONTRACT — Standard Form of Agreement Between Contractor and Subcontractor *(AIA Document A401)*	176
	Standard Form of Agreement Between Owner and Architect *(AIA Document B141)*	184
	Standard Form of Agreement Between Owner and Construction Manager *(AIA Document B801)*	195
	General Conditions of the Contract for Construction *(Comparison of the 1970 and 1976 Editions — AIA Document A201)*	205
	Standard Form of Agreement Between Architect and Engineer *(AIA Document C141)*	248
Appendix F —	National Society of Engineers Contracts	261
	Standard General Conditions of the Construction Contract. .	262
	Standard Form of Agreement Between Owner and Engineer for Professional Services	288
	Additional General Provisions (Architect-Engineer Contract)	317
Appendix G —	Contracts for Architectural and Engineering Services	326
Appendix H —	Contract Disputes Act of 1978	349

APPENDIX A

Partial Scheduling Clauses from *Dobson* v. *Rutgers* 157 N.J. Super. 357, 384 A. 2d 1121 (1978)

G4-F.4 PROJECT PLANNING, SCHEDULING AND CONTROL ("Critical Path Method"): As an aid to the General Contractor, and all other Contractors, to bring the completion of the project within the time allocated, the Owner has contracted for the services of a Critical Path Method Scheduling Consultant for project planning, scheduling, and control, and the *General Contractor shall incorporate and enforce combined schedule as his own.* Each Contractor agrees to cooperate and coordinate his own operations in order to meet effectively all scheduled task deadlines. [Emphasis supplied] .

G4-F.4 (a) [Stated in relevant part] (a.) Scope: The Critical Path Method (CPM) shall be used to control the timing and sequences of the project, as implemented by the CPM Consultant. All work shall be done in accordance with CPM Planning and Scheduling and it will be the responsibility of all Contractors to cooperate fully with the CPM Consultant. All costs incurred by the Contractors to implement correctly the CPM, other than the fees of the CPM Consultant, shall be borne by the Contractors and are part of their Contracts.

APPENDIX B

Scheduling Clauses from *Peter Kiewit Sons Co.* v. *Iowa Southern Utilities Co.*, 355 F. Supp. 376 (S.D. Iowa, 1973)

Article III of the Contract Agreement provided that:

> Time of completion is of the essence of the Contract Agreement, and that the Contractor shall proceed with the work in accordance with the specified schedule.

GC-34. BEGINNING, PROGRESS, AND TIME OF COMPLETION OF WORK. Unless otherwise specified the contractor shall begin work under this contract within ten (10) days after the date designated in a written order from the Owner to begin work. The rate of progress shall be such that the work will be completed in accordance with the terms of the contract on or before termination of the construction period named in the Contract Agreement. The Contractor shall furnish the Engineer a detailed schedule setting forth the procedure he proposes to follow and giving the dates he expects to start and to complete portions of the work. If in the opinion of the Engineer proper progress is not being maintained, changes shall be made in the Contractor's operations to ensure proper progress.

APPENDIX C

Southern Fireproofing Company v. *R.F. Ball Construction*
334 F. 2d 122 (8th Cir. 1964)

ARTICLE I. This Contract Agreement is subject to and shall be governed by all * * * General Conditions and Special Conditions of the Specifications and those contained in Contract between Owners and R.F. Ball Construction Company (A Joint Venture), where applicable.

ARTICLE IX. The Contractor shall prosecute the work rapidly, continuously and uninterruptedly and shall complete the entire work within *as required* so as not to interfere or delay R.F. Ball — Southwest Ball Construction Company in completing contract on the entire building *as required by Owner.*

The Contractor in agreeing to complete the work within the time mentioned, has taken into consideration and made allowance for the ordinary delays and hindrances incident to such work, whether growing out of delays of Common Carriers, delays in securing materials or workmen, changes, omissions, alterations, or otherwise, and is cognizant of the fact that R.F. Ball — Southwest Ball Construction Company in their contract with the Owner, is required to finish the entire work within a certain time, and failure on their part to do so renders them liable to heavy damages.

For the convenience of the Contractor, R.F. Ball — Southwest Ball Construction Company estimates that the building or buildings will be ready for the commencement of the work or its installation about *when notified,* 195...

But R.F. Ball — Southwest Ball Construction Company is not to be held responsible for any loss or damage incurred by the Contractor in the event the Contractor is unable to start or complete its work as herein contemplated.

Delays by the Contractor occasioned by failure on the part of Contractor to coordinate his work with that of other contractors on the job, or failure on the part of other contractors to coordinate their work with that of Contractor, shall be at and for the Contractor's responsibility; and such delays, when so occasioned, shall not relieve Contractor from any of its liabilities hereunder.

APPENDIX D

Scheduling Clause from *C. H. Leavell & Company* GSBCA No. 2901 70-2 BCA ¶8437 (1970)

2-34 Progress Charts and Construction Schedules 2-34 (d): In addition to the aforementioned requirements for a progress schedule, the Contractor shall, within 30 days after receipt of notice to proceed, prepare and submit for approval a network schedule of the project. This network shall be of the type generally referred to as the critical path method (CPM), critical path schedule (CPS), and critical path analysis (CPA), and other similar designations. The network analysis shall include as a minimum a graphic representation of all activities and events included in the construction of the project. The network shall show the major subdivisions of the work, the activities involved in each part, the relations of activities to each other as to dependency, concurrency, etc., the duration of each activity, and the critical path of activities for the overall job showing a total job duration equal to the contract time. A written statement of explanation is to be submitted with the network designation.

APPENDIX E

A.I.A. Contracts

THE AMERICAN INSTITUTE OF ARCHITECTS

AIA Document A101

Standard Form of Agreement Between Owner and Contractor

where the basis of payment is a

STIPULATED SUM

1977 EDITION

THIS DOCUMENT HAS IMPORTANT LEGAL CONSEQUENCES; CONSULTATION WITH AN ATTORNEY IS ENCOURAGED WITH RESPECT TO ITS COMPLETION OR MODIFICATION

Use only with the 1976 Edition of AIA Document A201, General Conditions of the Contract for Construction.

This document has been approved and endorsed by The Associated General Contractors of America.

AGREEMENT

made as of the day of in the year of Nineteen Hundred and

BETWEEN the Owner:

and the Contractor:

The Project:

The Architect:

The Owner and the Contractor agree as set forth below.

ARTICLE 1

THE CONTRACT DOCUMENTS

The Contract Documents consist of this Agreement, the Conditions of the Contract (General, Supplementary and other Conditions), the Drawings, the Specifications, all Addenda issued prior to and all Modifications issued after execution of this Agreement. These form the Contract, and all are as fully a part of the Contract as if attached to this Agreement or repeated herein. An enumeration of the Contract Documents appears in Article 7.

ARTICLE 2

THE WORK

The Contractor shall perform all the Work required by the Contract Documents for
(Here insert the caption descriptive of the Work as used on other Contract Documents.)

ARTICLE 3

TIME OF COMMENCEMENT AND SUBSTANTIAL COMPLETION

The Work to be performed under this Contract shall be commenced

and, subject to authorized adjustments, Substantial Completion shall be achieved not later than

(Here insert any special provisions for liquidated damages relating to failure to complete on time.)

ARTICLE 4

CONTRACT SUM

The Owner shall pay the Contractor in current funds for the performance of the Work, subject to additions and deductions by Change Order as provided in the Contract Documents, the Contract Sum of

The Contract Sum is determined as follows:
(State here the base bid or other lump sum amount, accepted alternates, and unit prices, as applicable.)

ARTICLE 5

PROGRESS PAYMENTS

Based upon Applications for Payment submitted to the Architect by the Contractor and Certificates for Payment issued by the Architect, the Owner shall make progress payments on account of the Contract Sum to the Contractor as provided in the Contract Documents for the period ending the day of the month as follows:

Not later than days following the end of the period covered by the Application for Payment percent (%) of the portion of the Contract Sum properly allocable to labor, materials and equipment incorporated in the Work and percent (%) of the portion of the Contract Sum properly allocable to materials and equipment suitably stored at the site or at some other location agreed upon in writing, for the period covered by the Application for Payment, less the aggregate of previous payments made by the Owner; and upon Substantial Completion of the entire Work, a sum sufficient to increase the total payments to percent (%) of the Contract Sum, less such amounts as the Architect shall determine for all incomplete Work and unsettled claims as provided in the Contract Documents.

(If not covered elsewhere in the Contract Documents, here insert any provision for limiting or reducing the amount retained after the Work reaches a certain stage of completion.)

Payments due and unpaid under the Contract Documents shall bear interest from the date payment is due at the rate entered below, or in the absence thereof, at the legal rate prevailing at the place of the Project.
(Here insert any rate of interest agreed upon.)

(Usury laws and requirements under the Federal Truth in Lending Act, similar state and local consumer credit laws and other regulations at the Owner's and Contractor's principal places of business, the location of the Project and elsewhere may affect the validity of this provision. Specific legal advice should be obtained with respect to deletion, modification, or other requirements such as written disclosures or waivers.)

ARTICLE 6

FINAL PAYMENT

Final payment, constituting the entire unpaid balance of the Contract Sum, shall be paid by the Owner to the Contractor when the Work has been completed, the Contract fully performed, and a final Certificate for Payment has been issued by the Architect.

ARTICLE 7

MISCELLANEOUS PROVISIONS

7.1 Terms used in this Agreement which are defined in the Conditions of the Contract shall have the meanings designated in those Conditions.

7.2 The Contract Documents, which constitute the entire agreement between the Owner and the Contractor, are listed in Article 1 and, except for Modifications issued after execution of this Agreement, are enumerated as follows:
(List below the Agreement, the Conditions of the Contract (General, Supplementary, and other Conditions), the Drawings, the Specifications, and any Addenda and accepted alternates, showing page or sheet numbers in all cases and dates where applicable.)

This Agreement entered into as of the day and year first written above.

OWNER	CONTRACTOR
______________________	______________________
______________________	______________________
______________________	______________________

THE AMERICAN INSTITUTE OF ARCHITECTS

AIA DOCUMENT A101/CM

CONSTRUCTION MANAGEMENT EDITION

Standard Form of Agreement Between Owner and Contractor

where the basis of payment is a

STIPULATED SUM

THIS DOCUMENT HAS IMPORTANT LEGAL CONSEQUENCES; CONSULTATION WITH AN ATTORNEY IS ENCOURAGED WITH RESPECT TO ITS COMPLETION OR MODIFICATION

Use only with the latest edition of AIA Document A201/CM, General Conditions of the Contract for Construction, Construction Management Edition.

AGREEMENT

made this day of in the year of Nineteen Hundred and

BETWEEN the Owner:

and the Contractor:

For the following Project:
(Include detailed description of Project location and scope)

the Construction Manager:

the Architect:

The Owner and the Contractor agree as set forth below.

ARTICLE 1

THE CONTRACT DOCUMENTS

The Contract Documents consist of this Agreement and any Exhibits attached hereto, Conditions of the Contract (General, Supplementary and other Conditions), Drawings, Specifications, all Addenda issued prior to execution of this Agreement and all Modifications issued subsequent thereto. These form the Contract, and all are as fully a part of the Contract as if attached to this Agreement or repeated herein. An enumeration of the Contract Documents appears in Article 8.

ARTICLE 2

THE WORK

The Contractor shall perform all the Work required by the Contract Documents for

(Here insert a precise description of the Work covered by this Contract and refer to numbers of Drawings and pages of Specifications including Addenda and accepted Alternates.)

ARTICLE 3

TIME OF COMMENCEMENT AND COMPLETION

(Here insert the specific provisions that are applicable to this Contract including any information pertaining to notice to proceed or other method of notification for commencement of Work starting and completion dates or duration and any provisions for liquidated damages relating to failure to complete on time.)

Time is of the essence of this Contract.

ARTICLE 4

CONTRACT SUM

The Owner shall pay the Contractor for the performance of the Work, subject to additions and deductions by Change Order as provided in the Conditions of the Contract, in current funds, the Contract Sum of

(State here the lump sum amount, unit prices, or both, as desired.)

ARTICLE 5

PROGRESS PAYMENTS

Based upon Applications for Payment submitted to the Construction Manager by the Contractor and Certificates for Payment issued by the Architect, the Owner shall make progress payments on account of the Contract Sum to the Contractor as provided in the Conditions of the Contract as follows:

On or about the day of each month per cent of the proportion of the Contract Sum properly allocable to labor, materials and equipment incorporated in the Work and per cent of the portion of the Contract Sum properly allocable to materials and equipment suitably stored at the site or at some other location agreed upon in writing by the parties up to days prior to the date on which the Application for Payment is submitted less the aggregate of previous payments in each case; and upon Substantial Completion of the entire Work, a sum sufficient to increase the total payments to per cent of the Contract Sum, less such retainages as the Architect shall determine for all incomplete Work and unsettled claims.

(If not covered elsewhere in the Contract Documents, here insert any provision for limiting or reducing the amount retained after the Work reaches a certain stage of completion.)

Any moneys not paid when due to either party under this Contract shall bear interest at the legal rate in force at the place of the Project.

ARTICLE 6

FINAL PAYMENT

Final payment, constituting the entire unpaid balance of the Contract Sum, shall be paid by the Owner to the Contractor days after Substantial Completion of the Work unless otherwise stipulated in the Certificate of Substantial Completion, provided the Work has then been completed, the Contract fully performed, and a final Certificate for Payment has been issued by the Architect.

ARTICLE 7

TEMPORARY FACILITIES AND SERVICES

(Here insert temporary facilities and services which are different from or in addition to those included elsewhere in the Contract Documents.)

ARTICLE 8

MISCELLANEOUS PROVISIONS

8.1 Terms used in this Agreement which are defined in the Conditions of the Contract shall have the meanings designated in those Conditions.

8.2 Working conditions:

(List below any special conditions affecting the Contract.)

8.3 The Contract Documents, which constitute the entire agreement between the Owner and the Contractor, are listed in Article 1 and, except for Modifications issued after execution of this Agreement, are enumerated as follows:

(List below the Agreement and any exhibits thereto, Conditions of the Contract (General, Supplementary, and other Conditions), Drawings, Specifications, Addenda and accepted Alternates, showing page or sheet numbers in all cases and dates where applicable.)

This Agreement executed the day and year first written above.

OWNER	CONTRACTOR
____________________	____________________
____________________	____________________

THE AMERICAN INSTITUTE OF ARCHITECTS

AIA Document A107

Abbreviated Form of Agreement Between Owner and Contractor

For CONSTRUCTION PROJECTS OF LIMITED SCOPE where
the Basis of Payment is a STIPULATED SUM

1978 EDITION

THIS DOCUMENT HAS IMPORTANT LEGAL CONSEQUENCES; CONSULTATION WITH AN ATTORNEY IS ENCOURAGED WITH RESPECT TO ITS COMPLETION OR MODIFICATION

This document includes abbreviated General Conditions and should not be used with other General Conditions.
It has been approved and endorsed by The Associated General Contractors of America.

AGREEMENT

made as of the day of in the year of Nineteen Hundred and

BETWEEN the Owner:

and the Contractor:

the Project:

the Architect:

The Owner and the Contractor agree as set forth below.

ARTICLE 1
THE WORK

1.1 The Contractor shall perform all the Work required by the Contract Documents for

(Here insert the caption descriptive of the Work as used on other Contract Documents.)

ARTICLE 2
TIME OF COMMENCEMENT AND SUBSTANTIAL COMPLETION

2.1 The Work to be performed under this Contract shall be commenced

and, subject to authorized adjustments, Substantial Completion shall be achieved not later than

(Here insert any special provisions for liquidated damages relating to failure to complete on time.)

ARTICLE 3
CONTRACT SUM

3.1 The Owner shall pay the Contractor in current funds for the performance of the Work, subject to additions and deductions by Change Order as provided in the Contract Documents, the Contract Sum of

3.2 The Contract Sum is determined as follows:

(State here the base bid or other lump sum amount, accepted alternates, and unit prices, as applicable.)

ARTICLE 4
PROGRESS PAYMENTS

4.1 Based upon Applications for Payment submitted to the Architect by the Contractor and Certificates for Payment issued by the Architect, the Owner shall make progress payments on account of the Contract Sum to the Contractor as provided in the Contract Documents for the period ending the day of the month as follows:

(Here insert payment procedures and provision for retainage, if any.)

4.2 Payments due and unpaid under the Contract Documents shall bear interest from the date payment is due at the rate entered below, or in the absence thereof, at the legal rate prevailing at the place of the Project.

(Here insert any rate of interest agreed upon.)

(Usury laws and requirements under the Federal Truth in Lending Act, similar state and local consumer credit laws and other regulations at the Owner's and Contractor's principal places of business, the location of the Project and elsewhere may affect the validity of this provision. Specific legal advice should be obtained with respect to deletion, modification, or other requirements such as written disclosure or waivers.)

ARTICLE 5
FINAL PAYMENT

5.1 Final payment, constituting the entire unpaid balance of the Contract Sum, shall be paid by the Owner to the Contractor when the Work has been completed, the Contract fully performed, and a final Certificate for Payment has been issued by the Architect.

ARTICLE 6
ENUMERATION OF CONTRACT DOCUMENTS

6.1 The Contract Documents, which constitute the entire agreement between the Owner and the Contractor, are listed in Article 7 and, except for Modifications issued after execution of this Agreement, are enumerated as follows:

(List below the Agreement, the Conditions of the Contract, [General, Supplementary, and other Conditions], the Drawings, the Specifications, and any Addenda and accepted alternates, showing page or sheet numbers in all cases and dates where applicable.)

GENERAL CONDITIONS

ARTICLE 7
CONTRACT DOCUMENTS

7.1 The Contract Documents consist of this Agreement with General Conditions, Supplementary and other Conditions, the Drawings, the Specifications, all Addenda issued prior to the execution of this Agreement, and all Modifications issued by the Architect after execution of the Contract such as Change Orders, written interpretations and written orders for minor changes in the Work. The intent of the Contract Documents is to include all items necessary for the proper execution and completion of the Work. The Contract Documents are complementary, and what is required by any one shall be as binding as if required by all. Work not covered in the Contract Documents will not be required unless it is consistent therewith and reasonably inferable therefrom as being necessary to produce the intended results.

7.2 Nothing contained in the Contract Documents shall create any contractual relationship between the Owner or the Architect and any Subcontractor or Sub-subcontractor.

7.3 By executing the Contract, the Contractor represents that he has visited the site and familiarized himself with the local conditions under which the Work is to be performed.

7.4 The Work comprises the completed construction required by the Contract Documents and includes all labor necessary to produce such construction, and all materials and equipment incorporated or to be incorporated in such construction.

ARTICLE 8
ARCHITECT

8.1 The Architect will provide administration of the Contract and will be the Owner's representative during construction and until final payment is due.

8.2 The Architect shall at all times have access to the Work wherever it is in preparation and progress.

8.3 The Architect will visit the site at intervals appropriate to the stage of construction to familiarize himself generally with the progress and quality of the Work and to determine in general if the Work is proceeding in accordance with the Contract Documents. However, the Architect will not be required to make exhaustive or continuous on-site inspections to check the quality or quantity of the Work. On the basis of his on-site observations as an architect, he will keep the Owner informed of the progress of the Work, and will endeavor to guard the Owner against defects and deficiencies in the Work of the Contractor. The Architect will not have control or charge of and will not be responsible for construction means, methods, techniques, sequences or procedures, or for safety precautions and programs in connection with the Work, and he will not be responsible for the Contractor's failure to carry out the Work in accordance with the Contract Documents.

8.4 Based on the Architect's observations and an evaluation of the Contractor's Applications for Payment, the Architect will determine the amounts owing to the Contractor and will issue Certificates for Payment in accordance with Article 15.

8.5 The Architect will be the interpreter of the requirements of the Contract Documents. He will make decisions on all claims, disputes or other matters in question between the Contractor and the Owner, but he will not be liable for the results of any interpretation or decision rendered in good faith. The Architect's decisions in matters relating to artistic effect will be final if consistent with the intent of the Contract Documents. All other decisions of the Architect, except those which have been waived by the making or acceptance of final payment, shall be subject to arbitration upon the written demand of either party.

8.6 The Architect will have authority to reject Work which does not conform to the Contract Documents.

8.7 The Architect will review and approve or take other appropriate action upon the Contractor's submittals such as Shop Drawings, Product Data and Samples, but only for conformance with the design concept of the Work and with the information given in the Contract Documents.

ARTICLE 9
OWNER

9.1 The Owner shall furnish all surveys and a legal description of the site.

9.2 Except as provided in Paragraph 10.5, the Owner shall secure and pay for necessary approvals, easements, assessments and charges required for the construction, use or occupancy of permanent structures or permanent changes in existing facilities.

9.3 The Owner shall forward all instructions to the Contractor through the Architect.

9.4 If the Contractor fails to correct defective Work or persistently fails to carry out the Work in accordance with the Contract Documents, the Owner, by a written order, may order the Contractor to stop the Work, or any portion thereof, until the cause for such order has been eliminated; however, this right of the Owner to stop the Work shall not give rise to any duty on the part of the Owner to exercise this right for the benefit of the Contractor or any other person or entity.

ARTICLE 10
CONTRACTOR

10.1 The Contractor shall supervise and direct the Work, using his best skill and attention and he shall be solely responsible for all construction means, methods, techniques, sequences and procedures and for coordinating all portions of the Work under the Contract.

10.2 Unless otherwise specifically provided in the Contract Documents, the Contractor shall provide and pay for all labor, materials, equipment, tools, construction equipment and machinery, water, heat, utilities, transportation, and other facilities and services necessary for the proper execution and completion of the Work, whether temporary or permanent and whether or not incorporated or to be incorporated in the Work.

10.3 The Contractor shall at all times enforce strict discipline and good order among his employees and shall not employ on the Work any unfit person or anyone not skilled in the task assigned to him.

10.4 The Contractor warrants to the Owner and the Architect that all materials and equipment incorporated in the Work will be new unless otherwise specified, and that all Work will be of good quality, free from faults and defects and in conformance with the Contract Documents. All Work not conforming to these requirements may be considered defective.

10.5 Unless otherwise provided in the Contract Documents, the Contractor shall pay all sales, consumer, use and other similar taxes which are legally enacted at the time bids are received, and shall secure and pay for the building permit and for all other permits and governmental fees, licenses and inspections necessary for the proper execution and completion of the Work.

10.6 The Contractor shall give all notices and comply with all laws, ordinances, rules, regulations, and lawful orders of any public authority bearing on the performance of the Work, and shall promptly notify the Architect if the Drawings and Specifications are at variance therewith.

10.7 The Contractor shall be responsible to the Owner for the acts and omissions of his employees, Subcontractors and their agents and employees, and other persons performing any of the Work under a contract with the Contractor.

10.8 The Contractor shall review, approve and submit all Shop Drawings, Product Data and Samples required by the Contract Documents. The Work shall be in accordance with approved submittals.

10.9 The Contractor at all times shall keep the premises free from accumulation of waste materials or rubbish caused by his operations. At the completion of the Work he shall remove all his waste materials and rubbish from and about the Project as well as his tools, construction equipment, machinery and surplus materials.

10.10 The Contractor shall pay all royalties and license fees. He shall defend all suits or claims for infringement of any patent rights and shall save the Owner harmless from loss on account thereof.

10.11 To the fullest extent permitted by law, the Contractor shall indemnify and hold harmless the Owner and the Architect and their agents and employees from and against all claims, damages, losses and expenses, including but not limited to attorneys' fees arising out of or resulting from the performance of the Work, provided that any such claim, damage, loss or expense (1) is attributable to bodily injury, sickness, disease or death, or to injury to or destruction of tangible property (other than the Work itself) including the loss of use resulting therefrom, and (2) is caused in whole or in part by any negligent act or omission of the Contractor, any Subcontractor, anyone directly or indirectly employed by any of them or anyone for whose acts any of them may be liable, regardless of whether or not it is caused in part by a party indemnified hereunder. Such obligation shall not be construed to negate, abridge, or otherwise reduce any other right or obligation of indemnity which would otherwise exist as to any party or person described in this Paragraph 10.11. In any and all claims against the Owner or the Architect or any of their agents or employees by any employee of the Contractor, any Subcontractor, anyone directly or indirectly employed by any of them or anyone for whose acts any of them may be liable, the indemnification obligation under this Paragraph 10.11 shall not be limited in any way by any limitation on the amount or type of damages, compensation or benefits payable by or for the Contractor or any Subcontractor under workers' or workmen's compensation acts, disability benefit acts or other employee benefit acts. The obligations of the Contractor under this Paragraph 10.11 shall not extend to the liability of the Architect, his agents or employees, arising out of (1) the preparation or approval of maps, drawings, opinions, reports, surveys, change orders, designs or specifications, or (2) the giving of or the failure to give directions or instructions by the Architect, his agents or employees provided such giving or failure to give is the primary cause of the injury or damage.

ARTICLE 11
SUBCONTRACTS

11.1 A Subcontractor is a person or entity who has a direct contract with the Contractor to perform any of the Work at the site.

11.2 Unless otherwise required by the Contract Documents or in the Bidding Documents, the Contractor, as soon as practicable after the award of the Contract, shall furnish to the Architect in writing the names of Subcontractors for each of the principal portions of the Work. The Contractor shall not employ any Subcontractor to whom the Architect or the Owner may have a reasonable objection. The Contractor shall not be required to contract with anyone to whom he has a reasonable objection. Contracts between the Contractor and the Subcontractors shall (1) require each Subcontractor, to the extent of the Work to be performed by the Subcontractor, to be bound to the Contractor by the terms of the Contract Documents, and to assume toward the Contractor all the obligations and responsibilities which the Contractor, by these Documents, assumes toward the Owner and the Architect, and (2) allow to the Subcontractor the benefit of all rights, remedies and redress afforded to the Contractor by these Contract Documents.

ARTICLE 12
WORK BY OWNER OR BY SEPARATE CONTRACTORS

12.1 The Owner reserves the right to perform work related to the Project with his own forces, and to award separate contracts in connection with other portions of the Project or other work on the site under these or similar Conditions of the Contract. If the Contractor claims that delay or additional cost is involved because of such action by the Owner, he shall make such claim as provided elsewhere in the Contract Documents.

12.2 The Contractor shall afford the Owner and separate contractors reasonable opportunity for the introduction and storage of their materials and equipment and the execution of their work, and shall connect and coordinate his Work with theirs as required by the Contract Documents.

12.3 Any costs caused by defective or ill-timed work shall be borne by the party responsible therefor.

ARTICLE 13
MISCELLANEOUS PROVISIONS

13.1 The Contract shall be governed by the law of the place where the Project is located.

13.2 All claims or disputes between the Contractor and the Owner arising out of, or relating to, the Contract Documents or the breach thereof shall be decided by arbitration in accordance with the Construction Industry Arbitration Rules of the American Arbitration Association then obtaining unless the parties mutually agree otherwise. Notice of the demand for arbitration shall be filed in writing with the other party to the Owner-Contractor Agreement and with the American Arbitration Association and shall be made within a reasonable time after the dispute has arisen. The award rendered by the arbitrators shall be final, and judgment may be entered upon it in accordance with applicable law in any court having jurisdiction thereof. Except by written consent of the person or entity sought to be joined, no arbitration arising out of or relating to the Contract Documents shall include, by consolidation, joinder or in any other manner, any person or entity not a party to the agreement under which such arbitration arises, unless it is shown at the time the demand for arbitration is filed that (1) such person or entity is substantially involved in a common question of fact or law, (2) the presence of such person or entity is required if complete relief is to be accorded in the arbitration, (3) the interest or responsibility of such person or entity in the matter is not insubstantial, and (4) such person or entity is not the Architect or any of his employees or consultants. The agreement herein among the parties to the Agreement and any other written agreement to arbitrate referred to herein shall be specifically enforceable under the prevailing arbitration law.

ARTICLE 14
TIME

14.1 All time limits stated in the Contract Documents are of the essence of the Contract. The Contractor shall expedite the Work and achieve Substantial Completion within the Contract Time.

14.2 The Date of Substantial Completion of the Work is the date certified by the Architect when construction is sufficiently complete so that the Owner can occupy or utilize the Work for the use for which it is intended.

14.3 If the Contractor is delayed at any time in the progress of the Work by changes ordered in the Work, by labor disputes, fire, unusual delay in transportation, adverse weather conditions not reasonably anticipatable, unavoidable casualties, or any causes beyond the Contractor's control, or by any other cause which the Architect determines may justify the delay, then the Contract Time shall be extended by Change Order for such reasonable time as the Architect may determine.

ARTICLE 15
PAYMENTS AND COMPLETION

15.1 Payments shall be made as provided in Article 4 and Article 5 of this Agreement.

15.2 Payments may be withheld on account of (1) defective work not remedied, (2) claims filed, (3) failure of the Contractor to make payments properly to Subcontractors or for labor, materials, or equipment, (4) damage to the Owner or another contractor, or (5) persistent failure to carry out the Work in accordance with the Contract Documents.

15.3 When the Architect agrees that the Work is substantially complete, he will issue a Certificate of Substantial Completion.

15.4 Final payment shall not be due until the Contractor has delivered to the Owner a complete release of all liens arising out of this Contract or receipts in full covering all labor, materials and equipment for which a lien could be filed, or a bond satisfactory to the Owner indemnifying him against any lien. If any lien remains unsatisfied after all payments are made, the Contractor shall refund to the Owner all moneys the latter may be compelled to pay in discharging such lien, including all costs and reasonable attorneys' fees.

15.5 The making of final payments shall constitute a waiver of all claims by the Owner except those arising from (1) unsettled liens, (2) faulty or defective Work appearing after Substantial Completion, (3) failure of the Work to comply with the requirements of the Contract Documents, or (4) terms of any special warranties required by the Contract Documents. The acceptance of final payment shall constitute a waiver of all claims by the Contractor except those previously made in writing and identified by the Contractor as unsettled at the time of the final Application for Payment.

ARTICLE 16
PROTECTION OF PERSONS AND PROPERTY

16.1 The Contractor shall be responsible for initiating, maintaining, and supervising all safety precautions and programs in connection with the Work. He shall take all reasonable precautions for the safety of, and shall provide all reasonable protection to prevent damage, injury or loss to (1) all employees on the Work and other persons who may be affected thereby, (2) all the Work and all materials and equipment to be incorporated therein, and (3) other property at the site or adjacent thereto. He shall give all notices and comply with all applicable laws, ordinances, rules, regulations and orders of any public authority bearing on the safety of persons and property and their protection from damage, injury or loss. The Contractor shall promptly remedy all damage or loss to any property caused in whole or in part by the Contractor, any Subcontractor, any Sub-subcontractor, or anyone directly or indirectly employed by any of them, or by anyone for whose acts any of them may be liable, except damage or loss attributable to the acts or omissions of the Owner or Architect or anyone directly or indirectly employed by either of them or by anyone for whose acts either of them may be liable, and not attributable to the fault or negligence of the Contractor. The foregoing obligations of the Contractor are in addition to his obligations under Paragraph 10.11.

ARTICLE 17
INSURANCE

17.1 Contractor's liability insurance shall be purchased and maintained by the Contractor to protect him from claims under workers' or workmen's compensation acts and other employee benefit acts, claims for damages because of bodily injury, including death, and from claims for damages, other than to the Work itself, to property which may arise out of or result from the Contractor's operations under this Contract, whether such operations be by himself or by any Subcontractor or anyone directly or indirectly employed by any of them. This insurance shall be written for not less than any limits of liability specified in the Contract Documents, or required by law, whichever is the greater, and shall include contractual liability insurance applicable to the Contractor's obligations under Paragraph 10.11. Certificates of such insurance shall be filed with the Owner prior to the commencement of the Work.

17.2 The Owner shall be responsible for purchasing and maintaining his own liability insurance and, at his option, may maintain such insurance as will protect him against claims which may arise from operations under the Contract.

17.3 Unless otherwise provided, the Owner shall purchase and maintain property insurance upon the entire Work at the site to the full insurable value thereof. This insurance shall include the interests of the Owner, the Contractor, Subcontractors and Sub-subcontractors in the Work and shall insure against the perils of fire and extended coverage and shall include "all risk" insurance for physical loss or damage including, without duplication of coverage, theft, vandalism, and malicious mischief.

17.4 Any loss insured under Paragraph 17.3 is to be adjusted with the Owner and made payable to the Owner as trustee for the insureds, as their interests may appear, subject to the requirements of any mortgagee clause.

17.5 The Owner shall file a copy of all policies with the Contractor before an exposure to loss may occur.

17.6 The Owner and Contractor waive all rights against each other for damages caused by fire or other perils to the extent covered by insurance obtained pursuant to this Article or any other property insurance applicable to the Work, except such rights as they may have to the proceeds of such insurance held by the Owner as trustee. The Contractor shall require similar waivers in favor of the Owner and the Contractor by Subcontractors and Sub-subcontractors.

ARTICLE 18
CHANGES IN THE WORK

18.1 The Owner, without invalidating the Contract, may order Changes in the Work consisting of additions, deletions, or modifications, the Contract Sum and the Contract Time being adjusted accordingly. All such changes in the Work shall be authorized by written Change Order signed by the Owner and the Architect.

18.2 The Contract Sum and the Contract Time may be changed only by Change Order.

18.3 The cost or credit to the Owner from a change in the Work shall be determined by mutual agreement.

ARTICLE 19
CORRECTION OF WORK

19.1 The Contractor shall promptly correct any Work rejected by the Architect as defective or as failing to conform to the Contract Documents whether observed before or after Substantial Completion and whether or not fabricated, installed or completed, and shall correct any Work found to be defective or nonconforming within a period of one year from the Date of Substantial Completion of the Contract or within such longer period of time as may be prescribed by law or by the terms of any applicable special warranty required by the Contract Documents. The provisions of this Article 19 apply to Work done by Subcontractors as well as to Work done by direct employees of the Contractor.

ARTICLE 20
TERMINATION OF THE CONTRACT

20.1 If the Architect fails to issue a Certificate for Payment for a period of thirty days through no fault of the Contractor, or if the Owner fails to make payment thereon for a period of thirty days, the Contractor may, upon seven additional days' written notice to the Owner and the Architect, terminate the Contract and recover from the Owner payment for all Work executed and for any proven loss sustained upon any materials, equipment, tools, and construction equipment and machinery, including reasonable profit and damages applicable to the Project.

20.2 If the Contractor defaults or persistently fails or neglects to carry out the Work in accordance with the Contract Documents or fails to perform any provision of the Contract, the Owner, after seven days' written notice to the Contractor and without prejudice to any other remedy he may have, may make good such deficiencies and may deduct the cost thereof including compensation for the Architect's additional services made necessary thereby, from the payment then or thereafter due the Contractor or, at his option, and upon certification by the Architect that sufficient cause exists to justify such action, may terminate the Contract and take possession of the site and of all materials, equipment, tools, and construction equipment and machinery thereon owned by the Contractor and may finish the Work by whatever method he may deem expedient, and if the unpaid balance of the Contract Sum exceeds the expense of finishing the Work, such excess shall be paid to the Contractor, but if such expense exceeds such unpaid balance, the Contractor shall pay the difference to the Owner.

ARTICLE 21
OTHER CONDITIONS OR PROVISIONS

This Agreement entered into as of the day and year first written above.

OWNER CONTRACTOR

THE AMERICAN INSTITUTE OF ARCHITECTS

AIA Document A111

Standard Form of Agreement Between Owner and Contractor

where the basis of payment is the

COST OF THE WORK PLUS A FEE

1978 EDITION

THIS DOCUMENT HAS IMPORTANT LEGAL CONSEQUENCES; CONSULTATION WITH AN ATTORNEY IS ENCOURAGED WITH RESPECT TO ITS COMPLETION OR MODIFICATION

Use only with the 1976 Edition of AIA Document A201, General Conditions of the Contract for Construction.

This document has been approved and endorsed by The Associated General Contactors of America

AGREEMENT

made as of the day of in the year of Nineteen Hundred and

BETWEEN the Owner:

and the Contractor:

the Project:

the Architect:

The Owner and the Contractor agree as set forth below.

ARTICLE 1

THE CONTRACT DOCUMENTS

1.1 The Contract Documents consist of this Agreement, the Conditions of the Contract (General, Supplementary and other Conditions), the Drawings, the Specifications, all Addenda issued prior to and all Modifications issued after execution of this Agreement. These form the Contract, and all are as fully a part of the Contract as if attached to this Agreement or repeated herein. An enumeration of the Contract Documents appears in Article 16. If anything in the Contract Documents is inconsistent with this Agreement, the Agreement shall govern.

ARTICLE 2

THE WORK

2.1 The Contractor shall perform all the Work required by the Contract Documents for

(Here insert the caption descriptive of the Work as used on other Contract Documents.)

ARTICLE 3

THE CONTRACTOR'S DUTIES AND STATUS

3.1 The Contractor accepts the relationship of trust and confidence established between him and the Owner by this Agreement. He covenants with the Owner to furnish his best skill and judgment and to cooperate with the Architect in furthering the interests of the Owner. He agrees to furnish efficient business administration and superintendence and to use his best efforts to furnish at all times an adequate supply of workmen and materials, and to perform the Work in the best way and in the most expeditious and economical manner consistent with the interests of the Owner.

ARTICLE 4

TIME OF COMMENCEMENT AND SUBSTANTIAL COMPLETION

4.1 The Work to be performed under this Contract shall be commenced

and, subject to authorized adjustments, Substantial Completion shall be achieved not later than

(Here insert any special provisions for liquidated damages relating to failure to complete on time.)

ARTICLE 5

COST OF THE WORK AND GUARANTEED MAXIMUM COST

5.1 The Owner agrees to reimburse the Contractor for the Cost of the Work as defined in Article 8. Such reimbursement shall be in addition to the Contractor's Fee stipulated in Article 6.

5.2 The maximum cost to the Owner, including the Cost of the Work and the Contractor's Fee, is guaranteed not to exceed the sum of dollars ($); such Guaranteed Maximum Cost shall be increased or decreased for Changes in the Work as provided in Article 7.

(Here insert any provision for distribution of any savings. Delete Paragraph 5.2 if there is no Guaranteed Maximum Cost.)

ARTICLE 6

CONTRACTOR'S FEE

6.1 In consideration of the performance of the Contract, the Owner agrees to pay the Contractor in current funds as compensation for his services a Contractor's Fee as follows:

6.2 For Changes in the Work, the Contractor's Fee shall be adjusted as follows:

6.3 The Contractor shall be paid percent (%) of the proportional amount of his Fee with each progress payment, and the balance of his Fee shall be paid at the time of final payment.

ARTICLE 7

CHANGES IN THE WORK

7.1 The Owner may make Changes in the Work as provided in the Contract Documents. The Contractor shall be reimbursed for Changes in the Work on the basis of Cost of the Work as defined in Article 8.

7.2 The Contractor's Fee for Changes in the Work shall be as set forth in Paragraph 6.2, or in the absence of specific provisions therein, shall be adjusted by negotiation on the basis of the Fee established for the original Work.

ARTICLE 8

COSTS TO BE REIMBURSED

8.1 The term Cost of the Work shall mean costs necessarily incurred in the proper performance of the Work and paid by the Contractor. Such costs shall be at rates not higher than the standard paid in the locality of the Work except with prior consent of the Owner, and shall include the items set forth below in this Article 8.

8.1.1 Wages paid for labor in the direct employ of the Contractor in the performance of the Work under applicable collective bargaining agreements, or under a salary or wage schedule agreed upon by the Owner and Contractor, and including such welfare or other benefits, if any, as may be payable with respect thereto.

8.1.2 Salaries of Contractor's personnel when stationed at the field office, in whatever capacity employed. Personnel engaged, at shops or on the road, in expediting the production or transportation of materials or equipment, shall be considered as stationed at the field office and their salaries paid for that portion of their time spent on this Work.

8.1.3 Cost of contributions, assessments or taxes incurred during the performance of the Work for such items as unemployment compensation and social security, insofar as such cost is based on wages, salaries, or other remuneration paid to employees of the Contractor and included in the Cost of the Work under Subparagraphs 8.1.1 and 8.1.2.

8.1.4 The portion of reasonable travel and subsistence expenses of the Contractor or of his officers or employees incurred while traveling in discharge of duties connected with the Work.

8.1.5 Cost of all materials, supplies and equipment incorporated in the Work, including costs of transportation thereof.

8.1.6 Payments made by the Contractor to Subcontractors for Work performed pursuant to subcontracts under this Agreement.

8.1.7 Cost, including transportation and maintenance, of all materials, supplies, equipment, temporary facilities and hand tools not owned by the workers, which are consumed in the performance of the Work, and cost less salvage value on such items used but not consumed which remain the property of the Contractor.

8.1.8 Rental charges of all necessary machinery and equipment, exclusive of hand tools, used at the site of the Work, whether rented from the Contractor or others, including installation, minor repairs and replacements, dismantling, removal, transportation and delivery costs thereof, at rental changes consistent with those prevailing in the area.

8.1.9 Cost of premiums for all bonds and insurance which the Contractor is required by the Contract Documents to purchase and maintain.

8.1.10 Sales, use or similar taxes related to the Work and for which the Contractor is liable imposed by any governmental authority.

8.1.11 Permit fees, royalties, damages for infringement of patents and costs of defending suits therefor, and deposits lost for causes other than the Contractor's negligence.

8.1.12 Losses and expenses, not compensated by insurance or otherwise, sustained by the Contractor in connection with the Work, provided they have resulted from causes other than the fault or neglect of the Contractor. Such losses shall include settlements made with the written consent and approval of the Owner. No such losses and expenses shall be included in the Cost of the Work for the purpose of determining the Contractor's Fee. If, however, such loss requires reconstruction and the Contractor is placed in charge thereof, he shall be paid for his services a Fee proportionate to that stated in Paragraph 6.1.

8.1.13 Minor expenses such as telegrams, long distance telephone calls, telephone service at the site, expressage, and similar petty cash items in connection with the Work.

8.1.14 Cost of removal of all debris.

8.1.15 Costs incurred due to an emergency affecting the safety of persons and property.

8.1.16 Other costs incurred in the performance of the Work if and to the extent approved in advance in writing by the Owner.

(Here insert modifications or limitations to any of the above Subparagraphs, such as equipment rental charges and small tool charges applicable to the Work.)

ARTICLE 9

COSTS NOT TO BE REIMBURSED

9.1 The term Cost of the Work shall not include any of the items set forth below in this Article 9.

9.1.1 Salaries or other compensation of the Contractor's personnel at the Contractor's principal office and branch offices.

9.1.2 Expenses of the Contractor's principal and branch offices other than the field office.

9.1.3 Any part of the Contractor's capital expenses, including interest on the Contractor's capital employed for the Work.

9.1.4 Except as specifically provided for in Subparagraph 8.1.8 or in modifications thereto, rental costs of machinery and equipment.

9.1.5 Overhead or general expenses of any kind, except as may be expressly included in Article 8.

9.1.6 Costs due to the negligence of the Contractor, any Subcontractor, anyone directly or indirectly employed by any of them, or for whose acts any of them may be liable, including but not limited to the correction of defective or nonconforming Work, disposal of materials and equipment wrongly supplied, or making good any damage to property.

9.1.7 The cost of any item not specifically and expressly included in the items described in Article 8.

9.1.8 Costs in excess of the Guaranteed Maximum Cost, if any, as set forth in Article 5 and adjusted pursuant to Article 7.

ARTICLE 10

DISCOUNTS, REBATES AND REFUNDS

10.1 All cash discounts shall accrue to the Contractor unless the Owner deposits funds with the Contractor with which to make payments, in which case the cash discounts shall accrue to the Owner. All trade discounts, rebates and refunds, and all returns from sale of surplus materials and equipment shall accrue to the Owner, and the Contractor shall make provisions so that they can be secured.

(Here insert any provisions relating to deposits by the Owner to permit the Contractor to obtain cash discounts.)

ARTICLE 11

SUBCONTRACTS AND OTHER AGREEMENTS

11.1 All portions of the Work that the Contractor's organization does not perform shall be performed under Subcontracts or by other appropriate agreement with the Contractor. The Contractor shall request bids from Subcontractors and shall deliver such bids to the Architect. The Owner will then determine, with the advice of the Contractor and subject to the reasonable objection of the Architect, which bids will be accepted.

11.2 All Subcontracts shall conform to the requirements of the Contract Documents. Subcontracts awarded on the basis of the cost of such work plus a fee shall also be subject to the provisions of this Agreement insofar as applicable.

ARTICLE 12

ACCOUNTING RECORDS

12.1 The Contractor shall check all materials, equipment and labor entering into the Work and shall keep such full and detailed accounts as may be necessary for proper financial management under this Agreement, and the system shall be satisfactory to the Owner. The Owner shall be afforded access to all the Contractor's records, books, correspondence, instructions, drawings, receipts, vouchers, memoranda and similar data relating to this Contract, and the Contractor shall preserve all such records for a period of three years, or for such longer period as may be required by law, after the final payment.

ARTICLE 13

APPLICATIONS FOR PAYMENT

13.1 The Contractor shall, at least ten days before each payment falls due, deliver to the Architect an itemized statement, notarized if required, showing in complete detail all moneys paid out or costs incurred by him on account of the Cost of the Work during the previous month for which he is to be reimbursed under Article 5 and the amount of the Contractor's Fee due as provided in Article 6, together with payrolls for all labor and such other data supporting the Contractor's right to payment for Subcontracts or materials as the Owner or the Architect may require.

ARTICLE 14

PAYMENTS TO THE CONTRACTOR

14.1 The Architect will review the Contractor's Applications for Payment and will promptly take appropriate action thereon as provided in the Contract Documents. Such amount as he may recommend for payment shall be payable by the Owner not later than the day of the month.

14.1.1 In taking action on the Contractor's Applications for Payment, the Architect shall be entitled to rely on the accuracy and completeness of the information furnished by the Contractor and shall not be deemed to represent that he has made audits of the supporting data, exhaustive or continuous on-site inspections or that he has made any examination to ascertain how or for what purposes the Contractor has used the moneys previously paid on account of the Contract.

14.2 Final payment, constituting the entire unpaid balance of the Cost of the Work and of the Contractor's Fee, shall be paid by the Owner to the Contractor days after Substantial Completion of the Work unless otherwise stipulated in the Certificate of Substantial Completion, provided the Work has been completed, the Contract fully performed, and final payment has been recommended by the Architect.

14.3 Payments due and unpaid under the Contract Documents shall bear interest from the date payment is due at the rate entered below, or in the absence thereof, at the legal rate prevailing at the place of the Project.

(Here insert any rate of interest agreed upon.)

(Usury laws and requirements under the Federal Truth in Lending Act, similar state and local consumer credit laws and other regulations at the Owner's and Contractor's principal places of business, the location of the Project and elsewhere may affect the validity of this provision. Specific legal advice should be obtained with respect to deletion, modification, or other requirements such as written disclosures or waivers.)

ARTICLE 15

TERMINATION OF CONTRACT

15.1 The Contract may be terminated by the Contractor as provided in the Contract Documents.

15.2 If the Owner terminates the Contract as provided in the Contract Documents, he shall reimburse the Contractor for any unpaid Cost of the Work due him under Article 5, plus (1) the unpaid balance of the Fee computed upon the Cost of the Work to the date of termination at the rate of the percentage named in Article 6, or (2) if the Contractor's Fee be stated as a fixed sum, such an amount as will increase the payments on account of his Fee to a sum which bears the same ratio to the said fixed sum as the Cost of the Work at the time of termination bears to the adjusted Guaranteed Maximum Cost, if any, otherwise to a reasonable estimated Cost of the Work when completed. The Owner shall also pay to the Contractor fair compensation, either by purchase or rental at the election of the Owner, for any equipment retained. In case of such termination of the Contract the Owner shall further assume and become liable for obligations, commitments and unsettled claims that the Contractor has previously undertaken or incurred in good faith in connection with said Work. The Contractor shall, as a condition of receiving the payments referred to in this Article 15, execute and deliver all such papers and take all such steps, including the legal assignment of his contractual rights, as the Owner may require for the purpose of fully vesting in himself the rights and benefits of the Contractor under such obligations or commitments.

ARTICLE 16

MISCELLANEOUS PROVISIONS

16.1 Terms used in this Agreement which are defined in the Contract Documents shall have the meanings designated in those Contract Documents.

16.2 The Contract Documents, which constitute the entire agreement between the Owner and the Contractor, are listed in Article 1 and, except for Modifications issued after execution of this Agreement, are enumerated as follows:

(List below the Agreement, the Conditions of the Contract, [General, Supplementary, and other Conditions], the Drawings, the Specifications, and any Addenda and accepted Alternates, showing page or sheet numbers in all cases and dates where applicable.)

This Agreement entered into as of the day and year first written above.

OWNER	CONTRACTOR
________________	________________
________________	________________
________________	________________

THE AMERICAN INSTITUTE OF ARCHITECTS

AIA Document A201

General Conditions of the Contract for Construction

THIS DOCUMENT HAS IMPORTANT LEGAL CONSEQUENCES; CONSULTATION WITH AN ATTORNEY IS ENCOURAGED WITH RESPECT TO ITS MODIFICATION

1976 EDITION

TABLE OF ARTICLES

1. CONTRACT DOCUMENTS
2. ARCHITECT
3. OWNER
4. CONTRACTOR
5. SUBCONTRACTORS
6. WORK BY OWNER OR BY SEPARATE CONTRACTORS
7. MISCELLANEOUS PROVISIONS
8. TIME
9. PAYMENTS AND COMPLETION
10. PROTECTION OF PERSONS AND PROPERTY
11. INSURANCE
12. CHANGES IN THE WORK
13. UNCOVERING AND CORRECTION OF WORK
14. TERMINATION OF THE CONTRACT

This document has been approved and endorsed by The Associated General Contractors of America.

INDEX

Acceptance of Defective or Non-Conforming Work . . 6.2.2, **13.3**
Acceptance of Work 9.5.5, 9.8.1, 9.9.1, 9.9.3
Access to Work 2.2.5, 6.2.1
Accident Prevention 2.2.4, 10
Acts and Omissions 2.2.4, 4.18.3, 7.4, 7.6.2, 8.3.1, 10.2.5
Additional Costs, Claims for 12.3
Administration of the Contract **2.2,** 4.3.3
All Risk Insurance 11.3.1
Allowances **4.8**
Applications for Payment 2.2.6, 9.2, **9.3,** 9.4, 9.5.3, 9.6.1, 9.8.2, 9.9.1, 9.9.3, 14.2.2
Approvals . . . . 2.2.14, 3.4, 4.3.3, 4.5, 4.12.4 through 4.12.6, 4.12.8, 4.18.3, 7.7, 9.3.2
Arbitration . . . . 2.2.7 through 2.2.13, 2.2.19, 6.2.5, **7.9,** 8.3.1, 11.3.7, 11.3.8
ARCHITECT **2**
Architect, **Definition** of **2.1**
Architect, Extent of Authority 2.2, 3.4, 4.12.8, 5.2, 6.3, 7.7.2, 8.1.3, 8.3.1, 9.2, 9.3.1, 9.4, 9.5.3, 9.6, 9.8, 9.9.1, 9.9.3, 12.1.1, 12.1.4, 12.3.1, 12.4.1, 13.1, 13.2.1, 13.2.5, 14.2
Architect, Limitations of Authority and Responsibility . . . . . 2.2.2 through 2.2.4, 2.2.10 through 2.2.14, 2.2.17, 2.2.18, 4.3.3, 4.12.6, 5.2.1, 9.4.2, 9.5.4, 9.5.5, 12.4
Architect's Additional Services . . 3.4, 7.7.2, 13.2.1, 13.2.5, 14.2.2
Architect's Approvals 2.2.14, 3.4, 4.5, 4.12.6, 4.12.8, 4.18.3
Architect's Authority to Reject Work 2.2.13, 4.5, 13.1.2, 13.2
Architect's Copyright 1.3
Architect's Decisions . . . . 2.2.7 through 2.2.13, 6.3, 7.7.2, 7.9.1, 8.3.1, 9.2, 9.4, 9.6.1, 9.8.1, 12.1.4, 12.3.1
Architect's Inspections 2.2.13, 2.2.16, 9.8.1, 9.9.1
Architect's Instructions 2.2.13, 2.2.15, 7.7.2, 12.4, 13.1
Architect's Interpretations . . . . 2.2.7 through 2.2.10, 12.3.2
Architect's On-Site Observations 2.2.3, 2.2.5, 2.2.6, 2.2.17, 7.7.1, 7.7.4, 9.4.2, 9.6.1, 9.9.1
Architect's Project Representative 2.2.17, 2.2.18
Architect's Relationship with Contractor 1.1.2, 2.2.4, 2.2.5, 2.2.10, 2.2.13, 4.3.3, 4.5, 4.7.3, 4.12.6, 4.18, 11.3.6
Architect's Relationship with Subcontractors 1.1.2, 2.2.13, 9.5.3, 9.5.4
Architect's Representations 9.4.2, 9.6.1, 9.9.1
Artistic Effect 1.2.3, 2.2.11, 2.2.12, 7.9.1
Attorneys' Fees 4.18.1, 6.2.5, 9.9.2
Award of Separate Contracts 6.1.1
Award of Subcontracts and Other Contracts for Portions of the Work **5.2**
Bonds, Lien 9.9.2
Bonds, Performance, Labor and Material Payment 7.5, 9.9.3
Building Permit 4.7
Certificate of Substantial Completion 9.8.1
Certificates of Inspection, Testing or Approval 7.7.3
Certificates of Insurance 9.3.2, 11.1.4
Certificates for Payment 2.2.6, 2.2.16, **9.4,** 9.5.1, 9.5.5, 9.6.1, 9.7.1, 9.8.2, 9.9.1, 9.9.3, 12.1.4, 14.2.2
Change Orders 1.1.1, 2.2.15, 3.4, 4.8.2.3, 5.2.3, 7.7.2, 8.3.1, 9.7, 9.9.3, 11.3.1, 11.3.5, 11.3.7, **12.1,** 13.1.2, 13.2.5, 13.3.1
Change Orders, Definition of 12.1.1
CHANGES IN THE WORK 2.2.15, 4.1.1, **12**
Claims for Additional Cost or Time 8.3.2, 8.3.3, 12.2.1, **12.3**
Claims for Damages 6.1.1, 6.2.5, **7.4,** 8.3, 9.6.1.1
Cleaning Up **4.15,** 6.3
Commencement of the Work, Conditions Relating to . . 3.2.1, 4.2, 4.7.1, 4.10, 5.2.1, 6.2.2, 7.5, 9.2, 11.1.4, 11.3.4
Commencement of the Work, Definition of 8.1.2
Communications 2.2.2, 3.2.6, 4.9.1, **4.16**
Completion, Conditions Relating to . . . 2.2.16, 4.11, 4.15, 9.4.2, 9.9, 13.2.2
COMPLETION, PAYMENTS AND **9**
Completion, Substantial 2.2.16, 8.1.1, 8.1.3, 8.2.2, 9.8, 13.2.2
Compliance with Laws 1.3, 2.1.1, 4.6, 4.7, 4.13, 7.1, 7.7, 10.2.2, 14
Concealed Conditions **12.2**
Consent, Written . . . 2.2.18, 4.14.2, 7.2, 7.6.2, 9.8.1, 9.9.2, 9.9.3, 11.3.9
Contract, Definition of **1.1.2**
Contract Administration 2.2, 4.3.3
Contract Award and Execution, Conditions Relating to 4.7.1, 4.10, 5.2, 7.5, 9.2, 11.1.4, 11.3.4
CONTRACT DOCUMENTS **1**
Contract Documents, Copies Furnished and Use of 1.3, 3.2.5, 5.3
Contract Documents, Definition of **1.1.1**
Contract Sum, Definition of **9.1.1**
Contract Termination **14**
Contract Time, Definition of 8.1.1
CONTRACTOR **4**
Contractor, **Definition** of **4.1,** 6.1.2
Contractor's Employees . . . . . . 4.3.2, 4.4.2, 4.8.1, 4.9, 4.18, 10.2.1 through 10.2.4, 10.2.6, 10.3, 11.1.1
Contractor's Liability Insurance **11.1**
Contractor's Relationship with Separate Contractors and Owner's Forces 3.2.7, 6
Contractor's Relationship with Subcontractors 1.2.4, 5.2, 5.3, 9.5.2, 11.3.3, 11.3.6
Contractor's Relationship with the Architect 1.1.2, 2.2.4, 2.2.5, 2.2.10, 2.2.13, 4.3.3, 4.5, 4.7.3, 4.12.6, 4.18, 11.3.6
Contractor's Representations 1.2.2, 4.5, 4.12.5, 9.3.3
Contractor's Responsibility for Those Performing the Work 4.3.2, 4.18, 10
Contractor's Review of Contract Documents 1.2.2, 4.2, 4.7.3
Contractor's Right to Stop the Work 9.7
Contractor's Right to Terminate the Contract 14.1
Contractor's Submittals 2.2.14, 4.10, 4.12, 5.2.1, 5.2.3, 9.2, 9.3.1, 9.8.1, 9.9.2, 9.9.3
Contractor's Superintendent 4.9, 10.2.6
Contractor's Supervision and Construction Procedures 1.2.4, 2.2.4, 4.3, 4.4, 10
Contractual Liability Insurance 11.1.3
Coordination and Correlation 1.2.2, 1.2.4, 4.3.1, 4.10.1, 4.12.5, 6.1.3, 6.2.1
Copies Furnished of Drawings and Specifications . . 1.3, 3.2.5, 5.3
Correction of Work 3.3, 3.4, 10.2.5, **13.2**
Cost, Definition of 12.1.4
Costs 3.4, 4.8.2, 4.15.2, 5.2.3, 6.1.1, 6.2.3, 6.2.5, 6.3, 7.7.1, 7.7.2, 9.7, 11.3.1, 11.3.5, 12.1.3, 12.1.4, 12.3, 13.1.2, 13.2, 14
Cutting and Patching of Work **4.14,** 6.2
Damage to the Work 6.2.4, 6.2.5, 9.6.1.5, 9.8.1, 10.2.1.2, 10.3, 11.3, 13.2.6
Damages, Claims for 6.1.1, 6.2.5, 7.4, 8.3.4, 9.6.1.2
Damages for Delay 6.1.1, 8.3.4, 9.7
Day, Definition of 8.1.4

Decisions of the Architect ......2.2.9 through 2.2.12, 6.3, 7.7.2, 7.9.1, 8.3.1, 9.2, 9.4, 9.6.1, 9.8.1, 12.1.4, 12.3.1, 14.2.1
Defective or Non-Conforming Work, Acceptance, Rejection and Correction of 2.2.3, 2.2.13, 3.3, 3.4, 4.5, 6.2.2, 6.2.3, 9.6.1.1, 9.9.4.2, 13
Definitions ....**1.1, 2.1, 3.1, 4.1,** 4.12.1 through 4.12.3, **5.1,** 6.1.2, **8.1,** 9.1.1, 12.1.1, 12.1.4
Delays and Extensions of Time **8.3**
Disputes 2.2.9, 2.2.12, 2.2.19, 6.2.5, 6.3, 7.9.1
Documents and Samples at the Site **4.11**
Drawings and Specifications, Use and Ownership of 1.1.1, 1.3, 3.2.5, 5.3
Emergencies **10.3**
Employees, Contractor's .....4.3.2, 4.4.2, 4.8.1, 4.9, 4.18, 10.2.1 through 10.2.4, 10.2.6, 10.3, 11.1.1
Equipment, Labor, Materials and 1.1.3, 4.4, 4.5, 4.12, 4.13, 4.15.1, 6.2.1, 9.3.2, 9.3.3, 11.3, 13.2.2, 13.2.5, 14
Execution and Progress of the Work 1.1.3, 1.2.3, 2.2.3, 2.2.4, 2.2.8, 4.2, 4.4.1, 4.5, 6.2.2, 7.9.3, 8.2, 8.3, 9.6.1, 10.2.3, 10.2.4, 14.2
Execution, Correlation and Intent of the Contract Documents **1.2,** 4.7.1
Extensions of Time 8.3, 12.1
Failure of Payment by Owner **9.7,** 14.1
Failure of Payment of Subcontractors ..9.5.2, 9.6.1.3, 9.9.2, 14.2.1
Final Completion and Final Payment ..2.2.12, 2.2.16, **9.9,** 13.3.1
Financial Arrangements, Owner's 3.2.1
Fire and Extended Coverage Insurance 11.3.1
Governing Law **7.1**
Guarantees (See Warranty and Warranties) 2.2.16, 4.5, 9.3.3, 9.8.1, 9.9.4, 13.2.2
Indemnification 4.17, **4.18,** 6.2.5, 9.9.2
Identification of Contract Documents 1.2.1
Identification of Subcontractors and Suppliers 5.2.1
Information and Services Required of the Owner **3.2,** 6, 9, 11.2, 11.3
Inspections 2.2.13, 2.2.16, 4.3.3, 7.7, 9.8.1, 9.9.1
Instructions to Bidders 1.1.1, 7.5
Instructions to the Contractor 2.2.2, 3.2.6, 4.8.1, 7.7.2, 12.1.2, 12.1.4
INSURANCE 9.8.1, **11**
Insurance, Contractor's Liability 11.1
Insurance, Loss of Use 11.4
Insurance, Owner's Liability 11.2
Insurance, Property 11.3
Insurance, Stored Materials 9.3.2, 11.3.1
Insurance Companies, Consent to Partial Occupancy 11.3.9
Insurance Companies, Settlement With 11.3.8
Intent of the Contract Documents ...1.2.3, 2.2.10, 2.2.13, 2.2.14, 12.4
Interest **7.8**
Interpretations, Written 1.1.1, 2.2.7, 2.2.8, 2.2.10, 12.4
Labor and Materials, Equipment 1.1.3, **4.4,** 4.5, 4.12, 4.13, 4.15.1, 6.2.1, 9.3.2, 9.3.3, 11.3, 13.2.2, 13.2.5, 14
Labor and Material Payment Bond 7.5
Labor Disputes 8.3.1
Laws and Regulations 1.3, 2.1., 4.6, 4.7, 4.13, 7.1, 7.7, 10.2.2, 14
Liens 9.3.3, 9.9.2, 9.9.4.1
Limitations of Authority 2.2.2, 2.2.17, 2.2.18, 11.3.8, 12.4.1
Limitations of Liability 2.2.10, 2.2.13, 2.2.14, 3.3, 4.2, 4.7.3, 4.12.6, 4.17, 4.18.3, 6.2.2, 7.6.2, 9.4.2, 9.9.4, 9.9.5, 10.2.5, 11.1.2, 11.3.6
Limitations of Time, General 2.2.8, 2.2.14, 3.2.4, 4.2, 4.7.3, 4.12.4, 4.15, 5.2.1, 5.2.3, 7.4, 7.7, 8.2, 9.5.2, 9.6, 9.8, 9.9, 11.3.4, 12.1.4, 12.4, 13.2.1, 13.2.2, 13.2.5
Limitations of Time, Specific 2.2.8, 2.2.12, 3.2.1, 3.4, 4.10, 5.3, 6.2.2, 7.9.2, 8.2, 8.3.2, 8.3.3, 9.2, 9.3.1, 9.4.1, 9.5.1, 9.7, 11.1.4, 11.3.1, 11.3.8, 11.3.9, 12.2, 12.3, 13.2.2, 13.2.5, 13.2.7, 14.1, 14.2.1
Limitations, Statutes of 7.9.2, 13.2.2, 13.2.7
Loss of Use Insurance **11.4**
Materials, Labor, Equipment and 1.1.3, 4.4, 4.5, 4.12, 4.13, 4.15.1, 6.2.1, 9.3.2, 9.3.3, 11.3.1, 13.2.2, 13.2.5, 14
Materials Suppliers 4.12.1, 5.2.1, 9.3.3
Means, Methods, Techniques, Sequences and Procedures of Construction 2.2.4, 4.3.1, 9.4.2
Minor Changes in the Work 1.1.1, 2.2.15, **12.4**
MISCELLANEOUS PROVISIONS **7**
Modifications, Definition of 1.1.1
Modifications to the Contract 1.1.1, 1.1.2, 2.2.2, 2.2.18, 4.7.3, 7.9.3, 12
Mutual Responsibility **6.2**
Non-Conforming Work, Acceptance of Defective or 13.3.1
Notice, Written 2.2.8, 2.2.12, 3.4, 4.2, 4.7.3, 4.7.4, 4.9, 4.12.6, 4.12.7, 4.17, 5.2.1, 7.3, 7.4, 7.7, 7.9.2, 8.1.2, 8.3.2, 8.3.3, 9.4.1, 9.6.1, 9.7, 9.9.1, 9.9.5, 10.2.6, 11.1.4, 11.3.1, 11.3.4, 11.3.5, 11.3.7, 11.3.8, 12.2, 12.3, 13.2.2, 13.2.5, 14
Notices, Permits, Fees and 4.7, 10.2.2
Notice of Testing and Inspections 7.7
Notice to Proceed 8.1.2
Observations, Architect's On-Site 2.2.3, 7.7.1, 7.7.4, 9.4.2
Observations, Contractor's 1.2.2, 4.2.1, 4.7.3
Occupancy 8.1.3, 9.5.5, 11.3.9
On-Site Inspections by the Architect 2.2.3, 2.2.16, 9.4.2, 9.8.1, 9.9.1
On-Site Observations by the Architect 2.2.3, 2.2.6, 2.2.17, 7.7.1, 7.7.4, 9.4.2, 9.6.1, 9.9.1
Orders, Written 3.3, 4.9, 12.1.4, 12.4.1, 13.1
OWNER **3**
Owner, **Definition** of **3.1**
Owner, Information and Services Required of the 3.2, 6.1.3, 6.2, 9, 11.2, 11.3
Owner's Authority 2.2.16, 4.8.1, 7.7.2, 9.3.1, 9.3.2, 9.8.1, 11.3.8, 12.1.2, 12.1.4
Owner's Financial Capability 3.2.1
Owner's Liability Insurance **11.2**
Owner's Relationship with Subcontractors 1.1.2, 9.5.4
Owner's Right to Carry Out the Work **3.4,** 13.2.4
Owner's Right to Clean Up 4.15.2, **6.3**
Owner's Right to Perform Work and to Award Separate Contracts **6.1**
Owner's Right to Terminate the Contract 14.2
Owner's Right to Stop the Work **3.3**
Ownership and Use of Documents 1.1.1, **1.3,** 3.2.5, 5.2.3
Patching of Work, Cutting and 4.14, 6.2.2
Patents, Royalties and 4.17.1
Payment Bond, Labor and Material 7.5
Payment, Applications for 2.2.6, 9.2, 9.3, 9.4, 9.5.3, 9.6.1, 9.8.2, 9.9.1, 9.9.3, 14.2.2
Payment, Certificates for 2.2.6, 2.2.16, 9.4, 9.5.1, 9.5.5, 9.6.1, 9.7.1, 9.8.2, 9.9.1, 9.9.3, 12.1.4, 14.2.2

Payment, Failure of 9.5.2, 9.6.1.3, 9.7, 9.9.2, 14
Payment, Final 2.2.12, 2.2.16, 9.9, 13.3.1
Payments, Progress 7.8, 7.9.3, 9.5.5, 9.8.2, 9.9.3, 12.1.4
PAYMENTS AND COMPLETION **9**
Payments to Subcontractors 9.5.2, 9.5.3, 9.5.4, 9.6.1.3, 11.3.3, 14.2.1
Payments Withheld **9.6**
Performance Bond and Labor and Material Payment Bond ..**7.5**
Permits, Fees and Notices 3.2.3, **4.7,** 4.13
PERSONS AND PROPERTY, PROTECTION OF **10**
Product Data, Definition of 4.12.2
Product Data, Shop Drawings, Samples and ...2.2.14, 4.2.1, 4.12
Progress and Completion 2.2.3, 7.9.3, **8.2**
Progress Payments 7.8, 7.9.3, 9.5.5, 9.8.2, 9.9.3, 12.1.4
Progress Schedule **4.10**
Project, Definition of **1.1.4**
Project Representative 2.2.17
Property Insurance **11.3**
PROTECTION OF PERSONS AND PROPERTY **10**
Regulations and Laws 1.3, 2.1.1, 4.6, 4.7, 4.13, 7.1, 10.2.2, 14
Rejection of Work 2.2.13, 4.5.1, 13.2
Releases of Waivers and Liens 9.9.2, 9.9.4
Representations 1.2.2, 4.5, 4.12.5, 9.4.2, 9.6.1, 9.9.1
Representatives 2.1, 2.2.2, 2.2.17, 2.2.18, 3.1, 4.1, 4.9, 5.1, 9.3.3
Responsibility for Those Performing the Work 2.2.4, 4.3.2, 6.1.3, 6.2, 9.8.1
Retainage 9.3.1, 9.5.2, 9.8.2, 9.9.2, 9.9.3
Review of Contract Documents
by the Contractor 1.2.2, **4.2,** 4.7.3
Reviews of Contractor's Submittals by
Owner and Architect 2.2.14, 4.10, 4.12, 5.2.1, 5.2.3, 9.2
Rights and Remedies 1.1.2, 2.2.12, 2.2.13, 3.3, 3.4, 5.3, 6.1, 6.3, **7.6,** 7.9, 8.3.1, 9.6.1, 9.7, 10.3, 12.1.2, 12.2, 13.2.2, 14
Royalties and Patents **4.17**
Safety of Persons and Property **10.2**
Safety Precautions and Programs 2.2.4, **10.1**
Samples, Definition of 4.12.3
Samples, Shop Drawings, Product Data and 2.2.14, 4.2, 4.12
Samples at the Site, Documents and 4.11
Schedule of Values **9.2**
Schedule, Progress 4.10
Separate Contracts and Contractors 4.14.2, 6, 11.3.6, 13.1.2
Shop Drawings, Definition of 4.12.1
Shop Drawings, Product Data and Samples ... 2.2.14, 4.2, **4.12**
Site, Use of 4.13, 6.2.1
Site Visits, Architect's 2.2.3, 2.2.5, 2.2.6, 2.2.17, 7.7.1, 7.7.4, 9.4.2, 9.6.1, 9.9.1
Site Inspections 1.2.2, 2.2.3, 2.2.16, 7.7, 9.8.1, 9.9.1
Special Inspection and Testing 2.2.13, 7.7
Specifications 1.1.1, 1.2.4, 1.3
Statutes of Limitations 7.9.2, 13.2.2, 13.2.7
Stopping the Work 3.3, 9.7.1, 10.3, 14.1
Stored Materials 6.2.1, 9.3.2, 10.2.1.2, 11.3.1, 13.2.5
SUBCONTRACTORS **5**
Subcontractors, **Definition** of **5.1**
Subcontractors, Work by 1.2.4, 2.2.4, 4.3.1, 4.3.2
Subcontractual Relations **5.3**
Submittals 1.3, 4.10, 4.12, 5.2.1, 5.2.3, 9.2, 9.3.1, 9.8.1, 9.9.2, 9.9.3
Subrogation, Waiver of 11.3.6
Substantial Completion 2.2.16, 8.1.1, 8.1.3, 8.2.2, **9.8,** 13.2.2
Substantial Completion, Definition of 8.1.3
Substitution of Subcontractors 5.2.3, 5.2.4
Substitution of the Architect 2.2.19
Substitutions of Materials 4.5, 12.1.4
Sub-subcontractors, Definition of 5.1.2
Subsurface Conditions 12.2.1
Successors and Assigns **7.2**
Supervision and Construction Procedures .1.2.4, 2.2.4, **4.3,** 4.4, 10
Superintendent, Contractor's **4.9,** 10.2.6
Surety, Consent of 9.9.2, 9.9.3
Surveys 3.2.2, 4.18.3
Taxes **4.6**
Termination by the Contractor **14.1**
Termination by the Owner **14.2**
Termination of the Architect 2.2.19
TERMINATION OF THE CONTRACT **14**
Tests 2.2.13, 4.3.3, **7.7,** 9.4.2
Time **8**
Time, **Definition** of **8.1**
Time, Delays and Extensions of 8.3, 12.1, 12.3, 13.2.7
Time Limits, Specific 2.2.8, 2.2.12, 3.2.1, 3.4, 4.10, 5.3, 6.2.2, 7.9.2, 8.2, 8.3.2, 8.3.3, 9.2, 9.3.1, 9.4.1, 9.5.1, 9.7, 11.1.4, 11.3.1, 11.3.8, 11.3.9, 12.2, 12.3, 13.2.2, 13.2.5, 13.2.7, 14.1, 14.2.1
Title to Work 9.3.2, 9.3.3
UNCOVERING AND CORRECTION OF WORK **13**
Uncovering of Work **13.1**
Unforseen Conditions 8.3, 12.2
Unit Prices 12.1.3, 12.1.5
Use of Documents 1.1.1, 1.3, 3.2.5, 5.3
Use of Site **4.13,** 6.2.1
Values, Schedule of 9.2
Waiver of Claims by the Contractor 7.6.2, 8.3.2, 9.9.5, 11.3.6
Waiver of Claims by the Owner 7.6.2, 9.9.4, 11.3.6, 11.4.1
Waiver of Liens 9.9.2
Warranty and Warranties 2.2.16, **4.5,** 9.3.3, 9.8.1, 9.9.4, 13.2.2
Weather Delays 8.3.1
Work, Definition of **1.1.3**
Work by Owner or by Separate Contractors **6**
Written Consent 2.2.18, 4.14.2, 7.2, 7.6.2, 9.8.1, 9.9.3, 9.9.4
Written Interpretations 1.1.1, 1.2.4, 2.2.8, 12.3.2
Written Notice 2.2.8, 2.2.12, 3.4, 4.2, 4.7.3, 4.7.4, 4.9, 4.12.6, 4.12.7, 4.17, 5.2.1, **7.3,** 7.4, 7.7, 7.9.2, 8.1.2, 8.3.2, 8.3.3, 9.4.1, 9.6.1, 9.7, 9.9.1, 9.9.5, 10.2.6, 11.1.4, 11.3.1, 11.3.4, 11.3.5, 11.3.7, 11.3.8, 12.2, 12.3, 13.2.2, 13.2.5, 14
Written Orders 3.3, 4.9, 12.1.4, 12.4.1, 13.1

GENERAL CONDITIONS OF THE CONTRACT FOR CONSTRUCTION

ARTICLE 1

CONTRACT DOCUMENTS

1.1 DEFINITIONS

1.1.1 THE CONTRACT DOCUMENTS

The Contract Documents consist of the Owner-Contractor Agreement, the Conditions of the Contract (General, Supplementary and other Conditions), the Drawings, the Specifications, and all Addenda issued prior to and all Modifications issued after execution of the Contract. A Modification is (1) a written amendment to the Contract signed by both parties, (2) a Change Order, (3) a written interpretation issued by the Architect pursuant to Subparagraph 2.2.8, or (4) a written order for a minor change in the Work issued by the Architect pursuant to Paragraph 12.4. The Contract Documents do not include Bidding Documents such as the Advertisement or Invitation to Bid, the Instructions to Bidders, sample forms, the Contractor's Bid or portions of Addenda relating to any of these, or any other documents, unless specifically enumerated in the Owner-Contractor Agreement.

1.1.2 THE CONTRACT

The Contract Documents form the Contract for Construction. This Contract represents the entire and integrated agreement between the parties hereto and supersedes all prior negotiations, representations, or agreements, either written or oral. The Contract may be amended or modified only by a Modification as defined in Subparagraph 1.1.1. The Contract Documents shall not be construed to create any contractual relationship of any kind between the Architect and the Contractor, but the Architect shall be entitled to performance of obligations intended for his benefit, and to enforcement thereof. Nothing contained in the Contract Documents shall create any contractual relationship between the Owner or the Architect and any Subcontractor or Sub-subcontractor.

1.1.3 THE WORK

The Work comprises the completed construction required by the Contract Documents and includes all labor necessary to produce such construction, and all materials and equipment incorporated or to be incorporated in such construction.

1.1.4 THE PROJECT

The Project is the total construction of which the Work performed under the Contract Documents may be the whole or a part.

1.2 EXECUTION, CORRELATION AND INTENT

1.2.1 The Contract Documents shall be signed in not less than triplicate by the Owner and Contractor. If either the Owner or the Contractor or both do not sign the Conditions of the Contract, Drawings, Specifications, or any of the other Contract Documents, the Architect shall identify such Documents.

1.2.2 By executing the Contract, the Contractor represents that he has visited the site, familiarized himself with the local conditions under which the Work is to be performed, and correlated his observations with the requirements of the Contract Documents.

1.2.3 The intent of the Contract Documents is to include all items necessary for the proper execution and completion of the Work. The Contract Documents are complementary, and what is required by any one shall be as binding as if required by all. Work not covered in the Contract Documents will not be required unless it is consistent therewith and is reasonably inferable therefrom as being necessary to produce the intended results. Words and abbreviations which have well-known technical or trade meanings are used in the Contract Documents in accordance with such recognized meanings.

1.2.4 The organization of the Specifications into divisions, sections and articles, and the arrangement of Drawings shall not control the Contractor in dividing the Work among Subcontractors or in establishing the extent of Work to be performed by any trade.

1.3 OWNERSHIP AND USE OF DOCUMENTS

1.3.1 All Drawings, Specifications and copies thereof furnished by the Architect are and shall remain his property. They are to be used only with respect to this Project and are not to be used on any other project. With the exception of one contract set for each party to the Contract, such documents are to be returned or suitably accounted for to the Architect on request at the completion of the Work. Submission or distribution to meet official regulatory requirements or for other purposes in connection with the Project is not to be construed as publication in derogation of the Architect's common law copyright or other reserved rights.

ARTICLE 2

ARCHITECT

2.1 DEFINITION

2.1.1 The Architect is the person lawfully licensed to practice architecture, or an entity lawfully practicing architecture identified as such in the Owner-Contractor Agreement, and is referred to throughout the Contract Documents as if singular in number and masculine in gender. The term Architect means the Architect or his authorized representative.

2.2 ADMINISTRATION OF THE CONTRACT

2.2.1 The Architect will provide administration of the Contract as hereinafter described.

2.2.2 The Architect will be the Owner's representative during construction and until final payment is due. The Architect will advise and consult with the Owner. The Owner's instructions to the Contractor shall be forwarded through the Architect. The Architect will have authority to act on behalf of the Owner only to the extent provided in the Contract Documents, unless otherwise modified by written instrument in accordance with Subparagraph 2.2.18.

2.2.3 The Architect will visit the site at intervals appropriate to the stage of construction to familiarize himself generally with the progress and quality of the Work and to determine in general if the Work is proceeding in accordance with the Contract Documents. However, the Architect will not be required to make exhaustive or continuous on-site inspections to check the quality or quantity of the Work. On the basis of his on-site observations as an architect, he will keep the Owner informed of the progress of the Work, and will endeavor to guard the Owner against defects and deficiencies in the Work of the Contractor.

2.2.4 The Architect will not be responsible for and will not have control or charge of construction means, methods, techniques, sequences or procedures, or for safety precautions and programs in connection with the Work, and he will not be responsible for the Contractor's failure to carry out the Work in accordance with the Contract Documents. The Architect will not be responsible for or have control or charge over the acts or omissions of the Contractor, Subcontractors, or any of their agents or employees, or any other persons performing any of the Work.

2.2.5 The Architect shall at all times have access to the Work wherever it is in preparation and progress. The Contractor shall provide facilities for such access so the Architect may perform his functions under the Contract Documents.

2.2.6 Based on the Architect's observations and an evaluation of the Contractor's Applications for Payment, the Architect will determine the amounts owing to the Contractor and will issue Certificates for Payment in such amounts, as provided in Paragraph 9.4.

2.2.7 The Architect will be the interpreter of the requirements of the Contract Documents and the judge of the performance thereunder by both the Owner and Contractor.

2.2.8 The Architect will render interpretations necessary for the proper execution or progress of the Work, with reasonable promptness and in accordance with any time limit agreed upon. Either party to the Contract may make written request to the Architect for such interpretations.

2.2.9 Claims, disputes and other matters in question between the Contractor and the Owner relating to the execution or progress of the Work or the interpretation of the Contract Documents shall be referred initially to the Architect for decision which he will render in writing within a reasonable time.

2.2.10 All interpretations and decisions of the Architect shall be consistent with the intent of and reasonably inferable from the Contract Documents and will be in writing or in the form of drawings. In his capacity as interpreter and judge, he will endeavor to secure faithful performance by both the Owner and the Contractor, will not show partiality to either, and will not be liable for the result of any interpretation or decision rendered in good faith in such capacity.

2.2.11 The Architect's decisions in matters relating to artistic effect will be final if consistent with the intent of the Contract Documents.

2.2.12 Any claim, dispute or other matter in question between the Contractor and the Owner referred to the Architect, except those relating to artistic effect as provided in Subparagraph 2.2.11 and except those which have been waived by the making or acceptance of final payment as provided in Subparagraphs 9.9.4 and 9.9.5, shall be subject to arbitration upon the written demand of either party. However, no demand for arbitration of any such claim, dispute or other matter may be made until the earlier of (1) the date on which the Architect has rendered a written decision, or (2) the tenth day after the parties have presented their evidence to the Architect or have been given a reasonable opportunity to do so, if the Architect has not rendered his written decision by that date. When such a written decision of the Architect states (1) that the decision is final but subject to appeal, and (2) that any demand for arbitration of a claim, dispute or other matter covered by such decision must be made within thirty days after the date on which the party making the demand receives the written decision, failure to demand arbitration within said thirty days' period will result in the Architect's decision becoming final and binding upon the Owner and the Contractor. If the Architect renders a decision after arbitration proceedings have been initiated, such decision may be entered as evidence but will not supersede any arbitration proceedings unless the decision is acceptable to all parties concerned.

2.2.13 The Architect will have authority to reject Work which does not conform to the Contract Documents. Whenever, in his opinion, he considers it necessary or advisable for the implementation of the intent of the Contract Documents, he will have authority to require special inspection or testing of the Work in accordance with Subparagraph 7.7.2 whether or not such Work be then fabricated, installed or completed. However, neither the Architect's authority to act under this Subparagraph 2.2.13, nor any decision made by him in good faith either to exercise or not to exercise such authority, shall give rise to any duty or responsibility of the Architect to the Contractor, any Subcontractor, any of their agents or employees, or any other person performing any of the Work.

2.2.14 The Architect will review and approve or take other appropriate action upon Contractor's submittals such as Shop Drawings, Product Data and Samples, but only for conformance with the design concept of the Work and with the information given in the Contract Documents. Such action shall be taken with reasonable promptness so as to cause no delay. The Architect's approval of a specific item shall not indicate approval of an assembly of which the item is a component.

2.2.15 The Architect will prepare Change Orders in accordance with Article 12, and will have authority to order minor changes in the Work as provided in Subparagraph 12.4.1.

2.2.16 The Architect will conduct inspections to determine the dates of Substantial Completion and final completion, will receive and forward to the Owner for the Owner's review written warranties and related documents required by the Contract and assembled by the Contractor, and will issue a final Certificate for Payment upon compliance with the requirements of Paragraph 9.9.

2.2.17 If the Owner and Architect agree, the Architect will provide one or more Project Representatives to assist the Architect in carrying out his responsibilities at the site. The duties, responsibilities and limitations of authority of any such Project Representative shall be as set forth in an exhibit to be incorporated in the Contract Documents.

2.2.18 The duties, responsibilities and limitations of authority of the Architect as the Owner's representative during construction as set forth in the Contract Documents will not be modified or extended without written consent of the Owner, the Contractor and the Architect.

2.2.19 In case of the termination of the employment of the Architect, the Owner shall appoint an architect against whom the Contractor makes no reasonable objection whose status under the Contract Documents shall be that of the former architect. Any dispute in connection with such appointment shall be subject to arbitration.

ARTICLE 3

OWNER

3.1 DEFINITION

3.1.1 The Owner is the person or entity identified as such in the Owner-Contractor Agreement and is referred to throughout the Contract Documents as if singular in number and masculine in gender. The term Owner means the Owner or his authorized representative.

3.2 INFORMATION AND SERVICES REQUIRED OF THE OWNER

3.2.1 The Owner shall, at the request of the Contractor, at the time of execution of the Owner-Contractor Agreement, furnish to the Contractor reasonable evidence that he has made financial arrangements to fulfill his obligations under the Contract. Unless such reasonable evidence is furnished, the Contractor is not required to execute the Owner-Contractor Agreement or to commence the Work.

3.2.2 The Owner shall furnish all surveys describing the physical characteristics, legal limitations and utility locations for the site of the Project, and a legal description of the site.

3.2.3 Except as provided in Subparagraph 4.7.1, the Owner shall secure and pay for necessary approvals, easements, assessments and charges required for the construction, use or occupancy of permanent structures or for permanent changes in existing facilities.

3.2.4 Information or services under the Owner's control shall be furnished by the Owner with reasonable promptness to avoid delay in the orderly progress of the Work.

3.2.5 Unless otherwise provided in the Contract Documents, the Contractor will be furnished, free of charge, all copies of Drawings and Specifications reasonably necessary for the execution of the Work.

3.2.6 The Owner shall forward all instructions to the Contractor through the Architect.

3.2.7 The foregoing are in addition to other duties and responsibilities of the Owner enumerated herein and especially those in respect to Work by Owner or by Separate Contractors, Payments and Completion, and Insurance in Articles 6, 9 and 11 respectively.

3.3 OWNER'S RIGHT TO STOP THE WORK

3.3.1 If the Contractor fails to correct defective Work as required by Paragraph 13.2 or persistently fails to carry out the Work in accordance with the Contract Documents, the Owner, by a written order signed personally or by an agent specifically so empowered by the Owner in writing, may order the Contractor to stop the Work, or any portion thereof, until the cause for such order has been eliminated; however, this right of the Owner to stop the Work shall not give rise to any duty on the part of the Owner to exercise this right for the benefit of the Contractor or any other person or entity, except to the extent required by Subparagraph 6.1.3.

3.4 OWNER'S RIGHT TO CARRY OUT THE WORK

3.4.1 If the Contractor defaults or neglects to carry out the Work in accordance with the Contract Documents and fails within seven days after receipt of written notice from the Owner to commence and continue correction of such default or neglect with diligence and promptness, the Owner may, after seven days following receipt by the Contractor of an additional written notice and without prejudice to any other remedy he may have, make good such deficiencies. In such case an appropriate Change Order shall be issued deducting from the payments then or thereafter due the Contractor the cost of correcting such deficiencies, including compensation for the Architect's additional services made necessary by such default, neglect or failure. Such action by the Owner and the amount charged to the Contractor are both subject to the prior approval of the Architect. If the payments then or thereafter due the Contractor are not sufficient to cover such amount, the Contractor shall pay the difference to the Owner.

ARTICLE 4

CONTRACTOR

4.1 DEFINITION

4.1.1 The Contractor is the person or entity identified as such in the Owner-Contractor Agreement and is referred to throughout the Contract Documents as if singular in number and masculine in gender. The term Contractor means the Contractor or his authorized representative.

4.2 REVIEW OF CONTRACT DOCUMENTS

4.2.1 The Contractor shall carefully study and compare the Contract Documents and shall at once report to the Architect any error, inconsistency or omission he may discover. The Contractor shall not be liable to the Owner or the Architect for any damage resulting from any such errors, inconsistencies or omissions in the Contract Documents. The Contractor shall perform no portion of the Work at any time without Contract Documents or, where required, approved Shop Drawings, Product Data or Samples for such portion of the Work.

4.3 SUPERVISION AND CONSTRUCTION PROCEDURES

4.3.1 The Contractor shall supervise and direct the Work, using his best skill and attention. He shall be solely responsible for all construction means, methods, techniques, sequences and procedures and for coordinating all portions of the Work under the Contract.

4.3.2 The Contractor shall be responsible to the Owner for the acts and omissions of his employees, Subcontractors and their agents and employees, and other persons performing any of the Work under a contract with the Contractor.

4.3.3 The Contractor shall not be relieved from his obligations to perform the Work in accordance with the Contract Documents either by the activities or duties of the Architect in his administration of the Contract, or by inspections, tests or approvals required or performed under Paragraph 7.7 by persons other than the Contractor.

4.4 LABOR AND MATERIALS

4.4.1 Unless otherwise provided in the Contract Documents, the Contractor shall provide and pay for all labor, materials, equipment, tools, construction equipment and machinery, water, heat, utilities, transportation, and other facilities and services necessary for the proper execution and completion of the Work, whether temporary or permanent and whether or not incorporated or to be incorporated in the Work.

4.4.2 The Contractor shall at all times enforce strict discipline and good order among his employees and shall not employ on the Work any unfit person or anyone not skilled in the task assigned to him.

4.5 WARRANTY

4.5.1 The Contractor warrants to the Owner and the Architect that all materials and equipment furnished under this Contract will be new unless otherwise specified, and that all Work will be of good quality, free from faults and defects and in conformance with the Contract Documents. All Work not conforming to these requirements, including substitutions not properly approved and authorized, may be considered defective. If required by the Architect, the Contractor shall furnish satisfactory evidence as to the kind and quality of materials and equipment. This warranty is not limited by the provisions of Paragraph 13.2.

4.6 TAXES

4.6.1 The Contractor shall pay all sales, consumer, use and other similar taxes for the Work or portions thereof provided by the Contractor which are legally enacted at the time bids are received, whether or not yet effective.

4.7 PERMITS, FEES AND NOTICES

4.7.1 Unless otherwise provided in the Contract Documents, the Contractor shall secure and pay for the building permit and for all other permits and governmental fees, licenses and inspections necessary for the proper execution and completion of the Work which are customarily secured after execution of the Contract and which are legally required at the time the bids are received.

4.7.2 The Contractor shall give all notices and comply with all laws, ordinances, rules, regulations and lawful orders of any public authority bearing on the performance of the Work.

4.7.3 It is not the responsibility of the Contractor to make certain that the Contract Documents are in accordance with applicable laws, statutes, building codes and regulations. If the Contractor observes that any of the Contract Documents are at variance therewith in any respect, he shall promptly notify the Architect in writing, and any necessary changes shall be accomplished by appropriate Modification.

4.7.4 If the Contractor performs any Work knowing it to be contrary to such laws, ordinances, rules and regulations, and without such notice to the Architect, he shall assume full responsibility therefor and shall bear all costs attributable thereto.

4.8 ALLOWANCES

4.8.1 The Contractor shall include in the Contract Sum all allowances stated in the Contract Documents. Items covered by these allowances shall be supplied for such amounts and by such persons as the Owner may direct, but the Contractor will not be required to employ persons against whom he makes a reasonable objection.

4.8.2 Unless otherwise provided in the Contract Documents:

.1 these allowances shall cover the cost to the Contractor, less any applicable trade discount, of the materials and equipment required by the allowance delivered at the site, and all applicable taxes;

.2 the Contractor's costs for unloading and handling on the site, labor, installation costs, overhead, profit and other expenses contemplated for the original allowance shall be included in the Contract Sum and not in the allowance;

.3 whenever the cost is more than or less than the allowance, the Contract Sum shall be adjusted accordingly by Change Order, the amount of which will recognize changes, if any, in handling costs on the site, labor, installation costs, overhead, profit and other expenses.

4.9 SUPERINTENDENT

4.9.1 The Contractor shall employ a competent superintendent and necessary assistants who shall be in attendance at the Project site during the progress of the Work. The superintendent shall represent the Contractor and all communications given to the superintendent shall be as binding as if given to the Contractor. Important communications shall be confirmed in writing. Other communications shall be so confirmed on written request in each case.

4.10 PROGRESS SCHEDULE

4.10.1 The Contractor, immediately after being awarded the Contract, shall prepare and submit for the Owner's and Architect's information an estimated progress schedule for the Work. The progress schedule shall be related to the entire Project to the extent required by the Contract Documents, and shall provide for expeditious and practicable execution of the Work.

4.11 DOCUMENTS AND SAMPLES AT THE SITE

4.11.1 The Contractor shall maintain at the site for the Owner one record copy of all Drawings, Specifications, Addenda, Change Orders and other Modifications, in good order and marked currently to record all changes made during construction, and approved Shop Drawings, Product Data and Samples. These shall be available to the Architect and shall be delivered to him for the Owner upon completion of the Work.

4.12 SHOP DRAWINGS, PRODUCT DATA AND SAMPLES

4.12.1 Shop Drawings are drawings, diagrams, schedules and other data specially prepared for the Work by the Contractor or any Subcontractor, manufacturer, supplier or distributor to illustrate some portion of the Work.

4.12.2 Product Data are illustrations, standard schedules, performance charts, instructions, brochures, diagrams and other information furnished by the Contractor to illustrate a material, product or system for some portion of the Work.

4.12.3 Samples are physical examples which illustrate materials, equipment or workmanship and establish standards by which the Work will be judged.

4.12.4 The Contractor shall review, approve and submit, with reasonable promptness and in such sequence as to cause no delay in the Work or in the work of the Owner or any separate contractor, all Shop Drawings, Product Data and Samples required by the Contract Documents.

4.12.5 By approving and submitting Shop Drawings, Product Data and Samples, the Contractor represents that he has determined and verified all materials, field measurements, and field construction criteria related thereto, or will do so, and that he has checked and coordinated the information contained within such submittals with the requirements of the Work and of the Contract Documents.

4.12.6 The Contractor shall not be relieved of responsibility for any deviation from the requirements of the Contract Documents by the Architect's approval of Shop Drawings, Product Data or Samples under Subparagraph 2.2.14 unless the Contractor has specifically informed the Architect in writing of such deviation at the time of submission and the Architect has given written approval to the specific deviation. The Contractor shall not be relieved from responsibility for errors or omissions in the Shop Drawings, Product Data or Samples by the Architect's approval thereof.

4.12.7 The Contractor shall direct specific attention, in writing or on resubmitted Shop Drawings, Product Data or Samples, to revisions other than those requested by the Architect on previous submittals.

4.12.8 No portion of the Work requiring submission of a Shop Drawing, Product Data or Sample shall be commenced until the submittal has been approved by the Architect as provided in Subparagraph 2.2.14. All such portions of the Work shall be in accordance with approved submittals.

4.13 USE OF SITE

4.13.1 The Contractor shall confine operations at the site to areas permitted by law, ordinances, permits and the Contract Documents and shall not unreasonably encumber the site with any materials or equipment.

4.14 CUTTING AND PATCHING OF WORK

4.14.1 The Contractor shall be responsible for all cutting, fitting or patching that may be required to complete the Work or to make its several parts fit together properly.

4.14.2 The Contractor shall not damage or endanger any portion of the Work or the work of the Owner or any separate contractors by cutting, patching or otherwise altering any work, or by excavation. The Contractor shall not cut or otherwise alter the work of the Owner or any separate contractor except with the written consent of the Owner and of such separate contractor. The Contractor shall not unreasonably withhold from the Owner or any separate contractor his consent to cutting or otherwise altering the Work.

4.15 CLEANING UP

4.15.1 The Contractor at all times shall keep the premises free from accumulation of waste materials or rubbish caused by his operations. At the completion of the Work he shall remove all his waste materials and rubbish from and about the Project as well as all his tools, construction equipment, machinery and surplus materials.

4.15.2 If the Contractor fails to clean up at the completion of the Work, the Owner may do so as provided in Paragraph 3.4 and the cost thereof shall be charged to the Contractor.

4.16 COMMUNICATIONS

4.16.1 The Contractor shall forward all communications to the Owner through the Architect.

4.17 ROYALTIES AND PATENTS

4.17.1 The Contractor shall pay all royalties and license fees. He shall defend all suits or claims for infringement of any patent rights and shall save the Owner harmless from loss on account thereof, except that the Owner shall be responsible for all such loss when a particular design, process or the product of a particular manufacturer or manufacturers is specified, but if the Contractor has reason to believe that the design, process or product specified is an infringement of a patent, he shall be responsible for such loss unless he promptly gives such information to the Architect.

4.18 INDEMNIFICATION

4.18.1 To the fullest extent permitted by law, the Contractor shall indemnify and hold harmless the Owner and the Architect and their agents and employees from and against all claims, damages, losses and expenses, including but not limited to attorneys' fees, arising out of or resulting from the performance of the Work, provided that any such claim, damage, loss or expense (1) is attributable to bodily injury, sickness, disease or death, or to injury to or destruction of tangible property (other than the Work itself) including the loss of use resulting therefrom, and (2) is caused in whole or in part by any negligent act or omission of the Contractor, any Subcontractor, anyone directly or indirectly employed by any of them or anyone for whose acts any of them may be liable, regardless of whether or not it is caused in part by a party indemnified hereunder. Such obligation shall not be construed to negate, abridge, or otherwise reduce any other right or obligation of indemnity which would otherwise exist as to any party or person described in this Paragraph 4.18.

4.18.2 In any and all claims against the Owner or the Architect or any of their agents or employees by any employee of the Contractor, any Subcontractor, anyone directly or indirectly employed by any of them or anyone for whose acts any of them may be liable, the indemnification obligation under this Paragraph 4.18 shall not be limited in any way by any limitation on the amount or type of damages, compensation or benefits payable by or for the Contractor or any Subcontractor under workers' or workmen's compensation acts, disability benefit acts or other employee benefit acts.

4.18.3 The obligations of the Contractor under this Paragraph 4.18 shall not extend to the liability of the Architect, his agents or employees, arising out of (1) the preparation or approval of maps, drawings, opinions, reports, surveys, change orders, designs or specifications, or (2) the giving of or the failure to give directions or instructions by the Architect, his agents or employees provided such giving or failure to give is the primary cause of the injury or damage.

ARTICLE 5

SUBCONTRACTORS

5.1 DEFINITION

5.1.1 A Subcontractor is a person or entity who has a direct contract with the Contractor to perform any of the Work at the site. The term Subcontractor is referred to throughout the Contract Documents as if singular in number and masculine in gender and means a Subcontractor or his authorized representative. The term Subcontractor does not include any separate contractor or his subcontractors.

5.1.2 A Sub-subcontractor is a person or entity who has a direct or indirect contract with a Subcontractor to perform any of the Work at the site. The term Sub-subcontractor is referred to throughout the Contract Documents as if singular in number and masculine in gender and means a Sub-subcontractor or an authorized representative thereof.

5.2 AWARD OF SUBCONTRACTS AND OTHER CONTRACTS FOR PORTIONS OF THE WORK

5.2.1 Unless otherwise required by the Contract Documents or the Bidding Documents, the Contractor, as soon as practicable after the award of the Contract, shall furnish to the Owner and the Architect in writing the names of the persons or entities (including those who are to furnish materials or equipment fabricated to a special design) proposed for each of the principal portions of the Work. The Architect will promptly reply to the Contractor in writing stating whether or not the Owner or the Architect, after due investigation, has reasonable objection to any such proposed person or entity. Failure of the Owner or Architect to reply promptly shall constitute notice of no reasonable objection.

5.2.2 The Contractor shall not contract with any such proposed person or entity to whom the Owner or the Architect has made reasonable objection under the provisions of Subparagraph 5.2.1. The Contractor shall not be required to contract with anyone to whom he has a reasonable objection.

5.2.3 If the Owner or the Architect has reasonable objection to any such proposed person or entity, the Contractor shall submit a substitute to whom the Owner or the Architect has no reasonable objection, and the Contract Sum shall be increased or decreased by the difference in cost occasioned by such substitution and an appropriate Change Order shall be issued; however, no increase in the Contract Sum shall be allowed for any such substitution unless the Contractor has acted promptly and responsively in submitting names as required by Subparagraph 5.2.1.

5.2.4 The Contractor shall make no substitution for any Subcontractor, person or entity previously selected if the Owner or Architect makes reasonable objection to such substitution.

5.3 SUBCONTRACTUAL RELATIONS

5.3.1 By an appropriate agreement, written where legally required for validity, the Contractor shall require each Subcontractor, to the extent of the Work to be performed by the Subcontractor, to be bound to the Contractor by the terms of the Contract Documents, and to assume toward the Contractor all the obligations and responsibilities which the Contractor, by these Documents, assumes toward the Owner and the Architect. Said agreement shall preserve and protect the rights of the Owner and the Architect under the Contract Documents with respect to the Work to be performed by the Subcontractor so that the subcontracting thereof will not prejudice such rights, and shall allow to the Subcontractor, unless specifically provided otherwise in the Contractor-Subcontractor agreement, the benefit of all rights, remedies and redress against the Contractor that the Contractor, by these Documents, has against the Owner. Where appropriate, the Contractor shall require each Subcontractor to enter into similar agreements with his Sub-subcontractors. The Contractor shall make available to each proposed Subcontractor, prior to the execution of the Subcontract, copies of the Contract Documents to which the Subcontractor will be bound by this Paragraph 5.3, and identify to the Subcontractor any terms and conditions of the proposed Subcontract which may be at variance with the Contract Documents. Each Subcontractor shall similarly make copies of such Documents available to his Sub-subcontractors.

ARTICLE 6

WORK BY OWNER OR BY SEPARATE CONTRACTORS

6.1 OWNER'S RIGHT TO PERFORM WORK AND TO AWARD SEPARATE CONTRACTS

6.1.1 The Owner reserves the right to perform work related to the Project with his own forces, and to award separate contracts in connection with other portions of the Project or other work on the site under these or similar Conditions of the Contract. If the Contractor claims that delay or additional cost is involved because of such action by the Owner, he shall make such claim as provided elsewhere in the Contract Documents.

6.1.2 When separate contracts are awarded for different portions of the Project or other work on the site, the term Contractor in the Contract Documents in each case shall mean the Contractor who executes each separate Owner-Contractor Agreement.

6.1.3 The Owner will provide for the coordination of the work of his own forces and of each separate contractor with the Work of the Contractor, who shall cooperate therewith as provided in Paragraph 6.2.

6.2 MUTUAL RESPONSIBILITY

6.2.1 The Contractor shall afford the Owner and separate contractors reasonable opportunity for the introduction and storage of their materials and equipment and the execution of their work, and shall connect and coordinate his Work with theirs as required by the Contract Documents.

6.2.2 If any part of the Contractor's Work depends for proper execution or results upon the work of the Owner or any separate contractor, the Contractor shall, prior to proceeding with the Work, promptly report to the Architect any apparent discrepancies or defects in such other work that render it unsuitable for such proper execution and results. Failure of the Contractor so to report shall constitute an acceptance of the Owner's or separate contractors' work as fit and proper to receive his Work, except as to defects which may subsequently become apparent in such work by others.

6.2.3 Any costs caused by defective or ill-timed work shall be borne by the party responsible therefor.

6.2.4 Should the Contractor wrongfully cause damage to the work or property of the Owner, or to other work on the site, the Contractor shall promptly remedy such damage as provided in Subparagraph 10.2.5.

6.2.5 Should the Contractor wrongfully cause damage to the work or property of any separate contractor, the Contractor shall upon due notice promptly attempt to settle with such other contractor by agreement, or otherwise to resolve the dispute. If such separate contractor sues or initiates an arbitration proceeding against the Owner on account of any damage alleged to have been caused by the Contractor, the Owner shall notify the Contractor who shall defend such proceedings at the Owner's expense, and if any judgment or award against the Owner arises therefrom the Contractor shall pay or satisfy it and shall reimburse the Owner for all attorneys' fees and court or arbitration costs which the Owner has incurred.

6.3 OWNER'S RIGHT TO CLEAN UP

6.3.1 If a dispute arises between the Contractor and separate contractors as to their responsibility for cleaning up as required by Paragraph 4.15, the Owner may clean up and charge the cost thereof to the contractors responsible therefor as the Architect shall determine to be just.

ARTICLE 7

MISCELLANEOUS PROVISIONS

7.1 GOVERNING LAW

7.1.1 The Contract shall be governed by the law of the place where the Project is located.

7.2 SUCCESSORS AND ASSIGNS

7.2.1 The Owner and the Contractor each binds himself, his partners, successors, assigns and legal representatives to the other party hereto and to the partners, successors, assigns and legal representatives of such other party in respect to all covenants, agreements and obligations contained in the Contract Documents. Neither party to the Contract shall assign the Contract or sublet it as a whole without the written consent of the other, nor shall the Contractor assign any moneys due or to become due to him hereunder, without the previous written consent of the Owner.

7.3 WRITTEN NOTICE

7.3.1 Written notice shall be deemed to have been duly served if delivered in person to the individual or member of the firm or entity or to an officer of the corporation for whom it was intended, or if delivered at or sent by registered or certified mail to the last business address known to him who gives the notice.

7.4 CLAIMS FOR DAMAGES

7.4.1 Should either party to the Contract suffer injury or damage to person or property because of any act or omission of the other party or of any of his employees, agents or others for whose acts he is legally liable, claim shall be made in writing to such other party within a reasonable time after the first observance of such injury or damage.

7.5 PERFORMANCE BOND AND LABOR AND MATERIAL PAYMENT BOND

7.5.1 The Owner shall have the right to require the Contractor to furnish bonds covering the faithful performance of the Contract and the payment of all obligations arising thereunder if and as required in the Bidding Documents or in the Contract Documents.

7.6 RIGHTS AND REMEDIES

7.6.1 The duties and obligations imposed by the Contract Documents and the rights and remedies available thereunder shall be in addition to and not a limitation of any duties, obligations, rights and remedies otherwise imposed or available by law.

7.6.2 No action or failure to act by the Owner, Architect or Contractor shall constitute a waiver of any right or duty afforded any of them under the Contract, nor shall any such action or failure to act constitute an approval of or acquiescence in any breach thereunder, except as may be specifically agreed in writing.

7.7 TESTS

7.7.1 If the Contract Documents, laws, ordinances, rules, regulations or orders of any public authority having jurisdiction require any portion of the Work to be inspected, tested or approved, the Contractor shall give the Architect timely notice of its readiness so the Architect may observe such inspection, testing or approval. The Contractor shall bear all costs of such inspections, tests or approvals conducted by public authorities. Unless otherwise provided, the Owner shall bear all costs of other inspections, tests or approvals.

7.7.2 If the Architect determines that any Work requires special inspection, testing, or approval which Subparagraph 7.7.1 does not include, he will, upon written authorization from the Owner, instruct the Contractor to order such special inspection, testing or approval, and the Contractor shall give notice as provided in Subparagraph 7.7.1. If such special inspection or testing reveals a failure of the Work to comply with the requirements of the Contract Documents, the Contractor shall bear all costs thereof, including compensation for the Architect's additional services made necessary by such failure; otherwise the Owner shall bear such costs, and an appropriate Change Order shall be issued.

7.7.3 Required certificates of inspection, testing or approval shall be secured by the Contractor and promptly delivered by him to the Architect.

7.7.4 If the Architect is to observe the inspections, tests or approvals required by the Contract Documents, he will do so promptly and, where practicable, at the source of supply.

7.8 INTEREST

7.8.1 Payments due and unpaid under the Contract Documents shall bear interest from the date payment is due at such rate as the parties may agree upon in writing or, in the absence thereof, at the legal rate prevailing at the place of the Project.

7.9 ARBITRATION

7.9.1 All claims, disputes and other matters in question between the Contractor and the Owner arising out of, or relating to, the Contract Documents or the breach thereof, except as provided in Subparagraph 2.2.11 with respect to the Architect's decisions on matters relating to artistic effect, and except for claims which have been waived by the making or acceptance of final payment as provided by Subparagraphs 9.9.4 and 9.9.5, shall be decided by arbitration in accordance with the Construction Industry Arbitration Rules of the American Arbitration Association then obtaining unless the parties mutually agree otherwise. No arbitration arising out of or relating to the Contract Documents shall include, by consolidation, joinder or in any other manner, the Architect, his employees or consultants except by written consent containing a specific reference to the Owner-Contractor Agreement and signed by the Architect, the Owner, the Contractor and any other person sought to be joined. No arbitration shall include by consolidation, joinder or in any other manner, parties other than the Owner, the Contractor and any other persons substantially involved in a common question of fact or law, whose presence is required if complete relief is to be accorded in the arbitration. No person other than the Owner or Contractor shall be included as an original third party or additional third party to an arbitration whose interest or responsibility is insubstantial. Any consent to arbitration involving an additional person or persons shall not constitute consent to arbitration of any dispute not described therein or with any person not named or described therein. The foregoing agreement to arbitrate and any other agreement to arbitrate with an additional person or persons duly consented to by the parties to the Owner-Contractor Agreement shall be specifically enforceable under the prevailing arbitration law. The award rendered by the arbitrators shall be final, and judgment may be entered upon it in accordance with applicable law in any court having jurisdiction thereof.

7.9.2 Notice of the demand for arbitration shall be filed in writing with the other party to the Owner-Contractor Agreement and with the American Arbitration Association, and a copy shall be filed with the Architect. The demand for arbitration shall be made within the time limits specified in Subparagraph 2.2.12 where applicable, and in all other cases within a reasonable time after the claim, dispute or other matter in question has arisen, and in no event shall it be made after the date when institution of legal or equitable proceedings based on such claim, dispute or other matter in question would be barred by the applicable statute of limitations.

7.9.3 Unless otherwise agreed in writing, the Contractor shall carry on the Work and maintain its progress during any arbitration proceedings, and the Owner shall continue to make payments to the Contractor in accordance with the Contract Documents.

ARTICLE 8

TIME

8.1 DEFINITIONS

8.1.1 Unless otherwise provided, the Contract Time is the period of time allotted in the Contract Documents for Substantial Completion of the Work as defined in Subparagraph 8.1.3, including authorized adjustments thereto.

8.1.2 The date of commencement of the Work is the date established in a notice to proceed. If there is no notice to proceed, it shall be the date of the Owner-Contractor Agreement or such other date as may be established therein.

8.1.3 The Date of Substantial Completion of the Work or designated portion thereof is the Date certified by the Architect when construction is sufficiently complete, in accordance with the Contract Documents, so the Owner can occupy or utilize the Work or designated portion thereof for the use for which it is intended.

8.1.4 The term day as used in the Contract Documents shall mean calendar day unless otherwise specifically designated.

8.2 PROGRESS AND COMPLETION

8.2.1 All time limits stated in the Contract Documents are of the essence of the Contract.

8.2.2 The Contractor shall begin the Work on the date of commencement as defined in Subparagraph 8.1.2. He shall carry the Work forward expeditiously with adequate forces and shall achieve Substantial Completion within the Contract Time.

8.3 DELAYS AND EXTENSIONS OF TIME

8.3.1 If the Contractor is delayed at any time in the progress of the Work by any act or neglect of the Owner or the Architect, or by any employee of either, or by any separate contractor employed by the Owner, or by changes ordered in the Work, or by labor disputes, fire, unusual delay in transportation, adverse weather conditions not reasonably anticipatable, unavoidable casualties, or any causes beyond the Contractor's control, or by delay authorized by the Owner pending arbitration, or by any other cause which the Architect determines may justify the delay, then the Contract Time shall be extended by Change Order for such reasonable time as the Architect may determine.

8.3.2 Any claim for extension of time shall be made in writing to the Architect not more than twenty days after the commencement of the delay; otherwise it shall be waived. In the case of a continuing delay only one claim is necessary. The Contractor shall provide an estimate of the probable effect of such delay on the progress of the Work.

8.3.3 If no agreement is made stating the dates upon which interpretations as provided in Subparagraph 2.2.8 shall be furnished, then no claim for delay shall be allowed on account of failure to furnish such interpretations until fifteen days after written request is made for them, and not then unless such claim is reasonable.

8.3.4 This Paragraph 8.3 does not exclude the recovery of damages for delay by either party under other provisions of the Contract Documents.

ARTICLE 9

PAYMENTS AND COMPLETION

9.1 CONTRACT SUM

9.1.1 The Contract Sum is stated in the Owner-Contractor Agreement and, including authorized adjustments thereto, is the total amount payable by the Owner to the Contractor for the performance of the Work under the Contract Documents.

9.2 SCHEDULE OF VALUES

9.2.1 Before the first Application for Payment, the Contractor shall submit to the Architect a schedule of values allocated to the various portions of the Work, prepared in such form and supported by such data to substantiate its accuracy as the Architect may require. This schedule, unless objected to by the Architect, shall be used only as a basis for the Contractor's Applications for Payment.

9.3 APPLICATIONS FOR PAYMENT

9.3.1 At least ten days before the date for each progress payment established in the Owner-Contractor Agreement, the Contractor shall submit to the Architect an itemized Application for Payment, notarized if required, supported by such data substantiating the Contractor's right to payment as the Owner or the Architect may require, and reflecting retainage, if any, as provided elsewhere in the Contract Documents.

9.3.2 Unless otherwise provided in the Contract Documents, payments will be made on account of materials or equipment not incorporated in the Work but delivered and suitably stored at the site and, if approved in advance by the Owner, payments may similarly be made for materials or equipment suitably stored at some other location agreed upon in writing. Payments for materials or equipment stored on or off the site shall be conditioned upon submission by the Contractor of bills of sale or such other procedures satisfactory to the Owner to establish the Owner's title to such materials or equipment or otherwise protect the Owner's interest, including applicable insurance and transportation to the site for those materials and equipment stored off the site.

9.3.3 The Contractor warrants that title to all Work, materials and equipment covered by an Application for Payment will pass to the Owner either by incorporation in the construction or upon the receipt of payment by the Contractor, whichever occurs first, free and clear of all liens, claims, security interests or encumbrances, hereinafter referred to in this Article 9 as "liens"; and that no Work, materials or equipment covered by an Application for Payment will have been acquired by the Contractor, or by any other person performing Work at the site or furnishing materials and equipment for the Project, subject to an agreement under which an interest therein or an encumbrance thereon is retained by the seller or otherwise imposed by the Contractor or such other person.

9.4 CERTIFICATES FOR PAYMENT

9.4.1 The Architect will, within seven days after the receipt of the Contractor's Application for Payment, either issue a Certificate for Payment to the Owner, with a copy to the Contractor, for such amount as the Architect determines is properly due, or notify the Contractor in writing his reasons for withholding a Certificate as provided in Subparagraph 9.6.1.

9.4.2 The issuance of a Certificate for Payment will constitute a representation by the Architect to the Owner, based on his observations at the site as provided in Subparagraph 2.2.3 and the data comprising the Application for Payment, that the Work has progressed to the point indicated; that, to the best of his knowledge, information and belief, the quality of the Work is in accordance with the Contract Documents (subject to an evaluation of the Work for conformance with the Contract Documents upon Substantial Completion, to the results of any subsequent tests required by or performed under the Contract Documents, to minor deviations from the Contract Documents correctable prior to completion, and to any specific qualifications stated in his Certificate); and that the Contractor is entitled to payment in the amount certified. However, by issuing a Certificate for Payment, the Architect shall not thereby be deemed to represent that he has made exhaustive or continuous on-site inspections to check the quality or quantity of the Work or that he has reviewed the construction means, methods, techniques, sequences or procedures, or that he has made any examination to ascertain how or for what purpose the Contractor has used the moneys previously paid on account of the Contract Sum.

9.5 PROGRESS PAYMENTS

9.5.1 After the Architect has issued a Certificate for Payment, the Owner shall make payment in the manner and within the time provided in the Contract Documents.

9.5.2 The Contractor shall promptly pay each Subcontractor, upon receipt of payment from the Owner, out of the amount paid to the Contractor on account of such Subcontractor's Work, the amount to which said Subcontractor is entitled, reflecting the percentage actually retained, if any, from payments to the Contractor on account of such Subcontractor's Work. The Contractor shall, by an appropriate agreement with each Subcontractor, require each Subcontractor to make payments to his Sub-subcontractors in similar manner.

9.5.3 The Architect may, on request and at his discretion, furnish to any Subcontractor, if practicable, information regarding the percentages of completion or the amounts applied for by the Contractor and the action taken thereon by the Architect on account of Work done by such Subcontractor.

9.5.4 Neither the Owner nor the Architect shall have any obligation to pay or to see to the payment of any moneys to any Subcontractor except as may otherwise be required by law.

9.5.5 No Certificate for a progress payment, nor any progress payment, nor any partial or entire use or occupancy of the Project by the Owner, shall constitute an acceptance of any Work not in accordance with the Contract Documents.

9.6 PAYMENTS WITHHELD

9.6.1 The Architect may decline to certify payment and may withhold his Certificate in whole or in part, to the extent necessary reasonably to protect the Owner, if in his opinion he is unable to make representations to the Owner as provided in Subparagraph 9.4.2. If the Architect is unable to make representations to the Owner as provided in Subparagraph 9.4.2 and to certify payment in the amount of the Application, he will notify the Contractor as provided in Subparagraph 9.4.1. If the Contractor and the Architect cannot agree on a revised amount, the Architect will promptly issue a Certificate for Payment for the amount for which he is able to make such representations to the Owner. The Architect may also decline to certify payment or, because of subsequently discovered evidence or subsequent observations, he may nullify the whole or any part of any Certificate for Payment previously issued, to such extent as may be necessary in his opinion to protect the Owner from loss because of:

.1 defective Work not remedied,

.2 third party claims filed or reasonable evidence indicating probable filing of such claims,

.3 failure of the Contractor to make payments properly to Subcontractors or for labor, materials or equipment,

.4 reasonable evidence that the Work cannot be completed for the unpaid balance of the Contract Sum,

.5 damage to the Owner or another contractor,

.6 reasonable evidence that the Work will not be completed within the Contract Time, or

.7 persistent failure to carry out the Work in accordance with the Contract Documents.

9.6.2 When the above grounds in Subparagraph 9.6.1 are removed, payment shall be made for amounts withheld because of them.

9.7 FAILURE OF PAYMENT

9.7.1 If the Architect does not issue a Certificate for Payment, through no fault of the Contractor, within seven days after receipt of the Contractor's Application for Payment, or if the Owner does not pay the Contractor within seven days after the date established in the Contract Documents any amount certified by the Architect or awarded by arbitration, then the Contractor may, upon seven additional days' written notice to the Owner and the Architect, stop the Work until payment of the amount owing has been received. The Contract Sum shall be increased by the amount of the Contractor's reasonable costs of shut-down, delay and start-up, which shall be effected by appropriate Change Order in accordance with Paragraph 12.3.

9.8 SUBSTANTIAL COMPLETION

9.8.1 When the Contractor considers that the Work, or a designated portion thereof which is acceptable to the Owner, is substantially complete as defined in Subparagraph 8.1.3, the Contractor shall prepare for submission to the Architect a list of items to be completed or corrected. The failure to include any items on such list does not alter the responsibility of the Contractor to complete all Work in accordance with the Contract Documents. When the Architect on the basis of an inspection determines that the Work or designated portion thereof is substantially complete, he will then prepare a Certificate of Substantial Completion which shall establish the Date of Substantial Completion, shall state the responsibilities of the Owner and the Contractor for security, maintenance, heat, utilities, damage to the Work, and insurance, and shall fix the time within which the Contractor shall complete the items listed therein. Warranties required by the Contract Documents shall commence on the Date of Substantial Completion of the Work or designated portion thereof unless otherwise provided in the Certificate of Substantial Completion. The Certificate of Substantial Completion shall be submitted to the Owner and the Contractor for their written acceptance of the responsibilities assigned to them in such Certificate.

9.8.2 Upon Substantial Completion of the Work or designated portion thereof and upon application by the Contractor and certification by the Architect, the Owner shall make payment, reflecting adjustment in retainage, if any, for such Work or portion thereof, as provided in the Contract Documents.

9.9 FINAL COMPLETION AND FINAL PAYMENT

9.9.1 Upon receipt of written notice that the Work is ready for final inspection and acceptance and upon receipt of a final Application for Payment, the Architect will promptly make such inspection and, when he finds the Work acceptable under the Contract Documents and the Contract fully performed, he will promptly issue a final Certificate for Payment stating that to the best of his knowledge, information and belief, and on the basis of his observations and inspections, the Work has been completed in accordance with the terms and conditions of the Contract Documents and that the entire balance found to be due the Contractor, and noted in said final Certificate, is due and payable. The Architect's final Certificate for Payment will constitute a further representation that the conditions precedent to the Contractor's being entitled to final payment as set forth in Subparagraph 9.9.2 have been fulfilled.

9.9.2 Neither the final payment nor the remaining retained percentage shall become due until the Contractor submits to the Architect (1) an affidavit that all payrolls, bills for materials and equipment, and other indebtedness connected with the Work for which the Owner or his property might in any way be responsible, have been paid or otherwise satisfied, (2) consent of surety, if any, to final payment and (3), if required by the Owner, other data establishing payment or satisfaction of all such obligations, such as receipts, releases and waivers of liens arising out of the Contract, to the extent and in such form as may be designated by the Owner. If any Subcontractor refuses to furnish a release or waiver required by the Owner, the Contractor may furnish a bond satisfactory to the Owner to indemnify him against any such lien. If any such lien remains unsatisfied after all payments are made, the Contractor shall refund to the Owner all moneys that the latter may be compelled to pay in discharging such lien, including all costs and reasonable attorneys' fees.

9.9.3 If, after Substantial Completion of the Work, final completion thereof is materially delayed through no fault of the Contractor or by the issuance of Change Orders affecting final completion, and the Architect so confirms, the Owner shall, upon application by the Contractor and certification by the Architect, and without terminating the Contract, make payment of the balance due for that portion of the Work fully completed and accepted. If the remaining balance for Work not fully completed or corrected is less than the retainage stipulated in the Contract Documents, and if bonds have been furnished as provided in Paragraph 7.5, the written consent of the surety to the payment of the balance due for that portion of the Work fully completed and accepted shall be submitted by the Contractor to the Architect prior to certification of such payment. Such payment shall be made under the terms and conditions governing final payment, except that it shall not constitute a waiver of claims.

9.9.4 The making of final payment shall constitute a waiver of all claims by the Owner except those arising from:

.1 unsettled liens,
.2 faulty or defective Work appearing after Substantial Completion,
.3 failure of the Work to comply with the requirements of the Contract Documents, or
.4 terms of any special warranties required by the Contract Documents.

9.9.5 The acceptance of final payment shall constitute a waiver of all claims by the Contractor except those previously made in writing and identified by the Contractor as unsettled at the time of the final Application for Payment.

ARTICLE 10

PROTECTION OF PERSONS AND PROPERTY

10.1 SAFETY PRECAUTIONS AND PROGRAMS

10.1.1 The Contractor shall be responsible for initiating, maintaining and supervising all safety precautions and programs in connection with the Work.

10.2 SAFETY OF PERSONS AND PROPERTY

10.2.1 The Contractor shall take all reasonable precautions for the safety of, and shall provide all reasonable protection to prevent damage, injury or loss to:

.1 all employees on the Work and all other persons who may be affected thereby;
.2 all the Work and all materials and equipment to be incorporated therein, whether in storage on or off the site, under the care, custody or control of the Contractor or any of his Subcontractors or Sub-subcontractors; and
.3 other property at the site or adjacent thereto, including trees, shrubs, lawns, walks, pavements, roadways, structures and utilities not designated for removal, relocation or replacement in the course of construction.

10.2.2 The Contractor shall give all notices and comply with all applicable laws, ordinances, rules, regulations and lawful orders of any public authority bearing on the safety of persons or property or their protection from damage, injury or loss.

10.2.3 The Contractor shall erect and maintain, as required by existing conditions and progress of the Work, all reasonable safeguards for safety and protection, including posting danger signs and other warnings against hazards, promulgating safety regulations and notifying owners and users of adjacent utilities.

10.2.4 When the use or storage of explosives or other hazardous materials or equipment is necessary for the execution of the Work, the Contractor shall exercise the utmost care and shall carry on such activities under the supervision of properly qualified personnel.

10.2.5 The Contractor shall promptly remedy all damage or loss (other than damage or loss insured under Paragraph 11.3) to any property referred to in Clauses 10.2.1.2 and 10.2.1.3 caused in whole or in part by the Contractor, any Subcontractor, any Sub-subcontractor, or anyone directly or indirectly employed by any of them, or by anyone for whose acts any of them may be liable and for which the Contractor is responsible under Clauses 10.2.1.2 and 10.2.1.3, except damage or loss attributable to the acts or omissions of the Owner or Architect or anyone directly or indirectly employed by either of them, or by anyone for whose acts either of them may be liable, and not attributable to the fault or negligence of the Contractor. The foregoing obligations of the Contractor are in addition to his obligations under Paragraph 4.18.

10.2.6 The Contractor shall designate a responsible member of his organization at the site whose duty shall be the prevention of accidents. This person shall be the Contractor's superintendent unless otherwise designated by the Contractor in writing to the Owner and the Architect.

10.2.7 The Contractor shall not load or permit any part of the Work to be loaded so as to endanger its safety.

10.3 EMERGENCIES

10.3.1 In any emergency affecting the safety of persons or property, the Contractor shall act, at his discretion, to prevent threatened damage, injury or loss. Any additional compensation or extension of time claimed by the Contractor on account of emergency work shall be determined as provided in Article 12 for Changes in the Work.

ARTICLE 11

INSURANCE

11.1 CONTRACTOR'S LIABILITY INSURANCE

11.1.1 The Contractor shall purchase and maintain such insurance as will protect him from claims set forth below which may arise out of or result from the Contractor's operations under the Contract, whether such operations be by himself or by any Subcontractor or by anyone directly or indirectly employed by any of them, or by anyone for whose acts any of them may be liable:

.1 claims under workers' or workmen's compensation, disability benefit and other similar employee benefit acts;

.2 claims for damages because of bodily injury, occupational sickness or disease, or death of his employees;

.3 claims for damages because of bodily injury, sickness or disease, or death of any person other than his employees;

.4 claims for damages insured by usual personal injury liability coverage which are sustained (1) by any person as a result of an offense directly or indirectly related to the employment of such person by the Contractor, or (2) by any other person;

.5 claims for damages, other than to the Work itself, because of injury to or destruction of tangible property, including loss of use resulting therefrom; and

.6 claims for damages because of bodily injury or death of any person or property damage arising out of the ownership, maintenance or use of any motor vehicle.

11.1.2 The insurance required by Subparagraph 11.1.1 shall be written for not less than any limits of liability specified in the Contract Documents, or required by law, whichever is greater.

11.1.3 The insurance required by Subparagraph 11.1.1 shall include contractual liability insurance applicable to the Contractor's obligations under Paragraph 4.18.

11.1.4 Certificates of Insurance acceptable to the Owner shall be filed with the Owner prior to commencement of the Work. These Certificates shall contain a provision that coverages afforded under the policies will not be cancelled until at least thirty days' prior written notice has been given to the Owner.

11.2 OWNER'S LIABILITY INSURANCE

11.2.1 The Owner shall be responsible for purchasing and maintaining his own liability insurance and, at his option, may purchase and maintain such insurance as will protect him against claims which may arise from operations under the Contract.

11.3 PROPERTY INSURANCE

11.3.1 Unless otherwise provided, the Owner shall purchase and maintain property insurance upon the entire Work at the site to the full insurable value thereof. This insurance shall include the interests of the Owner, the Contractor, Subcontractors and Sub-subcontractors in the Work and shall insure against the perils of fire and extended coverage and shall include "all risk" insurance for physical loss or damage including, without duplication of coverage, theft, vandalism and malicious mischief. If the Owner does not intend to purchase such insurance for the full insurable value of the entire Work, he shall inform the Contractor in writing prior to commencement of the Work. The Contractor may then effect insurance which will protect the interests of himself, his Subcontractors and the Sub-subcontractors in the Work, and by appropriate Change Order the cost thereof shall be charged to the Owner. If the Contractor is damaged by failure of the Owner to purchase or maintain such insurance and to so notify the Contractor, then the Owner shall bear all reasonable costs properly attributable thereto. If not covered under the all risk insurance or otherwise provided in the Contract Documents, the Contractor shall effect and maintain similar property insurance on portions of the Work stored off the site or in transit when such portions of the Work are to be included in an Application for Payment under Subparagraph 9.3.2.

11.3.2 The Owner shall purchase and maintain such boiler and machinery insurance as may be required by the Contract Documents or by law. This insurance shall include the interests of the Owner, the Contractor, Subcontractors and Sub-subcontractors in the Work.

11.3.3 Any loss insured under Subparagraph 11.3.1 is to be adjusted with the Owner and made payable to the Owner as trustee for the insureds, as their interests may appear, subject to the requirements of any applicable mortgagee clause and of Subparagraph 11.3.8. The Contractor shall pay each Subcontractor a just share of any insurance moneys received by the Contractor, and by appropriate agreement, written where legally required for validity, shall require each Subcontractor to make payments to his Sub-subcontractors in similar manner.

11.3.4 The Owner shall file a copy of all policies with the Contractor before an exposure to loss may occur.

11.3.5 If the Contractor requests in writing that insurance for risks other than those described in Subparagraphs 11.3.1 and 11.3.2 or other special hazards be included in the property insurance policy, the Owner shall, if possible, include such insurance, and the cost thereof shall be charged to the Contractor by appropriate Change Order.

11.3.6 The Owner and Contractor waive all rights against (1) each other and the Subcontractors, Sub-subcontractors, agents and employees each of the other, and (2) the Architect and separate contractors, if any, and their subcontractors, sub-subcontractors, agents and employees, for damages caused by fire or other perils to the extent covered by insurance obtained pursuant to this Paragraph 11.3 or any other property insurance applicable to the Work, except such rights as they may have to the proceeds of such insurance held by the Owner as trustee. The foregoing waiver afforded the Architect, his agents and employees shall not extend to the liability imposed by Subparagraph 4.18.3. The Owner or the Contractor, as appropriate, shall require of the Architect, separate contractors, Subcontractors and Sub-subcontractors by appropriate agreements, written where legally required for validity, similar waivers each in favor of all other parties enumerated in this Subparagraph 11.3.6.

11.3.7 If required in writing by any party in interest, the Owner as trustee shall, upon the occurrence of an insured loss, give bond for the proper performance of his duties. He shall deposit in a separate account any money so received, and he shall distribute it in accordance with such agreement as the parties in interest may reach, or in accordance with an award by arbitration in which case the procedure shall be as provided in Paragraph 7.9. If after such loss no other special agreement is made, replacement of damaged work shall be covered by an appropriate Change Order.

11.3.8 The Owner as trustee shall have power to adjust and settle any loss with the insurers unless one of the parties in interest shall object in writing within five days after the occurrence of loss to the Owner's exercise of this power, and if such objection be made, arbitrators shall be chosen as provided in Paragraph 7.9. The Owner as trustee shall, in that case, make settlement with the insurers in accordance with the directions of such arbitrators. If distribution of the insurance proceeds by arbitration is required, the arbitrators will direct such distribution.

11.3.9 If the Owner finds it necessary to occupy or use a portion or portions of the Work prior to Substantial Completion thereof, such occupancy or use shall not commence prior to a time mutually agreed to by the Owner and Contractor and to which the insurance company or companies providing the property insurance have consented by endorsement to the policy or policies. This insurance shall not be cancelled or lapsed on account of such partial occupancy or use. Consent of the Contractor and of the insurance company or companies to such occupancy or use shall not be unreasonably withheld.

11.4 LOSS OF USE INSURANCE

11.4.1 The Owner, at his option, may purchase and maintain such insurance as will insure him against loss of use of his property due to fire or other hazards, however caused. The Owner waives all rights of action against the Contractor for loss of use of his property, including consequential losses due to fire or other hazards however caused, to the extent covered by insurance under this Paragraph 11.4.

ARTICLE 12

CHANGES IN THE WORK

12.1 CHANGE ORDERS

12.1.1 A Change Order is a written order to the Contractor signed by the Owner and the Architect, issued after execution of the Contract, authorizing a change in the Work or an adjustment in the Contract Sum or the Contract Time. The Contract Sum and the Contract Time may be changed only by Change Order. A Change Order signed by the Contractor indicates his agreement therewith, including the adjustment in the Contract Sum or the Contract Time.

12.1.2 The Owner, without invalidating the Contract, may order changes in the Work within the general scope of the Contract consisting of additions, deletions or other revisions, the Contract Sum and the Contract Time being adjusted accordingly. All such changes in the Work shall be authorized by Change Order, and shall be performed under the applicable conditions of the Contract Documents.

12.1.3 The cost or credit to the Owner resulting from a change in the Work shall be determined in one or more of the following ways:

.1 by mutual acceptance of a lump sum properly itemized and supported by sufficient substantiating data to permit evaluation;

.2 by unit prices stated in the Contract Documents or subsequently agreed upon;

.3 by cost to be determined in a manner agreed upon by the parties and a mutually acceptable fixed or percentage fee; or

.4 by the method provided in Subparagraph 12.1.4.

12.1.4 If none of the methods set forth in Clauses 12.1.3.1, 12.1.3.2 or 12.1.3.3 is agreed upon, the Contractor, provided he receives a written order signed by the Owner, shall promptly proceed with the Work involved. The cost of such Work shall then be determined by the Architect on the basis of the reasonable expenditures and savings of those performing the Work attributable to the change, including, in the case of an increase in the Contract Sum, a reasonable allowance for overhead and profit. In such case, and also under Clauses 12.1.3.3 and 12.1.3.4 above, the Contractor shall keep and present, in such form as the Architect may prescribe, an itemized accounting together with appropriate supporting data for inclusion in a Change Order. Unless otherwise provided in the Contract Documents, cost shall be limited to the following: cost of materials, including sales tax and cost of delivery; cost of labor, including social security, old age and unemployment insurance, and fringe benefits required by agreement or custom; workers' or workmen's compensation insurance; bond premiums; rental value of equipment and machinery; and the additional costs of supervision and field office personnel directly attributable to the change. Pending final determination of cost to the Owner, payments on account shall be made on the Architect's Certificate for Payment. The amount of credit to be allowed by the Contractor to the Owner for any deletion or change which results in a net decrease in the Contract Sum will be the amount of the actual net cost as confirmed by the Architect. When both additions and credits covering related Work or substitutions are involved in any one change, the allowance for overhead and profit shall be figured on the basis of the net increase, if any, with respect to that change.

12.1.5 If unit prices are stated in the Contract Documents or subsequently agreed upon, and if the quantities originally contemplated are so changed in a proposed Change Order that application of the agreed unit prices to the quantities of Work proposed will cause substantial inequity to the Owner or the Contractor, the applicable unit prices shall be equitably adjusted.

12.2 CONCEALED CONDITIONS

12.2.1 Should concealed conditions encountered in the performance of the Work below the surface of the ground or should concealed or unknown conditions in an existing structure be at variance with the conditions indicated by the Contract Documents, or should unknown physical conditions below the surface of the ground or should concealed or unknown conditions in an existing structure of an unusual nature, differing materially from those ordinarily encountered and generally recognized as inherent in work of the character provided for in this Contract, be encountered, the Contract Sum shall be equitably adjusted by Change Order upon claim by either party made within twenty days after the first observance of the conditions.

12.3 CLAIMS FOR ADDITIONAL COST

12.3.1 If the Contractor wishes to make a claim for an increase in the Contract Sum, he shall give the Architect written notice thereof within twenty days after the occurrence of the event giving rise to such claim. This notice shall be given by the Contractor before proceeding to execute the Work, except in an emergency endangering life or property in which case the Contractor shall proceed in accordance with Paragraph 10.3. No such claim shall be valid unless so made. If the Owner and the Contractor cannot agree on the amount of the adjustment in the Contract Sum, it shall be determined by the Architect. Any change in the Contract Sum resulting from such claim shall be authorized by Change Order.

12.3.2 If the Contractor claims that additional cost is involved because of, but not limited to, (1) any written interpretation pursuant to Subparagraph 2.2.8, (2) any order by the Owner to stop the Work pursuant to Paragraph 3.3 where the Contractor was not at fault, (3) any written order for a minor change in the Work issued pursuant to Paragraph 12.4, or (4) failure of payment by the Owner pursuant to Paragraph 9.7, the Contractor shall make such claim as provided in Subparagraph 12.3.1.

12.4 MINOR CHANGES IN THE WORK

12.4.1 The Architect will have authority to order minor changes in the Work not involving an adjustment in the Contract Sum or an extension of the Contract Time and not inconsistent with the intent of the Contract Documents. Such changes shall be effected by written order, and shall be binding on the Owner and the Contractor. The Contractor shall carry out such written orders promptly.

ARTICLE 13

UNCOVERING AND CORRECTION OF WORK

13.1 UNCOVERING OF WORK

13.1.1 If any portion of the Work should be covered contrary to the request of the Architect or to requirements specifically expressed in the Contract Documents, it must, if required in writing by the Architect, be uncovered for his observation and shall be replaced at the Contractor's expense.

13.1.2 If any other portion of the Work has been covered which the Architect has not specifically requested to observe prior to being covered, the Architect may request to see such Work and it shall be uncovered by the Contractor. If such Work be found in accordance with the Contract Documents, the cost of uncovering and replacement shall, by appropriate Change Order, be charged to the Owner. If such Work be found not in accordance with the Contract Documents, the Contractor shall pay such costs unless it be found that this condition was caused by the Owner or a separate contractor as provided in Article 6, in which event the Owner shall be responsible for the payment of such costs.

13.2 CORRECTION OF WORK

13.2.1 The Contractor shall promptly correct all Work rejected by the Architect as defective or as failing to conform to the Contract Documents whether observed before or after Substantial Completion and whether or not fabricated, installed or completed. The Contractor shall bear all costs of correcting such rejected Work, including compensation for the Architect's additional services made necessary thereby.

13.2.2 If, within one year after the Date of Substantial Completion of the Work or designated portion thereof or within one year after acceptance by the Owner of designated equipment or within such longer period of time as may be prescribed by law or by the terms of any applicable special warranty required by the Contract Documents, any of the Work is found to be defective or not in accordance with the Contract Documents, the Contractor shall correct it promptly after receipt of a written notice from the Owner to do so unless the Owner has previously given the Contractor a written acceptance of such condition. This obligation shall survive termination of the Contract. The Owner shall give such notice promptly after discovery of the condition.

13.2.3 The Contractor shall remove from the site all portions of the Work which are defective or non-conforming and which have not been corrected under Subparagraphs 4.5.1, 13.2.1 and 13.2.2, unless removal is waived by the Owner.

13.2.4 If the Contractor fails to correct defective or non-conforming Work as provided in Subparagraphs 4.5.1, 13.2.1 and 13.2.2, the Owner may correct it in accordance with Paragraph 3.4.

13.2.5 If the Contractor does not proceed with the correction of such defective or non-conforming Work within a reasonable time fixed by written notice from the Architect, the Owner may remove it and may store the materials or equipment at the expense of the Contractor. If the Contractor does not pay the cost of such removal and storage within ten days thereafter, the Owner may upon ten additional days' written notice sell such Work at auction or at private sale and shall account for the net proceeds thereof, after deducting all the costs that should have been borne by the Contractor, including compensation for the Architect's additional services made necessary thereby. If such proceeds of sale do not cover all costs which the Contractor should have borne, the difference shall be charged to the Contractor and an appropriate Change Order shall be issued. If the payments then or thereafter due the Contractor are not sufficient to cover such amount, the Contractor shall pay the difference to the Owner.

13.2.6 The Contractor shall bear the cost of making good all work of the Owner or separate contractors destroyed or damaged by such correction or removal.

13.2.7 Nothing contained in this Paragraph 13.2 shall be construed to establish a period of limitation with respect to any other obligation which the Contractor might have under the Contract Documents, including Paragraph 4.5 hereof. The establishment of the time period of one year after the Date of Substantial Completion or such longer period of time as may be prescribed by law or by the terms of any warranty required by the Contract Documents relates only to the specific obligation of the Contractor to correct the Work, and has no relationship to the time within which his obligation to comply with the Contract Documents may be sought to be enforced, nor to the time within which proceedings may be commenced to establish the Contractor's liability with respect to his obligations other than specifically to correct the Work.

13.3 ACCEPTANCE OF DEFECTIVE OR NON-CONFORMING WORK

13.3.1 If the Owner prefers to accept defective or non-conforming Work, he may do so instead of requiring its removal and correction, in which case a Change Order will be issued to reflect a reduction in the Contract Sum where appropriate and equitable. Such adjustment shall be effected whether or not final payment has been made.

ARTICLE 14

TERMINATION OF THE CONTRACT

14.1 TERMINATION BY THE CONTRACTOR

14.1.1 If the Work is stopped for a period of thirty days under an order of any court or other public authority having jurisdiction, or as a result of an act of government, such as a declaration of a national emergency making materials unavailable, through no act or fault of the Contractor or a Subcontractor or their agents or employees or any other persons performing any of the Work under a contract with the Contractor, or if the Work should be stopped for a period of thirty days by the Contractor because the Architect has not issued a Certificate for Payment as provided in Paragraph 9.7 or because the Owner has not made payment thereon as provided in Paragraph 9.7, then the Contractor may, upon seven additional days' written notice to the Owner and the Architect, terminate the Contract and recover from the Owner payment for all Work executed and for any proven loss sustained upon any materials, equipment, tools, construction equipment and machinery, including reasonable profit and damages.

14.2 TERMINATION BY THE OWNER

14.2.1 If the Contractor is adjudged a bankrupt, or if he makes a general assignment for the benefit of his creditors, or if a receiver is appointed on account of his insolvency, or if he persistently or repeatedly refuses or fails, except in cases for which extension of time is provided, to supply enough properly skilled workmen or proper materials, or if he fails to make prompt payment to Subcontractors or for materials or labor, or persistently disregards laws, ordinances, rules, regulations or orders of any public authority having jurisdiction, or otherwise is guilty of a substantial violation of a provision of the Contract Documents, then the Owner, upon certification by the Architect that sufficient cause exists to justify such action, may, without prejudice to any right or remedy and after giving the Contractor and his surety, if any, seven days' written notice, terminate the employment of the Contractor and take possession of the site and of all materials, equipment, tools, construction equipment and machinery thereon owned by the Contractor and may finish the Work by whatever method he may deem expedient. In such case the Contractor shall not be entitled to receive any further payment until the Work is finished.

14.2.2 If the unpaid balance of the Contract Sum exceeds the costs of finishing the Work, including compensation for the Architect's additional services made necessary thereby, such excess shall be paid to the Contractor. If such costs exceed the unpaid balance, the Contractor shall pay the difference to the Owner. The amount to be paid to the Contractor or to the Owner, as the case may be, shall be certified by the Architect, upon application, in the manner provided in Paragraph 9.4, and this obligation for payment shall survive the termination of the Contract.

THE AMERICAN INSTITUTE OF ARCHITECTS

AIA Documents A201 and A201/SC

General and Federal Supplementary Conditions of the Contract for Construction

TABLE OF ARTICLES

GENERAL CONDITIONS

AIA Document A201 — 1976 Edition

1. CONTRACT DOCUMENTS
2. ARCHITECT
3. OWNER
4. CONTRACTOR
5. SUBCONTRACTORS
6. WORK BY OWNER OR BY SEPARATE CONTRACTORS
7. MISCELLANEOUS PROVISIONS
8. TIME
9. PAYMENTS AND COMPLETION
10. PROTECTION OF PERSONS AND PROPERTY
11. INSURANCE
12. CHANGES IN THE WORK
13. UNCOVERING AND CORRECTION OF WORK
14. TERMINATION OF THE CONTRACT

FEDERAL SUPPLEMENTARY CONDITIONS

AIA Document A201/SC — 1977 Edition

15. MODIFICATIONS OF THE GENERAL CONDITIONS
16. ADDITIONAL CONDITIONS
17. INSURANCE REQUIREMENTS

This document may be used for U.S. Department of Health, Education and Welfare Federally Assisted Construction Projects.

NOTICE

1. AIA Documents A201 and A201/SC are intended for joint use in Project Manuals prepared for U.S. Department of Health, Education and Welfare Federally Assisted Construction Projects. This COVER should be detached and discarded, and the Documents placed in their proper sequence in the Project Manual.
2. On Projects for which A201/SC may be used, AIA Document A701, Instructions to Bidders, January 1974 Edition, should not be used without modifying the following inconsistent provisions:
   a. A201/SC, Paragraph 15.3, is in conflict with, and takes precedence over A701, Paragraph 7.1.
   b. A201/SC, Paragraph 15.4, is in conflict with, and takes precedence over A701, Paragraph 8.1.
3. In A201/SC the Owner must provide the information required on insurance coverage to be included in Article 17.
4. Attach any wage determination decision required by Subparagraph 16.6.1.

THE AMERICAN INSTITUTE OF ARCHITECTS

AIA Document A201/SC

Federal Supplementary Conditions of the Contract For Construction

THIS DOCUMENT HAS IMPORTANT LEGAL CONSEQUENCES; CONSULTATION WITH AN ATTORNEY IS ENCOURAGED WITH RESPECT TO ITS COMPLETION OR MODIFICATION

1977 EDITION

TABLE OF ARTICLES

15. MODIFICATIONS OF THE GENERAL CONDITIONS

.1 Modification of Paragraph 1.2, Execution, Correlation and Intent
.2 Modification of Paragraph 4.8, Allowances
.3 Modification of Paragraph 5.2, Award of Subcontracts and Other Contracts for Portions of the Work
.4 Modification of Paragraph 7.5, Performance Bond and Labor and Material Payment Bond
.5 Modification of Paragraph 10.1, Safety Precautions and Programs
.6 Modification of Paragraph 11.1, Contractor's Liability Insurance

16. ADDITIONAL CONDITIONS

.1 Substitution of Materials and Equipment
.2 Federal Inspection
.3 Lands and Rights-of-Way
.4 Equal Opportunity
.5 Certification of Nonsegregated Facilities
.6 Minimum Wages
.7 Payrolls and Basic Records
.8 Apprentices and Trainees
.9 Compliance with Copeland Regulations
.10 Contract Work Hours and Safety Standards Act — Overtime Compensation and Safety Standards
.11 Withholding of Funds
.12 Subcontracts
.13 Contract Termination — Debarment
.14 Use and Occupancy of Project Prior to Acceptance by the Owner

17. INSURANCE REQUIREMENTS

This document may be used for U.S. Department of Health, Education and Welfare Federally Assisted Construction Projects.

SUPPLEMENTARY CONDITIONS OF THE CONTRACT FOR CONSTRUCTION

ARTICLE 15

MODIFICATIONS OF THE GENERAL CONDITIONS

15.1 MODIFICATION OF PARAGRAPH 1.2, EXECUTION, CORRELATION AND INTENT

15.1.1 Add the following to Subparagraph 1.2.1:

Such Documents shall be enumerated in the Owner-Contractor Agreement.

15.2 MODIFICATION OF PARAGRAPH 4.8, ALLOWANCES

Substitute the following for Subparagraph 4.8.1:

4.8.1 The Contractor shall include in the Contract Sum all allowances stated in the Contract Documents. Items covered by these allowances shall be supplied for such amounts as the Owner may direct and by such persons as the Owner shall have determined by competitive bidding through public advertising. The Contractor shall purchase the items covered by these allowances by the award of Subcontract to the lowest responsive and responsible bidder.

15.3 MODIFICATION OF PARAGRAPH 5.2, AWARD OF SUBCONTRACTS AND OTHER CONTRACTS FOR PORTIONS OF THE WORK

Substitute the following for Subparagraph 5.2.2:

5.2.2 The Contractor shall not contract with any person or entity declared ineligible under Federal laws or regulations from participating in Federally assisted construction projects or to whom the Owner or the Architect has made reasonable objection under the provisions of Subparagraph 5.2.1. The Contractor shall not be required to contract with anyone to whom he has a reasonable objection.

15.4 MODIFICATION OF PARAGRAPH 7.5, PERFORMANCE BOND AND LABOR AND MATERIAL PAYMENT BOND

Substitute the following for Subparagraph 7.5.1:

7.5.1 The Contractor shall furnish a Performance Bond in an amount equal to one hundred percent (100%) of the Contract Sum as security for the faithful performance of this Contract and also a Labor and Material Payment Bond in an amount not less than one hundred percent (100%) of the Contract Sum or in a penal sum not less than that prescribed by State, Territorial or local law, as security for the payment of persons performing labor on the Project under this Contract and furnishing materials in connection with this Contract. The Performance Bond and the Labor and Material Payment Bond may be in one or in separate instruments in accordance with local law and shall be delivered to the Owner not later than the date of execution of the Contract.

15.5 MODIFICATION OF PARAGRAPH 10.1, SAFETY PRECAUTIONS AND PROGRAMS

Add the following Subparagraph 10.1.2:

10.1.2 Attention is invited to the regulations issued by the Secretary of Labor pursuant to Section 107 of the Contract Work Hours and Safety Standards Act (40 U.S.C. 333) entitled "Safety and Health Regulations for Construction" (29 CFR Part 1926). The Contractor shall be required to comply with those regulations to the extent that any resulting Contract involves construction.

15.6 MODIFICATION OF ARTICLE 11, INSURANCE

Refer to Article 17 for the limits of and any modifications to the insurance requirements of this Article.

ARTICLE 16

ADDITIONAL CONDITIONS

16.1 SUBSTITUTION OF MATERIALS AND EQUIPMENT

16.1.1 Whenever a material, article or piece of equipment is identified on the Drawings or in the Specifications by reference to manufacturers' or vendors' names, trade names, catalog numbers, or the like, it is so identified for the purpose of establishing a standard, and any material, article, or piece of equipment of other manufacturers or vendors which will perform adequately the duties imposed by the general design will be considered equally acceptable provided the material, article, or piece of equipment so proposed is, in the opinion of the Architect, of equal substance, appearance and function. It shall not be purchased or installed by the Contractor without the Architect's written approval.

16.2 FEDERAL INSPECTION

16.2.1 The authorized representatives and agents of the Federal Government shall be permitted to inspect all Work, materials, payrolls, records of personnel, invoices of materials, and other relevant data and records.

16.3 LANDS AND RIGHTS-OF-WAY

16.3.1 Prior to the start of construction, the Owner shall obtain all lands and rights-of-way necessary for the execution and completion of Work to be performed under this Contract.

16.4 EQUAL OPPORTUNITY

16.4.1 During the performance of this Contract the Contractor agrees as follows:

.1 The Contractor will not discriminate against any employee or applicant for employment because of race, color, sex, religion, national origin or age. The

Contractor will take affirmative action to insure that applicants are employed and that employees are treated during employment without regard to their race, color, sex, religion, national origin or age. Such action shall include, but not be limited to, the following: employment, upgrading, demotion or transfer; recruitment or recruitment advertising; layoff or termination; rates of pay or other forms of compensation; and selection for training, including apprenticeship. The Contractor agrees to post in conspicuous places, available to employees and applicants for employment, notices to be provided by an appropriate agency of the Federal Government setting forth the requirements of these nondiscrimination provisions.

.2 The Contractor will state, in all solicitations or advertisements for employees placed by or on behalf of the Contractor, that all qualified applicants will receive consideration for employment without regard to race, color, sex, religion, national origin or age.

.3 The Contractor will send to each labor union or representative of workers with which he has a collective bargaining agreement or other contract or understanding a notice to be provided by the Owner, advising the labor union or workers' representative of the Contractor's commitments under Section 202 of Executive Order No. 11246 of September 24, 1965 as amended, and shall post copies of the notice in conspicuous places available to employees and applicants for employment.

.4 The Contractor will comply with all provisions of Executive Order No. 11246 of September 24, 1965, as amended, and of the rules, regulations and relevant orders of the Secretary of Labor.

.5 The Contractor will furnish all information and reports required by Executive Order No. 11246 of September 24, 1965 as amended, and by the rules, regulations and orders of the Secretary of Labor, or pursuant thereto, and shall permit access to his books, records and accounts by an appropriate agency of the Federal Government and by the Secretary of Labor for purposes of investigation to ascertain compliance with such rules, regulations and orders.

.6 In the event of the Contractor's noncompliance with the Equal Opportunity conditions of this Contract or with any of such rules, regulations or orders, this Contract may be cancelled, terminated or suspended in whole or in part, and the Contractor may be declared ineligible for further Government contracts or Federally assisted contracts in accordance with procedures authorized in Executive Order No. 11246 of September 24, 1965, as amended, and such other sanctions may be imposed and remedies invoked as provided in said Executive Order, or by rule, regulation or order of the Secretary of Labor, or as otherwise provided by law.

.7 The Contractor will include all clauses 16.4.1.1 to 16.4.1.7 inclusive in every Subcontract or purchase order unless exempted by rules, regulations or orders of the Secretary of Labor issued pursuant to Section 204 of Executive Order 11246 of September 24, 1965, as amended, so that such provisions will be binding upon each Subcontractor or vendor. The Contractor will take such action with respect to any Subcontractor or vendor as the appropriate agency of the Federal Government may direct as a means of enforcing such provisions, including sanctions for noncompliance, provided, however, that in the event the Contractor becomes involved in, or is threatened with, litigation with a Subcontractor or vendor as a result of such direction by the appropriate agency of the Federal Government, the Contractor may request the United States to enter into such litigation to protect the interests of the United States.

16.4.2 Exemptions to the requirements of the above Equal Opportunity conditions are construction Contracts and Subcontracts not exceeding $10,000, and Contracts and Subcontracts with regard to Work performed outside the United States by employees who were not recruited in the United States.

16.4.3 Unless otherwise provided, the above Equal Opportunity provisions are not required to be inserted in sub-subcontracts except for Sub-subcontracts involving the performance of construction Work at the site of construction, in which case the provisions must be inserted in all Sub-subcontracts.

16.5 CERTIFICATION OF NONSEGREGATED FACILITIES

(Applicable to Contracts and Subcontracts exceeding $10,000 which are not exempt from the provisions of Paragraph 16.4, "Equal Opportunity" of this Article 16.)

16.5.1 By entering into an agreement related to the Work described in the Contract Documents the Contractor or Subcontractor certifies that he does not maintain or provide for his employees any segregated facilities at any of his establishments, and that he does not permit his employees to perform their services at any location under his control where segregated facilities are maintained. The Contractor or Subcontractor further certifies that he will not maintain or provide for his employees any segregated facilities at any of his establishments and that he will not permit his employees to perform their services at any location under his control where segregated facilities are maintained. The Contractor or Subcontractor agrees that a breach of this certification is a violation of Paragraph 16.4, "Equal Opportunity." As used herein, the term "segregated facilities" means any waiting rooms, work areas, rest rooms and washrooms, restaurants and other eating areas, time clocks, locker rooms and other storage or dressing areas, parking lots, drinking fountains, recreation or entertainment areas, transportation, and housing facilities provided for employees on the basis of race, color, religion, age or national origin, because of habit, local custom, or otherwise. The Contractor further agrees that (except where he has obtained identical certifications from proposed Subcontractors for specific time periods) he will obtain identical certifications from proposed Subcontractors prior to the award of Subcontracts exceeding $10,000 which are not exempt from the provisions of Paragraph 16.4 "Equal Opportunity;" that he will retain such certifications in his files; and that he will forward the following notice to such proposed Subcontractors (except where the proposed Subcontractors have submitted identical certifications for specific time periods):

"NOTICE TO PROSPECTIVE SUBCONTRACTORS OF REQUIREMENT FOR CERTIFICATION OF NONSEGREGATED FACILITIES

A Certification of Nonsegregated Facilities, as required by the May 9, 1967, order (32 Federal Register 7439, May 19, 1967) on Elimination of Segregated Facilities, by the Secretary of Labor, must be submitted prior to the award of a Subcontract exceeding $10,000 which is not exempt from the provisions of Paragraph 16.4 'Equal Opportunity.' The Certification may be submitted either for each Subcontract or for all Subcontracts during a period, i.e. quarterly, semiannually or annually."

16.5.2 The penalty for making false statements in Certifications required by Paragraph 16.5.1 is prescribed in 18 USC 1001.

16.6 MINIMUM WAGES

16.6.1 All mechanics and laborers employed or working directly upon the site of the Work shall be paid unconditionally and not less often than once a week, and without subsequent deduction or rebate on any account [except such payroll deductions as are permitted by the Copeland Regulations (29 CFR Part 3)], the full amounts due at time of payment computed at wage rates not less than the aggregate of the basic hourly rates and the rates of payments, contributions, or costs for any fringe benefits contained in the wage determination decision of the Secretary of Labor, which is attached hereto and made a part hereof, regardless of any contractual relationship which may be alleged to exist between the Contractor or Subcontractor and such laborers and mechanics, and the wage decision shall be posted by the Contractor at the site of the Work in a prominent place where it can easily be seen by the workers.

16.6.2 The Owner shall require that any class of laborers or mechanics, including apprentices and trainees, which is not listed in the wage determination and which is to be employed under the Contract shall be classified or reclassified conformably to the wage determination and a report of the action taken shall be sent to the appropriate Federal agency. If the interested parties cannot agree on the proper classification or reclassification of a particular class of laborers or mechanics, including apprentices and trainees, to be used, the Owner shall submit the question together with his recommendation through the appropriate Federal agency to the Secretary of Labor for final determination.

16.6.3 The Owner shall require, whenever the minimum wage rate prescribed in the Contract for a class of laborers or mechanics includes a fringe benefit which is not expressed as an hourly wage rate and the Contractor is obliged to pay a cash equivalent of such a fringe benefit, an hourly cash equivalent thereof to be established. In the event interested parties cannot agree upon a cash equivalent of the fringe benefit, the question, accompanied by the recommendation of the Owner shall be referred to the Secretary of Labor for determination.

16.6.4 If the Contractor does not make payments to a trustee or other third person, he may consider as part of the wages of any laborer or mechanic the amount of any costs reasonably anticipated in providing benefits under a plan or program of a type expressly listed in the wage determination decision of the Secretary of Labor which is a part of this Contract: provided, however, the Secretary of Labor has found, upon the written request of the Contractor, that the applicable standards of the Davis-Bacon Act have been met. The Secretary of Labor may require the Contractor to set aside in a separate account assets for the meeting of obligations under the plan or program.

16.7 PAYROLLS AND BASIC RECORDS

16.7.1 The Contractor shall maintain payrolls and basic records relating thereto during the course of the Work and shall preserve them for a period of three years thereafter for all laborers and mechanics, including apprentices and trainees, working at the site of the Work. Such records shall contain the name and address of each employee, his correct classification, rate of pay (including rates of contributions for, or costs assumed to provide, fringe benefits), daily and weekly number of hours worked, deductions made and actual wages paid. Whenever the Contractor has obtained approval from the Secretary of Labor as provided in Paragraph 16.6.4 the Contractor shall maintain records which show that the commitment to provide such benefits is enforceable, that the plan or program is financially responsible, and that the plan or program has been communicated in writing to the laborers or mechanics affected, and records which show the costs anticipated or the actual cost incurred in providing such benefits.

16.7.2 The Contractor shall submit weekly a copy of all payrolls to the Owner. The prime Contractor shall be responsible for the submission of copies of payrolls of all Subcontractors. Each such copy shall be accompanied by a statement signed by the Contractor indicating that the payrolls are correct and complete, that the wage rates contained therein are not less than those determined by the Secretary of Labor, and that the classification set forth for each laborer or mechanic, including apprentices and trainees, conform to the work he performed. Submission of the "Weekly Statement of Compliance" required under this Contract and the Copeland Regulations of the Secretary of Labor (29 CFR Part 3) shall satisfy the requirement for submission of the above statement. The Contractor shall submit also a copy of any approval by the Secretary of Labor with respect to fringe benefits which is required by Paragraph 16.6.4.

16.7.3 Contractors employing apprentices or trainees under approved programs shall include a notation on the first weekly certified payrolls submitted to the Owner that their employment is pursuant to an approved program and shall identify the program.

16.7.4 The Contractor will make the records required under the labor standards clauses of the Contract available for inspection by authorized representatives of the Owner, the appropriate Federal agency and the U.S. Department of Labor, and shall permit such representatives to interview employees during working hours on the job.

16.8 APPRENTICES AND TRAINEES

16.8.1 Apprentices will be permitted to work at less than the predetermined rate for the work they perform when they are employed and individually registered in a bona fide apprenticeship program registered with the U.S. Department of Labor, Employment Training Administration (formerly Manpower), Bureau of Apprenticeship and Training, or with a State apprenticeship agency recognized by the Bureau, or if a person is employed in his first 90 days of probationary employment as an apprentice in such an apprenticeship program who is not individually registered in the program but who has been certified by the Bureau of Apprenticeship and Training or a State apprenticeship agency (where appropriate) to be eligible for probationary employment as an apprentice. The allowable ratio of apprentices to journeymen in any craft classification shall not be greater than the ratio permitted to the Contractor as to his entire work force under the registered program. Any employee listed on a payroll at an apprentice wage rate, who is not a trainee as defined in Subparagraph 16.8.2 of this Paragraph or is not registered or otherwise employed as stated above, shall be paid the wage rate determined by the Secretary of Labor for the classification of work he actually performed. The Contractor will be required to furnish to the Owner or a representative of the U.S. Department of Labor written evidence of the registration of his program and apprentices, as well as the appropriate ratios and wage rates (expressed in percentages of the journeyman hourly rates), for the area of construction prior to using any apprentices on the Contract work. The wage rate paid apprentices shall be not less than the appropriate percentage of the journeyman's rate contained in the applicable wage determination.

16.8.2 Trainees will not be permitted to work at less than the predetermined rate for the work performed unless they are employed pursuant to and individually registered in a program which has received prior approval, evidenced by formal certification, by the U.S. Department of Labor, Employment Training Administration (formerly Manpower), Bureau of Apprenticeship and Training. The ratio of trainees to journeymen shall not be greater than permitted under the plan approved by the Bureau of Apprenticeship and Training. Every trainee must be paid not less than the rate specified in the approved program for his level of progress. Any employee listed in the payroll at a trainee rate who is not registered and participating in a training plan approved by the Bureau of Apprenticeship and Training shall be paid not less than the wage rate determined by the Secretary of Labor for the classification of work he actually performed. The Contractor will be required to furnish to the Owner or a representative of the U.S. Department of Labor, written evidence of the certification of his program, the registration of the trainees, and the ratios and wage rates prescribed in that program. In the event the Bureau of Apprenticeship and Training withdraws approval of a training program, the Contractor will no longer be permitted to utilize trainees at less than the applicable predetermined rate for the work performed until an acceptable program is approved.

16.8.3 The utilization of apprentices, trainees and journeymen under this part shall be in conformity with the equal employment opportunity requirements of Executive Order 11246, as amended, and 29 CFR Part 30.

16.9 COMPLIANCE WITH COPELAND REGULATIONS

16.9.1 The Contractor shall comply with the Copeland Regulations of the Secretary of Labor (29 CFR Part 3), which are incorporated herein by reference. In addition, the Weekly Statement of Compliance required by these regulations shall also contain a statement that the fringe benefits paid are equal to or greater than those set forth in the minimum wage decision.

16.10 CONTRACT WORK HOURS AND SAFETY STANDARDS ACT—OVERTIME COMPENSATION AND SAFETY STANDARDS (40 USC 327-330)

16.10.1 The Contractor shall not require or permit any laborer or mechanic, including apprentices and trainees, in any work week in which he is employed on any work under this Contract to work in excess of 8 hours in any calendar day or in excess of 40 hours in such work week on work subject to the provisions of the Contract Work Hours and Safety Standards Act unless such laborer or mechanic, including apprentices and trainees, receives compensation at a rate not less than one and one-half times his basic rate of pay for all such hours worked in excess of 8 hours in any calendar day or in excess of 40 hours in such work week, whichever is the greater number of overtime hours. The "basic rate of pay" as used in this provision shall be the amount paid per hour, exclusive of the Contractor's contribution or cost for fringe benefits, and any cash payment made in lieu of providing fringe benefits, or the basic hourly rate contained in the wage determination, whichever is greater.

16.10.2 In the event of any violation of the provisions of Subparagraph 16.10.1 the Contractor shall be liable to any affected employee for any amounts due, and to the United States for liquidated damages. Such liquidated damages shall be computed with respect to each individual laborer or mechanic, including apprentices and trainees, employed in violation of the provisions of Subparagraph 16.10.1, in the sum of $10 for each calendar day on which such employee was required or permitted to be employed on such work in excess of 8 hours or in excess of the standard work week of 40 hours without payment of the overtime wages required by Subparagraph 16.10.1.

16.10.3 The Contractor shall not require or permit any laborer or mechanic, including apprentices and trainees, employed in the performance of this Contract to work in surroundings or under conditions which are unsanitary, hazardous, or dangerous to his health as determined under construction safety and health standards promulgated by the Secretary of Labor by regulation (29 CFR Part 1926, 36 FR 7340, April 17, 1971) pursuant to Section 107 of the Contract Work Hours and Safety Standards Act.

16.11 WITHHOLDING OF FUNDS

16.11.1 The Owner may withhold or cause to be withheld from the Contractor as much of the accrued payments or advances as may be considered necessary (a) to pay the laborers and mechanics, including apprentices and trainees, employed by the Contractor or any Subcontractor on the Work the full amount of wages required by the Contract, and (b) to satisfy any liability of any Contractor for liquidated damages under Paragraph 16.10 hereof entitled "Contract Work Hours and Safety Standards Act—Overtime Compensation and Safety Standards (40 USC 327-330)."

16.11.2 If the Contractor or any Subcontractor fails to pay any laborer or mechanic, including apprentices and trainees, employed or working on the site of the Work, all or part of the wages required by the Contract, the Owner may, after written notice to the prime Contractor take such action as may be necessary to cause suspension of any further payments or advances until such violations have ceased.

16.12 SUBCONTRACTS

16.12.1 The Contractor will insert in all Subcontracts, Paragraphs 16.6 through 16.13 inclusive, entitled: "Minimum Wages," "Payrolls and Basic Records," "Apprentices and Trainees," "Compliance with Copeland Regulations," "Contract Work Hours and Safety Standards Act—Overtime Compensation and Safety Standards (40 USC 327-330)," "Withholding of Funds," "Subcontracts," and "Contract Termination—Debarment," and shall further require all Subcontractors to incorporate physically these same Paragraphs in all Sub-subcontracts.

16.12.2 The term "Contractor" as used in such Paragraphs in any Subcontract shall be deemed to refer to the Subcontractor except when the phrase "prime Contractor" is used.

16.13 CONTRACT TERMINATION—DEBARMENT

16.13.1 A breach of Paragraphs 16.6 through 16.13 inclusive, respectively entitled "Minimum Wages," "Payrolls and Basic Records," "Apprentices and Trainees," "Compliance with Copeland Regulations," "Contract Work Hours and Safety Standards (40 USC 327-330)," "Withholding of Funds," and "Subcontracts" may be grounds for termination of the Contract and for debarment as provided in 29 CFR 5.6.

16.14 USE AND OCCUPANCY OF PROJECT PRIOR TO ACCEPTANCE BY THE OWNER

16.14.1 The Contractor agrees to use and occupancy of a portion or unit of the Project before formal acceptance by the Owner under the following conditions:

.1 A Certificate of Substantial Completion shall be prepared and executed as provided in Subparagraph 9.8.1 of the accompanying General Conditions of the Contract for Construction, except that when, in the opinion of the Architect, the Contractor is chargeable with unwarranted delay in completing Work or other Contract requirements the signature of the Contractor will not be required. The Certificate of Substantial Completion shall be accompanied by a written endorsement of the Contractor's insurance carrier and surety permitting occupancy by the Owner during the remaining period of Project Work.

.2 Occupancy by the Owner shall not be construed by the Contractor as being an acceptance of that part of the Project to be occupied.

.3 The Contractor shall not be held responsible for any damage to the occupied part of the Project resulting from the Owner's occupancy.

.4 Occupancy by the Owner shall not be deemed to constitute a waiver of existing claims in behalf of the Owner or Contractor against each other.

.5 If the Project consists of more than one building, and one of the buildings is to be occupied, the Owner, prior to occupancy of that building, shall secure permanent property insurance on the building to be occupied and necessary permits which may be required for use and occupancy.

16.14.2 With the exception of Clause 16.14.1.5, use and occupancy by the Owner prior to Project acceptance does not relieve the Contractor of his responsibility to maintain all insurance and bonds required of the Contractor under the Contract until the Project is completed and accepted by the Owner.

16.15 ENUMERATION OF THE DRAWINGS, THE SPECIFICATIONS AND ADDENDA

16.15.1 Following are the Drawings, Specifications and Addenda which form a part of this Contract, as set forth in Subparagraph 1.1.1 of the accompanying General Conditions of the Contract for Construction.

(List below all Drawings, Specifications and Addenda, showing page or sheet numbers and dates where applicable.)

DRAWINGS

SPECIFICATIONS: DATED ______________

ADDENDA:

ARTICLE 17

INSURANCE REQUIREMENTS

17.1 The insurance required by Subparagraph 11.1.1 shall be written for not less than the following:

1. Workers' Compensation
   - (a) State — Statutory
   - (b) Applicable Federal (e.g., Longshoremen's) — Statutory
   - (c) Employer's Liability — $________________

2. Comprehensive General Liability
   (Including Premises—Operations; Independent Contractor's Protective; Products and Completed Operation Broad Form Property Damage)
   - (a) Bodily Injury
     - (1) Each Occurrence — $________________
     - (2) Annual Aggregate — $________________
   - (b) Property Damage
     - (1) Each Occurence — $________________
     - (2) Annual Aggregate — $________________

4. Personal Injury
   Annual Aggregate — $________________

5. Completed Operations and Products Liability shall be maintained for ____________( ) years after final payment.

6. Property Damage Liability Insurance shall include coverage for the following hazards:
   - ☐ X
   - ☐ C
   - ☐ U

7. Comprehensive Automobile Liability
   - (a) Bodily Injury
     - (1) Each Person — $________________
     - (2) Each Occurrence — $________________
   - (b) Property Damage
     - (1) Each Occurrence — $________________

8. If an exposure exists, Aircraft Liability (owned and non-owned) and Watercraft Liability (owned and non-owned), with limits approved by the Owner shall be provided.

9. The Contractor shall carry insurance in addition to that specifically named above as follows:

Coverage	*Amount*
	$________________

17.2 Revise the first sentence of Subparagraph 11.3.1 as follows:

11.3.1 Until the Work is completed and accepted by the Owner, the ________________________* shall purchase and maintain property insurance upon the entire Work at the site to the full insurable value thereof.

*(Insert Owner or Contractor. Occasionally the Owner may require the Contractor to furnish the property insurance.)

THE AMERICAN INSTITUTE OF ARCHITECTS

AIA Document A401

SUBCONTRACT

Standard Form of Agreement Between Contractor and Subcontractor

1978 EDITION

Use with the latest edition of the appropriate AIA Documents as follows:

A101, Owner-Contractor Agreement — Stipulated Sum
A107, Abbreviated Owner-Contractor Agreement with General Conditions
A111, Owner-Contractor Agreement — Cost plus Fee
A201, General Conditions of the Contract for Construction.

THIS DOCUMENT HAS IMPORTANT LEGAL CONSEQUENCES; CONSULTATION WITH AN ATTORNEY IS ENCOURAGED WITH RESPECT TO ITS COMPLETION OR MODIFICATION

This document has been approved and endorsed by the American Subcontractors Association and the Associated Specialty Contractors, Inc.

AGREEMENT

made as of the day of in the year Nineteen Hundred and

BETWEEN the Contractor:

and the Subcontractor:

The Project:

The Owner:

The Architect:

The Contractor and Subcontractor agree as set forth below.

ARTICLE 1
THE CONTRACT DOCUMENTS

1.1 The Contract Documents for this Subcontract consist of this Agreement and any Exhibits attached hereto, the Agreement between the Owner and Contractor dated as of , the Conditions of the Contract between the Owner and Contractor (General, Supplementary and other Conditions), the Drawings, the Specifications, all Addenda issued prior to and all Modifications issued after execution of the Agreement between the Owner and Contractor and agreed upon by the parties to this Subcontract. These form the Subcontract, and are as fully a part of the Subcontract as if attached to this Agreement or repeated herein.

1.2 Copies of the above documents which are applicable to the Work under this Subcontract shall be furnished to the Subcontractor upon his request. An enumeration of the applicable Contract Documents appears in Article 15.

ARTICLE 2
THE WORK

2.1 The Subcontractor shall perform all the Work required by the Contract Documents for

(Here insert a precise description of the Work covered by this Subcontract and refer to numbers of Drawings and pages of Specifications including Addenda, Modifications and accepted Alternates.)

ARTICLE 3
TIME OF COMMENCEMENT AND SUBSTANTIAL COMPLETION

3.1 The Work to be performed under this Subcontract shall be commenced and, subject to authorized adjustments, shall be substantially completed not later than

(Here insert the specific provisions that are applicable to this Subcontract including any information pertaining to notice to proceed or other method of modification for commencement of Work, starting and completion dates, or duration, and any provisions for liquidated damages relating to failure to complete on time.)

3.2 Time is of the essence of this Subcontract.

3.3 No extension of time will be valid without the Contractor's written consent after claim made by the Subcontractor in accordance with Paragraph 11.10.

ARTICLE 4
THE CONTRACT SUM

4.1 The Contractor shall pay the Subcontractor in current funds for the performance of the Work, subject to additions and deductions authorized pursuant to Paragraph 11.9, the Contract Sum of

dollars ($).

The Contract Sum is determined as follows:

(State here the base bid or other lump sum amount, accepted alternates, and unit prices, as applicable.)

ARTICLE 5
PROGRESS PAYMENTS

5.1 The Contractor shall pay the Subcontractor monthly progress payments in accordance with Paragraph 12.4 of this Subcontract.

5.2 Applications for monthly progress payments shall be in writing and in accordance with Paragraph 11.8, shall state the estimated percentage of the Work in this Subcontract that has been satisfactorily completed and shall be submitted to the Contractor on or before the day of each month.

(Here insert details on (1) payment procedures and date of monthly applications, or other procedure if on other than a monthly basis, (2) the basis on which payment will be made on account of materials and equipment suitably stored at the site or other location agreed upon in writing, and (3) any provisions consistent with the Contract Documents for limiting or reducing the amount retained after the Work reaches a certain stage of completion.)

5.3 When the Subcontractor's Work or a designated portion thereof is substantially complete and in accordance with the Contract Documents, the Contractor shall, upon application by the Subcontractor, make prompt application for payment of such Work. Within thirty days following issuance by the Architect of the Certificate for Payment covering such substantially completed Work, the Contractor shall, to the full extent provided in the Contract Documents, make payment to the Subcontractor of the entire unpaid balance of the Contract Sum or of that portion of the Contract Sum attributable to the substantially completed Work, less any portion of the funds for the Subcontractor's Work withheld in accordance with the Certificate to cover costs of items to be completed or corrected by the Subcontractor.

(Delete the above Paragraph if the Contract Documents do not provide for, and the Subcontractor agrees to forego, release of retainage for the Subcontractor's Work prior to completion of the entire Project.)

5.4 Progress payments or final payment due and unpaid under this Subcontract shall bear interest from the date payment is due at the rate entered below or, in the absence thereof, at the legal rate prevailing at the place of the Project.

(Here insert any rate of interest agreed upon.)

(Usury laws and requirements under the Federal Truth in Lending Act, similar state and local consumer credit laws and other regulations at the Owner's, Contractor's and Subcontractor's principal places of business, the location of the Project and elsewhere may affect the validity of this provision. Specific legal advice should be obtained with respect to deletion, modification, or other requirements such as written disclosures or waivers.)

ARTICLE 6
FINAL PAYMENT

6.1 Final payment, constituting the entire unpaid balance of the Contract Sum, shall be due when the Work described in this Subcontract is fully completed and performed in accordance with the Contract Documents and is satisfactory to the Architect, and shall be payable as follows, in accordance with Article 5 and with Paragraph 12.4 of this Subcontract:

(Here insert the relevant conditions under which, or time in which, final payment will become payable.)

6.2 Before issuance of the final payment, the Subcontractor, if required, shall submit evidence satisfactory to the Contractor that all payrolls, bills for materials and equipment, and all known indebtedness connected with the Subcontractor's Work have been satisfied.

ARTICLE 7
PERFORMANCE BOND AND LABOR AND MATERIAL PAYMENT BOND

(Here insert any requirement for the furnishing of bonds by the Subcontractor.)

ARTICLE 8
TEMPORARY FACILITIES AND SERVICES

8.1 Unless otherwise provided in this Subcontract, the Contractor shall furnish and make available at no cost to the Subcontractor the following temporary facilities and services:

ARTICLE 9
INSURANCE

9.1 Prior to starting work, the Subcontractor shall obtain the required insurance from a responsible insurer, and shall furnish satisfactory evidence to the Contractor that the Subcontractor has complied with the requirements of this Article 9. Similarly, the Contractor shall furnish to the Subcontractor satisfactory evidence of insurance required of the Contractor by the Contract Documents.

9.2 The Contractor and Subcontractor waive all rights against each other and against the Owner, the Architect, separate contractors and all other subcontractors for damages caused by fire or other perils to the extent covered by property insurance provided under the General Conditions, except such rights as they may have to the proceeds of such insurance.

(Here insert any insurance requirements and Subcontractor's responsibility for obtaining, maintaining and paying for necessary insurance with limits equaling or exceeding those specified in the Contract Documents and inserted below, or required by law. If applicable, this shall include fire insurance and extended coverage, public liability, property damage, employer's liability, and workers' or workmen's compensation insurance for the Subcontractor and his employees. The insertion should cover provisions for notice of cancellation, allocation of insurance proceeds, and other aspects of insurance.)

ARTICLE 10
WORKING CONDITIONS

(Here insert any applicable arrangements concerning working conditions and labor matters for the Project.)

GENERAL CONDITIONS

ARTICLE 11
SUBCONTRACTOR

11.1 RIGHTS AND RESPONSIBILITIES

11.1.1 The Subcontractor shall be bound to the Contractor by the terms of this Agreement and, to the extent that provisions of the Contract Documents between the Owner and Contractor apply to the Work of the Subcontractor as defined in this Agreement, the Subcontractor shall assume toward the Contractor all the obligations and responsibilities which the Contractor, by those Documents, assumes toward the Owner and the Architect, and shall have the benefit of all rights, remedies and redress against the Contractor which the Contractor, by those Documents, has against the Owner, insofar as applicable to this Subcontract, provided that where any provision of the Contract Documents between the Owner and Contractor is inconsistent with any provision of this Agreement, this Agreement shall govern.

11.1.2 The Subcontractor shall not assign this subcontract without the written consent of the Contractor, nor subcontract the whole of this Subcontract without the written consent of the Contractor, nor further subcontract portions of this Subcontract without written notification to the Contractor when such notification is requested by the Contractor. The Subcontractor shall not assign any amounts due or to become due under this Subcontract without written notice to the Contractor.

11.2 EXECUTION AND PROGRESS OF THE WORK

11.2.1 The Subcontractor agrees that the Contractor's equipment will be available to the Subcontractor only at the Contractor's discretion and on mutually satisfactory terms.

11.2.2 The Subcontractor shall cooperate with the Contractor in scheduling and performing his Work to avoid conflict or interference with the work of others.

11.2.3 The Subcontractor shall promptly submit shop drawings and samples required in order to perform his Work efficiently, expeditiously and in a manner that will not cause delay in the progress of the Work of the Contractor or other subcontractors.

11.2.4 The Subcontractor shall furnish periodic progress reports on the Work as mutually agreed, including information on the status of materials and equipment under this Subcontract which may be in the course of preparation or manufacture.

11.2.5 The Subcontractor agrees that all Work shall be done subject to the final approval of the Architect. The Architect's decisions in matters relating to artistic effect shall be final if consistent with the intent of the Contract Documents.

11.2.6 The Subcontractor shall pay for all materials, equipment and labor used in, or in connection with, the performance of this Subcontract through the period covered by previous payments received from the Contractor, and shall furnish satisfactory evidence, when requested by the Contractor, to verify compliance with the above requirements.

11.3 LAWS, PERMITS, FEES AND NOTICES

11.3.1 The Subcontractor shall give all notices and comply with all laws, ordinances, rules, regulations and orders of any public authority bearing on the performance of the Work under this Subcontract. The Subcontractor shall secure and pay for all permits and governmental fees, licenses and inspections necessary for the proper execution and completion of the Subcontractor's Work, the furnishing of which is required of the Contractor by the Contract Documents.

11.3.2 The Subcontractor shall comply with Federal, State and local tax laws, social security acts, unemployment compensation acts and workers' or workmen's compensation acts insofar as applicable to the performance of this Subcontract.

11.4 WORK OF OTHERS

11.4.1 In carrying out his Work, the Subcontractor shall take necessary precautions to protect properly the finished work of other trades from damage caused by his operations.

11.4.2 The Subcontractor shall cooperate with the Contractor and other subcontractors whose work might interfere with the Subcontractor's Work, and shall participate in the preparation of coordinated drawings in areas of congestion as required by the Contract Documents, specifically noting and advising the Contractor of any such interference.

11.5 SAFETY PRECAUTIONS AND PROCEDURES

11.5.1 The Subcontractor shall take all reasonable safety precautions with respect to his Work, shall comply with all safety measures initiated by the Contractor and with all applicable laws, ordinances, rules, regulations and orders of any public authority for the safety of persons or property in accordance with the requirements of the Contract Documents. The Subcontractor shall report within three days to the Contractor any injury to any of the Subcontractor's employees at the site.

11.6 CLEANING UP

11.6.1 The Subcontractor shall at all times keep the premises free from accumulation of waste materials or rubbish arising out of the operations of this Subcontract. Unless otherwise provided, the Subcontractor shall not be held responsible for unclean conditions caused by other contractors or subcontractors.

11.7 WARRANTY

11.7.1 The Subcontractor warrants to the Owner, the Architect and the Contractor that all materials and equipment furnished shall be new unless otherwise specified, and that all Work under this Subcontract shall be of good quality, free from faults and defects and in conformance with the Contract Documents. All Work not conforming to these requirements, including substitutions not properly approved and authorized, may be considered defective. The warranty provided in this Paragraph 11.7 shall be in addition to and not in limitation of any other warranty or remedy required by law or by the Contract Documents.

11.8 APPLICATIONS FOR PAYMENT

11.8.1 The Subcontractor shall submit to the Contractor applications for payment at such times as stipulated in Article 5 to enable the Contractor to apply for payment.

11.8.2 If payments are made on the valuation of Work done, the Subcontractor shall, before the first application, submit to the Contractor a schedule of values of the various parts of the Work aggregating the total sum of this Subcontract, made out in such detail as the Subcontractor and Contractor may agree upon or as required by the Owner, and supported by such evidence as to its correctness as the Contractor may direct. This schedule, when approved by the Contractor, shall be used only as a basis for Applications for Payment, unless it be found to be in error. In applying for payment, the Subcontractor shall submit a statement based upon this schedule.

11.8.3 If payments are made on account of materials or equipment not incorporated in the Work but delivered and suitably stored at the site or at some other location agreed upon in writing, such payments shall be in accordance with the Terms and Conditions of the Contract Documents.

11.9 CHANGES IN THE WORK

11.9.1 The Subcontractor may be ordered in writing by the Contractor, without invalidating this Subcontract, to make changes in the Work within the general scope of this Subcontract consisting of additions, deletions or other revisions, the Contract Sum and the Contract Time being adjusted accordingly. The Subcontractor, prior to the commencement of such changed or revised Work, shall submit promptly to the Contractor written copies of any claim for adjustment to the Contract Sum and Contract Time for such revised Work in a manner consistent with the Contract Documents.

11.10 CLAIMS OF THE SUBCONTRACTOR

11.10.1 The Subcontractor shall make all claims promptly to the Contractor for additional cost, extensions of time, and damages for delays or other causes in accordance with the Contract Documents. Any such claim which will affect or become part of a claim which the Contractor is required to make under the Contract Documents within a specified time period or in a specified manner shall be made in sufficient time to permit the Contractor to satisfy the requirements of the Contract Documents. Such claims shall be received by the Contractor not less than two working days preceding the time by which the Contractor's claim must be made. Failure of the Subcontractor to make such a timely claim shall bind the Subcontractor to the same consequences as those to which the Contractor is bound.

11.11 INDEMNIFICATION

11.11.1 To the fullest extent permitted by law, the Subcontractor shall indemnify and hold harmless the Owner, the Architect and the Contractor and all of their agents and employees from and against all claims, damages, losses and expenses, including but not limited to attorney's fees, arising out of or resulting from the performance of the Subcontractor's Work under this Subcontract, provided that any such claim, damage, loss, or expense is attributable to bodily injury, sickness, disease, or death, or to injury to or destruction of tangible property (other than the Work itself) including the loss of use resulting therefrom, to the extent caused in whole or in part by any negligent act or omission of the Subcontractor or anyone directly or indirectly employed by him or anyone for whose acts he may be liable, regardless of whether it is caused in part by a party indemnified hereunder. Such obligation shall not be construed to negate, or abridge, or otherwise reduce any other right or obligation of indemnity which would otherwise exist as to any party or person described in this Paragraph 11.11.

11.11.2 In any and all claims against the Owner, the Architect, or the Contractor or any of their agents or employees by any employee of the Subcontractor, anyone directly or indirectly employed by him or anyone for whose acts he may be liable, the indemnification obligation under this Paragraph 11.11 shall not be limited in any way by any limitation on the amount or type of damages, compensation or benefits payable by or for the Subcontractor under workers' or workmen's compensation acts, disability benefit acts or other employee benefit acts.

11.11.3 The obligations of the Subcontractor under this Paragraph 11.11 shall not extend to the liability of the Architect, his agents or employees arising out of (1) the preparation or approval of maps, drawings, opinions, reports, surveys, Change Orders, designs or specifications, or (2) the giving of or the failure to give directions or instructions by the Architect, his agents or employees provided such giving or failure to give is the primary cause of the injury or damage.

11.12 SUBCONTRACTOR'S REMEDIES

11.12.1 If the Contractor does not pay the Subcontractor through no fault of the Subcontractor, within seven days from the time payment should be made as provided in Paragraph 12.4, the Subcontractor may, without prejudice to any other remedy he may have, upon seven additional days' written notice to the Contractor, stop his Work until payment of the amount owing has been received. The Contract Sum shall, by appropriate adjustment, be increased by the amount of the Subcontractor's reasonable costs of shutdown, delay and start-up.

ARTICLE 12
CONTRACTOR

12.1 RIGHTS AND RESPONSIBILITIES

12.1.1 The Contractor shall be bound to the Subcontractor by the terms of this Agreement, and to the extent that provisions of the Contract Documents between the Owner and the Contractor apply to the Work of the Subcontractor as defined in this Agreement, the Contractor shall assume toward the Subcontractor all the obligations and responsibilities that the Owner, by those Documents, assumes toward the Contractor, and shall have the benefit of all rights, remedies and redress against the Subcontractor which the Owner, by those Documents, has against the Contractor. Where any provision of the

Contract Documents between the Owner and the Contractor is inconsistent with any provisions of this Agreement, this Agreement shall govern.

12.2 SERVICES PROVIDED BY THE CONTRACTOR

12.2.1 The Contractor shall cooperate with the Subcontractor in scheduling and performing his Work to avoid conflicts or interference in the Subcontractor's Work, and shall expedite written responses to submittals made by the Subcontractor in accordance with Paragraphs 11.2, 11.9 and 11.10. As soon as practicable after execution of this Agreement, the Contractor shall provide the Subcontractor a copy of the estimated progress schedule of the Contractor's entire Work which the Contractor has prepared and submitted for the Owner's and the Architect's information, together with such additional scheduling details as will enable the Subcontractor to plan and perform his Work properly. The Subcontractor shall be notified promptly of any subsequent changes in the progress schedule and the additional scheduling details.

12.2.2 The Contractor shall provide suitable areas for storage of the Subcontractor's materials and equipment during the course of the Work. Any additional costs to the Subcontractor resulting from the relocation of such facilities at the direction of the Contractor shall be reimbursed by the Contractor.

12.3 COMMUNICATIONS

12.3.1 The Contractor shall promptly notify the Subcontractor of all modifications to the Contract between the Owner and the Contractor which affect this Subcontract and which were issued or entered into subsequent to the execution of this Subcontract.

12.3.2 The Contractor shall not give instructions or orders directly to employees or workmen of the Subcontractor except to persons designated as authorized representatives of the Subcontractor.

12.4 PAYMENTS TO THE SUBCONTRACTOR

12.4.1 Unless otherwise provided in the Contract Documents, the Contractor shall pay the Subcontractor each progress payment and the final payment under this Subcontract within three working days after he receives payment from the Owner, except as provided in Subparagraph 12.4.3. The amount of each progress payment to the Subcontractor shall be the amount to which the Subcontractor is entitled, reflecting the percentage of completion allowed to the Contractor for the Work of this Subcontractor applied to the Contract Sum of this Subcontract, and the percentage actually retained, if any, from payments to the Contractor on account of such Subcontractor's Work, plus, to the extent permitted by the Contract Documents, the amount allowed for materials and equipment suitably stored by the Subcontractor, less the aggregate of previous payments to the Subcontractor.

12.4.2 The Contractor shall permit the Subcontractor to request directly from the Architect information regarding the percentages of completion or the amount certified on account of Work done by the Subcontractor.

12.4.3 If the Architect does not issue a Certificate for Payment or the Contractor does not receive payment for any cause which is not the fault of the Subcontractor, the Contractor shall pay the Subcontractor, on demand, a progress payment computed as provided in Subparagraph 12.4.1 or the final payment as provided in Article 6.

12.5 CLAIMS BY THE CONTRACTOR

12.5.1 The Contractor shall make no demand for liquidated damages for delay in any sum in excess of such amount as may be specifically named in this Subcontract, and liquidated damages shall be assessed against this Subcontractor only for his negligent acts and his failure to act in accordance with the terms of this Agreement, and in no case for delays or causes arising outside the scope of this Subcontract, or for which other subcontractors are responsible.

12.5.2 Except as may be indicated in this Agreement, the Contractor agrees that no claim for payment for services rendered or materials and equipment furnished by the Contractor to the Subcontractor shall be valid without prior notice to the Subcontractor and unless written notice thereof is given by the Contractor to the Subcontractor not later than the tenth day of the calendar month following that in which the claim originated.

12.6 CONTRACTOR'S REMEDIES

12.6.1 If the Subcontractor defaults or neglects to carry out the Work in accordance with this Agreement and fails within three working days after receipt of written notice from the Contractor to commence and continue correction of such default or neglect with diligence and promptness, the Contractor may, after three days following receipt by the Subcontractor of an additional written notice, and without prejudice to any other remedy he may have, make good such deficiencies and may deduct the cost thereof from the payments then or thereafter due the Subcontractor, provided, however, that if such action is based upon faulty workmanship or materials and equipment, the Architect shall first have determined that the workmanship or materials and equipment are not in accordance with the Contract Documents.

ARTICLE 13
ARBITRATION

13.1 All claims, disputes and other matters in question arising out of, or relating to, this Subcontract, or the breach thereof, shall be decided by arbitration, which shall be conducted in the same manner and under the same procedure as provided in the Contract Documents with respect to disputes between the Owner and the Contractor, except that a decision by the Architect shall not be a condition precedent to arbitration. If the Contract Documents do not provide for arbitration or fail to specify the manner and procedure for arbitration, it shall be conducted in accordance with the Construction Industry Arbitration Rules of the American Arbitration Association then obtaining unless the parties mutually agree otherwise.

13.2 Except by written consent of the person or entity sought to be joined, no arbitration arising out of or relating to the Contract Documents shall include, by consolidation, joinder or in any other manner, any person or entity not a party to the Agreement under which such arbitration arises, unless it is shown at the time the demand for arbitration is filed that (1) such person or entity is substantially involved in a common question of fact or law,

(2) the presence of such person or entity is required if complete relief is to be accorded in the arbitration, (3) the interest or responsibility of such person or entity in the matter is not insubstantial, and (4) such person or entity is not the Architect, his employee or his consultant. This agreement to arbitrate and any other written agreement to arbitrate with an additional person or persons referred to herein shall be specifically enforceable under the prevailing arbitration law.

13.3 The Contractor shall permit the Subcontractor to be present and to submit evidence in any arbitration proceeding involving his rights.

13.4 The Contractor shall permit the Subcontractor to exercise whatever rights the Contractor may have under the Contract Documents in the choice of arbitrators in any dispute, if the sole cause of the dispute is the Work, materials, equipment, rights or responsibilities of the Subcontractor; or if the dispute involves the Subcontractor and any other subcontractor or subcontractors jointly, the Contractor shall permit them to exercise such rights jointly.

13.5 The award rendered by the arbitrators shall be final, and judgment may be entered upon it in accordance with applicable law in any court having jurisdiction thereof.

13.6 This Article shall not be deemed a limitation of any rights or remedies which the Subcontractor may have under any Federal or State mechanics' lien laws or under any applicable labor and material payment bonds unless such rights or remedies are expressly waived by him.

ARTICLE 14
TERMINATION

14.1 TERMINATION BY THE SUBCONTRACTOR

14.1.1 If the Work is stopped for a period of thirty days through no fault of the Subcontractor because the Contractor has not made payments thereon as provided in this Agreement, then the Subcontractor may without prejudice to any other remedy he may have, upon seven additional days' written notice to the Contractor, terminate this Subcontract and recover from the Contractor payment for all Work executed and for any proven loss resulting from the stoppage of the Work, including reasonable overhead, profit and damages.

14.2 TERMINATION BY THE CONTRACTOR

14.2.1 If the Subcontractor persistently or repeatedly fails or neglects to carry out the Work in accordance with the Contract Documents or otherwise to perform in accordance with this Agreement and fails within seven days after receipt of written notice to commence and continue correction of such default or neglect with diligence and promptness, the Contractor may, after seven days following receipt by the Subcontractor of an additional written notice and without prejudice to any other remedy he may have, terminate the Subcontract and finish the Work by whatever method he may deem expedient. If the unpaid balance of the Contract Sum exceeds the expense of finishing the Work, such excess shall be paid to the Subcontractor, but if such expense exceeds such unpaid balance, the Subcontractor shall pay the difference to the Contractor.

ARTICLE 15
MISCELLANEOUS PROVISIONS

15.1 Terms used in this Agreement which are defined in the Conditions of the Contract shall have the meanings designated in those Conditions.

15.2 The Contract Documents, which constitute the entire Agreement between the Owner and the Contractor, are listed in Article 1, and the documents which are applicable to this Subcontract, except for Addenda and Modifications issued after execution of this Subcontract, are enumerated as follows:

(List below the Agreement, the Conditions of the Contract [General, Supplementary, and other Conditions], the Drawings, the Specifications, and any Addenda and accepted Alternates, showing page or sheet numbers in all cases and dates where applicable. Continue on succeeding pages as required.)

This Agreement entered into as of the day and year first written above.

CONTRACTOR	SUBCONTRACTOR
______________________	______________________
______________________	______________________
______________________	______________________

THE AMERICAN INSTITUTE OF ARCHITECTS

AIA Document B141

Standard Form of Agreement Between Owner and Architect

1977 EDITION

THIS DOCUMENT HAS IMPORTANT LEGAL CONSEQUENCES; CONSULTATION WITH AN ATTORNEY IS ENCOURAGED WITH RESPECT TO ITS COMPLETION OR MODIFICATION

AGREEMENT

made as of the day of in the year of Nineteen Hundred and

BETWEEN the Owner:

and the Architect:

For the following Project:
(Include detailed description of Project location and scope.)

The Owner and the Architect agree as set forth below.

TERMS AND CONDITIONS OF AGREEMENT BETWEEN OWNER AND ARCHITECT

ARTICLE 1

ARCHITECT'S SERVICES AND RESPONSIBILITIES

BASIC SERVICES

The Architect's Basic Services consist of the five phases described in Paragraphs 1.1 through 1.5 and include normal structural, mechanical and electrical engineering services and any other services included in Article 15 as part of Basic Services.

1.1 SCHEMATIC DESIGN PHASE

1.1.1 The Architect shall review the program furnished by the Owner to ascertain the requirements of the Project and shall review the understanding of such requirements with the Owner.

1.1.2 The Architect shall provide a preliminary evaluation of the program and the Project budget requirements, each in terms of the other, subject to the limitations set forth in Subparagraph 3.2.1.

1.1.3 The Architect shall review with the Owner alternative approaches to design and construction of the Project.

1.1.4 Based on the mutually agreed upon program and Project budget requirements, the Architect shall prepare, for approval by the Owner, Schematic Design Documents consisting of drawings and other documents illustrating the scale and relationship of Project components.

1.1.5 The Architect shall submit to the Owner a Statement of Probable Construction Cost based on current area, volume or other unit costs.

1.2 DESIGN DEVELOPMENT PHASE

1.2.1 Based on the approved Schematic Design Documents and any adjustments authorized by the Owner in the program or Project budget, the Architect shall prepare, for approval by the Owner, Design Development Documents consisting of drawings and other documents to fix and describe the size and character of the entire Project as to architectural, structural, mechanical and electrical systems, materials and such other elements as may be appropriate.

1.2.2 The Architect shall submit to the Owner a further Statement of Probable Construction Cost.

1.3 CONSTRUCTION DOCUMENTS PHASE

1.3.1 Based on the approved Design Development Documents and any further adjustments in the scope or quality of the Project or in the Project budget authorized by the Owner, the Architect shall prepare, for approval by the Owner, Construction Documents consisting of Drawings and Specifications setting forth in detail the requirements for the construction of the Project.

1.3.2 The Architect shall assist the Owner in the preparation of the necessary bidding information, bidding forms, the Conditions of the Contract, and the form of Agreement between the Owner and the Contractor.

1.3.3 The Architect shall advise the Owner of any adjustments to previous Statements of Probable Construction Cost indicated by changes in requirements or general market conditions.

1.3.4 The Architect shall assist the Owner in connection with the Owner's responsibility for filing documents required for the approval of governmental authorities having jurisdiction over the Project.

1.4 BIDDING OR NEGOTIATION PHASE

1.4.1 The Architect, following the Owner's approval of the Construction Documents and of the latest Statement of Probable Construction Cost, shall assist the Owner in obtaining bids or negotiated proposals, and assist in awarding and preparing contracts for construction.

1.5 CONSTRUCTION PHASE—ADMINISTRATION OF THE CONSTRUCTION CONTRACT

1.5.1 The Construction Phase will commence with the award of the Contract for Construction and, together with the Architect's obligation to provide Basic Services under this Agreement, will terminate when final payment to the Contractor is due, or in the absence of a final Certificate for Payment or of such due date, sixty days after the Date of Substantial Completion of the Work, whichever occurs first.

1.5.2 Unless otherwise provided in this Agreement and incorporated in the Contract Documents, the Architect shall provide administration of the Contract for Construction as set forth below and in the edition of AIA Document A201, General Conditions of the Contract for Construction, current as of the date of this Agreement.

1.5.3 The Architect shall be a representative of the Owner during the Construction Phase, and shall advise and consult with the Owner. Instructions to the Contractor shall be forwarded through the Architect. The Architect shall have authority to act on behalf of the Owner only to the extent provided in the Contract Documents unless otherwise modified by written instrument in accordance with Subparagraph 1.5.16.

1.5.4 The Architect shall visit the site at intervals appropriate to the stage of construction or as otherwise agreed by the Architect in writing to become generally familiar with the progress and quality of the Work and to determine in general if the Work is proceeding in accordance with the Contract Documents. However, the Architect shall not be required to make exhaustive or continuous on-site inspections to check the quality or quantity of the Work. On the basis of such on-site observations as an architect, the Architect shall keep the Owner informed of the progress and quality of the Work, and shall endeavor to guard the Owner against defects and deficiencies in the Work of the Contractor.

1.5.5 The Architect shall not have control or charge of and shall not be responsible for construction means, methods, techniques, sequences or procedures, or for safety precautions and programs in connection with the Work, for the acts or omissions of the Contractor, Subcontractors or any other persons performing any of the Work, or for the failure of any of them to carry out the Work in accordance with the Contract Documents.

1.5.6 The Architect shall at all times have access to the Work wherever it is in preparation or progress.

1.5.7 The Architect shall determine the amounts owing to the Contractor based on observations at the site and on evaluations of the Contractor's Applications for Payment, and shall issue Certificates for Payment in such amounts, as provided in the Contract Documents.

1.5.8 The issuance of a Certificate for Payment shall constitute a representation by the Architect to the Owner, based on the Architect's observations at the site as provided in Subparagraph 1.5.4 and on the data comprising the Contractor's Application for Payment, that the Work has progressed to the point indicated; that, to the best of the Architect's knowledge, information and belief, the quality of the Work is in accordance with the Contract Documents (subject to an evaluation of the Work for conformance with the Contract Documents upon Substantial Completion, to the results of any subsequent tests required by or performed under the Contract Documents, to minor deviations from the Contract Documents correctable prior to completion, and to any specific qualifications stated in the Certificate for Payment); and that the Contractor is entitled to payment in the amount certified. However, the issuance of a Certificate for Payment shall not be a representation that the Architect has made any examination to ascertain how and for what purpose the Contractor has used the moneys paid on account of the Contract Sum.

1.5.9 The Architect shall be the interpreter of the requirements of the Contract Documents and the judge of the performance thereunder by both the Owner and Contractor. The Architect shall render interpretations necessary for the proper execution or progress of the Work with reasonable promptness on written request of either the Owner or the Contractor, and shall render written decisions, within a reasonable time, on all claims, disputes and other matters in question between the Owner and the Contractor relating to the execution or progress of the Work or the interpretation of the Contract Documents.

1.5.10 Interpretations and decisions of the Architect shall be consistent with the intent of and reasonably inferable from the Contract Documents and shall be in written or graphic form. In the capacity of interpreter and judge, the Architect shall endeavor to secure faithful performance by both the Owner and the Contractor, shall not show partiality to either, and shall not be liable for the result of any interpretation or decision rendered in good faith in such capacity.

1.5.11 The Architect's decisions in matters relating to artistic effect shall be final if consistent with the intent of the Contract Documents. The Architect's decisions on any other claims, disputes or other matters, including those in question between the Owner and the Contractor, shall be subject to arbitration as provided in this Agreement and in the Contract Documents.

1.5.12 The Architect shall have authority to reject Work which does not conform to the Contract Documents. Whenever, in the Architect's reasonable opinion, it is necessary or advisable for the implementation of the intent of the Contract Documents, the Architect will have authority to require special inspection or testing of the Work in accordance with the provisions of the Contract Documents, whether or not such Work be then fabricated, installed or completed.

1.5.13 The Architect shall review and approve or take other appropriate action upon the Contractor's submittals such as Shop Drawings, Product Data and Samples, but only for conformance with the design concept of the Work and with the information given in the Contract Documents. Such action shall be taken with reasonable promptness so as to cause no delay. The Architect's approval of a specific item shall not indicate approval of an assembly of which the item is a component.

1.5.14 The Architect shall prepare Change Orders for the Owner's approval and execution in accordance with the Contract Documents, and shall have authority to order minor changes in the Work not involving an adjustment in the Contract Sum or an extension of the Contract Time which are not inconsistent with the intent of the Contract Documents.

1.5.15 The Architect shall conduct inspections to determine the Dates of Substantial Completion and final completion, shall receive and forward to the Owner for the Owner's review written warranties and related documents required by the Contract Documents and assembled by the Contractor, and shall issue a final Certificate for Payment.

1.5.16 The extent of the duties, responsibilities and limitations of authority of the Architect as the Owner's representative during construction shall not be modified or extended without written consent of the Owner, the Contractor and the Architect.

1.6 PROJECT REPRESENTATION BEYOND BASIC SERVICES

1.6.1 If the Owner and Architect agree that more extensive representation at the site than is described in Paragraph 1.5 shall be provided, the Architect shall provide one or more Project Representatives to assist the Architect in carrying out such responsibilities at the site.

1.6.2 Such Project Representatives shall be selected, employed and directed by the Architect, and the Architect shall be compensated therefor as mutually agreed between the Owner and the Architect as set forth in an exhibit appended to this Agreement, which shall describe the duties, responsibilities and limitations of authority of such Project Representatives.

1.6.3 Through the observations by such Project Representatives, the Architect shall endeavor to provide further protection for the Owner against defects and deficiencies in the Work, but the furnishing of such project representation shall not modify the rights, responsibilities or obligations of the Architect as described in Paragraph 1.5.

1.7 ADDITIONAL SERVICES

The following Services are not included in Basic Services unless so identified in Article 15. They shall be provided if authorized or confirmed in writing by the Owner, and they shall be paid for by the Owner as provided in this Agreement, in addition to the compensation for Basic Services.

1.7.1 Providing analyses of the Owner's needs, and programming the requirements of the Project.

1.7.2 Providing financial feasibility or other special studies.

1.7.3 Providing planning surveys, site evaluations, environmental studies or comparative studies of prospective sites, and preparing special surveys, studies and submissions required for approvals of governmental authorities or others having jurisdiction over the Project.

1.7.4 Providing services relative to future facilities, systems and equipment which are not intended to be constructed during the Construction Phase.

1.7.5 Providing services to investigate existing conditions or facilities or to make measured drawings thereof, or to verify the accuracy of drawings or other information furnished by the Owner.

1.7.6 Preparing documents of alternate, separate or sequential bids or providing extra services in connection with bidding, negotiation or construction prior to the completion of the Construction Documents Phase, when requested by the Owner.

1.7.7 Providing coordination of Work performed by separate contractors or by the Owner's own forces.

1.7.8 Providing services in connection with the work of a construction manager or separate consultants retained by the Owner.

1.7.9 Providing Detailed Estimates of Construction Cost, analyses of owning and operating costs, or detailed quantity surveys or inventories of material, equipment and labor.

1.7.10 Providing interior design and other similar services required for or in connection with the selection, procurement or installation of furniture, furnishings and related equipment.

1.7.11 Providing services for planning tenant or rental spaces.

1.7.12 Making revisions in Drawings, Specifications or other documents when such revisions are inconsistent with written approvals or instructions previously given, are required by the enactment or revision of codes, laws or regulations subsequent to the preparation of such documents or are due to other causes not solely within the control of the Architect.

1.7.13 Preparing Drawings, Specifications and supporting data and providing other services in connection with Change Orders to the extent that the adjustment in the Basic Compensation resulting from the adjusted Construction Cost is not commensurate with the services required of the Architect, provided such Change Orders are required by causes not solely within the control of the Architect.

1.7.14 Making investigations, surveys, valuations, inventories or detailed appraisals of existing facilities, and services required in connection with construction performed by the Owner.

1.7.15 Providing consultation concerning replacement of any Work damaged by fire or other cause during construction, and furnishing services as may be required in connection with the replacement of such Work.

1.7.16 Providing services made necessary by the default of the Contractor, or by major defects or deficiencies in the Work of the Contractor, or by failure of performance of either the Owner or Contractor under the Contract for Construction.

1.7.17 Preparing a set of reproducible record drawings showing significant changes in the Work made during construction based on marked-up prints, drawings and other data furnished by the Contractor to the Architect.

1.7.18 Providing extensive assistance in the utilization of any equipment or system such as initial start-up or testing, adjusting and balancing, preparation of operation and maintenance manuals, training personnel for operation and maintenance, and consultation during operation.

1.7.19 Providing services after issuance to the Owner of the final Certificate for Payment, or in the absence of a final Certificate for Payment, more than sixty days after the Date of Substantial Completion of the Work.

1.7.20 Preparing to serve or serving as an expert witness in connection with any public hearing, arbitration proceeding or legal proceeding.

1.7.21 Providing services of consultants for other than the normal architectural, structural, mechanical and electrical engineering services for the Project.

1.7.22 Providing any other services not otherwise included in this Agreement or not customarily furnished in accordance with generally accepted architectural practice.

1.8 TIME

1.8.1 The Architect shall perform Basic and Additional Services as expeditiously as is consistent with professional skill and care and the orderly progress of the Work. Upon request of the Owner, the Architect shall submit for the Owner's approval a schedule for the performance of the Architect's services which shall be adjusted as required as the Project proceeds, and shall include allowances for periods of time required for the Owner's review and approval of submissions and for approvals of authorities having jurisdiction over the Project. This schedule, when approved by the Owner, shall not, except for reasonable cause, be exceeded by the Architect.

ARTICLE 2

THE OWNER'S RESPONSIBILITIES

2.1 The Owner shall provide full information regarding requirements for the Project including a program, which shall set forth the Owner's design objectives, constraints and criteria, including space requirements and relationships, flexibility and expandability, special equipment and systems and site requirements.

2.2 If the Owner provides a budget for the Project it shall include contingencies for bidding, changes in the Work during construction, and other costs which are the responsibility of the Owner, including those described in this Article 2 and in Subparagraph 3.1.2. The Owner shall, at the request of the Architect, provide a statement of funds available for the Project, and their source.

2.3 The Owner shall designate, when necessary, a representative authorized to act in the Owner's behalf with respect to the Project. The Owner or such authorized representative shall examine the documents submitted by the Architect and shall render decisions pertaining thereto promptly, to avoid unreasonable delay in the progress of the Architect's services.

2.4 The Owner shall furnish a legal description and a certified land survey of the site, giving, as applicable, grades and lines of streets, alleys, pavements and adjoining property; rights-of-way, restrictions, easements, encroachments, zoning, deed restrictions, boundaries and contours of the site; locations, dimensions and complete data pertaining to existing buildings, other improvements and trees; and full information concerning available service and utility lines both public and private, above and below grade, including inverts and depths.

2.5 The Owner shall furnish the services of soil engineers or other consultants when such services are deemed necessary by the Architect. Such services shall include test borings, test pits, soil bearing values, percolation tests, air and water pollution tests, ground corrosion and resistivity tests, including necessary operations for determining subsoil, air and water conditions, with reports and appropriate professional recommendations.

2.6 The Owner shall furnish structural, mechanical, chemical and other laboratory tests, inspections and reports as required by law or the Contract Documents.

2.7 The Owner shall furnish all legal, accounting and insurance counseling services as may be necessary at any time for the Project, including such auditing services as the Owner may require to verify the Contractor's Applications for Payment or to ascertain how or for what purposes the Contractor uses the moneys paid by or on behalf of the Owner.

2.8 The services, information, surveys and reports required by Paragraphs 2.4 through 2.7 inclusive shall be furnished at the Owner's expense, and the Architect shall be entitled to rely upon the accuracy and completeness thereof.

2.9 If the Owner observes or otherwise becomes aware of any fault or defect in the Project or nonconformance with the Contract Documents, prompt written notice thereof shall be given by the Owner to the Architect.

2.10 The Owner shall furnish required information and services and shall render approvals and decisions as expeditiously as necessary for the orderly progress of the Architect's services and of the Work.

ARTICLE 3

CONSTRUCTION COST

3.1 DEFINITION

3.1.1 The Construction Cost shall be the total cost or estimated cost to the Owner of all elements of the Project designed or specified by the Architect.

3.1.2 The Construction Cost shall include at current market rates, including a reasonable allowance for overhead and profit, the cost of labor and materials furnished by the Owner and any equipment which has been designed, specified, selected or specially provided for by the Architect.

3.1.3 Construction Cost does not include the compensation of the Architect and the Architect's consultants, the cost of the land, rights-of-way, or other costs which are the responsibility of the Owner as provided in Article 2.

3.2 RESPONSIBILITY FOR CONSTRUCTION COST

3.2.1 Evaluations of the Owner's Project budget, Statements of Probable Construction Cost and Detailed Estimates of Construction Cost, if any, prepared by the Architect, represent the Architect's best judgment as a design professional familiar with the construction industry. It is recognized, however, that neither the Architect nor the Owner has control over the cost of labor, materials or equipment, over the Contractor's methods of determining bid prices, or over competitive bidding, market or negotiating conditions. Accordingly, the Architect cannot and does not warrant or represent that bids or negotiated prices will not vary from the Project budget proposed, established or approved by the Owner, if any, or from any Statement of Probable Construction Cost or other cost estimate or evaluation prepared by the Architect.

3.2.2 No fixed limit of Construction Cost shall be established as a condition of this Agreement by the furnishing, proposal or establishment of a Project budget under Subparagraph 1.1.2 or Paragraph 2.2 or otherwise, unless such fixed limit has been agreed upon in writing and signed by the parties hereto. If such a fixed limit has been established, the Architect shall be permitted to include contingencies for design, bidding and price escalation, to determine what materials, equipment, component systems and types of construction are to be included in the Contract Documents, to make reasonable adjustments in the scope of the Project and to include in the Contract Documents alternate bids to adjust the Construction Cost to the fixed limit. Any such fixed limit shall be increased in the amount of any increase in the Contract Sum occurring after execution of the Contract for Construction.

3.2.3 If the Bidding or Negotiation Phase has not commenced within three months after the Architect submits the Construction Documents to the Owner, any Project budget or fixed limit of Construction Cost shall be adjusted to reflect any change in the general level of prices in the construction industry between the date of submission of the Construction Documents to the Owner and the date on which proposals are sought.

3.2.4 If a Project budget or fixed limit of Construction Cost (adjusted as provided in Subparagraph 3.2.3) is exceeded by the lowest bona fide bid or negotiated proposal, the Owner shall (1) give written approval of an increase in such fixed limit, (2) authorize rebidding or renegotiating of the Project within a reasonable time, (3) if the Project is abandoned, terminate in accordance with Paragraph 10.2, or (4) cooperate in revising the Project scope and quality as required to reduce the Construction Cost. In the case of (4), provided a fixed limit of Construction Cost has been established as a condition of this Agreement, the Architect, without additional charge, shall modify the Drawings and Specifications as necessary to comply with the fixed limit. The providing of such service shall be the limit of the Architect's responsibility arising from the establishment of such fixed limit, and having done so, the Architect shall be entitled to compensation for all services performed, in accordance with this Agreement, whether or not the Construction Phase is commenced.

ARTICLE 4

DIRECT PERSONNEL EXPENSE

4.1 Direct Personnel Expense is defined as the direct salaries of all the Architect's personnel engaged on the Project, and the portion of the cost of their mandatory and customary contributions and benefits related thereto, such as employment taxes and other statutory employee benefits, insurance, sick leave, holidays, vacations, pensions and similar contributions and benefits.

ARTICLE 5

REIMBURSABLE EXPENSES

5.1 Reimbursable Expenses are in addition to the Compensation for Basic and Additional Services and include actual expenditures made by the Architect and the Architect's employees and consultants in the interest of the Project for the expenses listed in the following Subparagraphs:

5.1.1 Expense of transportation in connection with the Project; living expenses in connection with out-of-town travel; long distance communications, and fees paid for securing approval of authorities having jurisdiction over the Project.

5.1.2 Expense of reproductions, postage and handling of Drawings, Specifications and other documents, excluding reproductions for the office use of the Architect and the Architect's consultants.

5.1.3 Expense of data processing and photographic production techniques when used in connection with Additional Services.

5.1.4 If authorized in advance by the Owner, expense of overtime work requiring higher than regular rates.

5.1.5 Expense of renderings, models and mock-ups requested by the Owner.

5.1.6 Expense of any additional insurance coverage or limits, including professional liability insurance, requested by the Owner in excess of that normally carried by the Architect and the Architect's consultants.

ARTICLE 6

PAYMENTS TO THE ARCHITECT

6.1 PAYMENTS ON ACCOUNT OF BASIC SERVICES

6.1.1 An initial payment as set forth in Paragraph 14.1 is the minimum payment under this Agreement.

6.1.2 Subsequent payments for Basic Services shall be made monthly and shall be in proportion to services performed within each Phase of services, on the basis set forth in Article 14.

6.1.3 If and to the extent that the Contract Time initially established in the Contract for Construction is exceeded or extended through no fault of the Architect, compensation for any Basic Services required for such extended period of Administration of the Construction Contract shall be computed as set forth in Paragraph 14.4 for Additional Services.

6.1.4 When compensation is based on a percentage of Construction Cost, and any portions of the Project are deleted or otherwise not constructed, compensation for such portions of the Project shall be payable to the extent services are performed on such portions, in accordance with the schedule set forth in Subparagraph 14.2.2, based on (1) the lowest bona fide bid or negotiated proposal or, (2) if no such bid or proposal is received, the most recent Statement of Probable Construction Cost or Detailed Estimate of Construction Cost for such portions of the Project.

6.2 PAYMENTS ON ACCOUNT OF ADDITIONAL SERVICES

6.2.1 Payments on account of the Architect's Additional Services as defined in Paragraph 1.7 and for Reimbursable Expenses as defined in Article 5 shall be made monthly upon presentation of the Architect's statement of services rendered or expenses incurred.

6.3 PAYMENTS WITHHELD

6.3.1 No deductions shall be made from the Architect's compensation on account of penalty, liquidated damages or other sums withheld from payments to contractors, or on account of the cost of changes in the Work other than those for which the Architect is held legally liable.

6.4 PROJECT SUSPENSION OR TERMINATION

6.4.1 If the Project is suspended or abandoned in whole or in part for more than three months, the Architect shall be compensated for all services performed prior to receipt of written notice from the Owner of such suspension or abandonment, together with Reimbursable Expenses then due and all Termination Expenses as defined in Paragraph 10.4. If the Project is resumed after being suspended for more than three months, the Architect's compensation shall be equitably adjusted.

ARTICLE 7

ARCHITECT'S ACCOUNTING RECORDS

7.1 Records of Reimbursable Expenses and expenses pertaining to Additional Services and services performed on the basis of a Multiple of Direct Personnel Expense shall be kept on the basis of generally accepted accounting principles and shall be available to the Owner or the Owner's authorized representative at mutually convenient times.

ARTICLE 8

OWNERSHIP AND USE OF DOCUMENTS

8.1 Drawings and Specifications as instruments of service are and shall remain the property of the Architect whether the Project for which they are made is executed or not. The Owner shall be permitted to retain copies, including reproducible copies, of Drawings and Specifications for information and reference in connection with the Owner's use and occupancy of the Project. The Drawings and Specifications shall not be used by the Owner on other projects, for additions to this Project, or for completion of this Project by others provided the Architect is not in default under this Agreement, except by agreement in writing and with appropriate compensation to the Architect.

8.2 Submission or distribution to meet official regulatory requirements or for other purposes in connection with the Project is not to be construed as publication in derogation of the Architect's rights.

ARTICLE 9

ARBITRATION

9.1 All claims, disputes and other matters in question between the parties to this Agreement, arising out of or relating to this Agreement or the breach thereof, shall be decided by arbitration in accordance with the Construction Industry Arbitration Rules of the American Arbitration Association then obtaining unless the parties mutually agree otherwise. No arbitration, arising out of or relating to this Agreement, shall include, by consolidation, joinder or in any other manner, any additional person not a party to this Agreement except by written consent containing a specific reference to this Agreement and signed by the Architect, the Owner, and any other person sought to be joined. Any consent to arbitration involving an additional person or persons shall not constitute consent to arbitration of any dispute not described therein or with any person not named or described therein. This Agreement to arbitrate and any agreement to arbitrate with an additional person or persons duly consented to by the parties to this Agreement shall be specifically enforceable under the prevailing arbitration law.

9.2 Notice of the demand for arbitration shall be filed in writing with the other party to this Agreement and with the American Arbitration Association. The demand shall be made within a reasonable time after the claim, dispute or other matter in question has arisen. In no event shall the demand for arbitration be made after the date when institution of legal or equitable proceedings based on such claim, dispute or other matter in question would be barred by the applicable statute of limitations.

9.3 The award rendered by the arbitrators shall be final, and judgment may be entered upon it in accordance with applicable law in any court having jurisdiction thereof.

ARTICLE 10

TERMINATION OF AGREEMENT

10.1 This Agreement may be terminated by either party upon seven days' written notice should the other party fail substantially to perform in accordance with its terms through no fault of the party initiating the termination.

10.2 This Agreement may be terminated by the Owner upon at least seven days' written notice to the Architect in the event that the Project is permanently abandoned.

10.3 In the event of termination not the fault of the Architect, the Architect shall be compensated for all services performed to termination date, together with Reimbursable Expenses then due and all Termination Expenses as defined in Paragraph 10.4.

10.4 Termination Expenses include expenses directly attributable to termination for which the Architect is not otherwise compensated, plus an amount computed as a percentage of the total Basic and Additional Compensation earned to the time of termination, as follows:

.1 20 percent if termination occurs during the Schematic Design Phase; or

.2 10 percent if termination occurs during the Design Development Phase; or

.3 5 percent if termination occurs during any subsequent phase.

ARTICLE 11

MISCELLANEOUS PROVISIONS

11.1 Unless otherwise specified, this Agreement shall be governed by the law of the principal place of business of the Architect.

11.2 Terms in this Agreement shall have the same meaning as those in AIA Document A201, General Conditions of the Contract for Construction, current as of the date of this Agreement.

11.3 As between the parties to this Agreement: as to all acts or failures to act by either party to this Agreement, any applicable statute of limitations shall commence to run and any alleged cause of action shall be deemed to have accrued in any and all events not later than the relevant Date of Substantial Completion of the Work, and as to any acts or failures to act occurring after the relevant Date of Substantial Completion, not later than the date of issuance of the final Certificate for Payment.

11.4 The Owner and the Architect waive all rights against each other and against the contractors, consultants, agents and employees of the other for damages covered by any property insurance during construction as set forth in the edition of AIA Document A201, General Conditions, current as of the date of this Agreement. The Owner and the Architect each shall require appropriate similar waivers from their contractors, consultants and agents.

ARTICLE 12

SUCCESSORS AND ASSIGNS

12.1 The Owner and the Architect, respectively, bind themselves, their partners, successors, assigns and legal representatives to the other party to this Agreement and to the partners, successors, assigns and legal representatives of such other party with respect to all covenants of this Agreement. Neither the Owner nor the Architect shall assign, sublet or transfer any interest in this Agreement without the written consent of the other.

ARTICLE 13

EXTENT OF AGREEMENT

13.1 This Agreement represents the entire and integrated agreement between the Owner and the Architect and supersedes all prior negotiations, representations or agreements, either written or oral. This Agreement may be amended only by written instrument signed by both Owner and Architect.

ARTICLE 14

BASIS OF COMPENSATION

The Owner shall compensate the Architect for the Scope of Services provided, in accordance with Article 6, Payments to the Architect, and the other Terms and Conditions of this Agreement, as follows:

14.1 AN INITIAL PAYMENT of dollars ($)

shall be made upon execution of this Agreement and credited to the Owner's account as follows:

14.2 BASIC COMPENSATION

14.2.1 FOR BASIC SERVICES, as described in Paragraphs 1.1 through 1.5, and any other services included in Article 15 as part of Basic Services, Basic Compensation shall be computed as follows:

(Here insert basis of compensation, including fixed amounts, multiples or percentages, and identify Phases to which particular methods of compensation apply, if necessary.)

14.2.2 Where compensation is based on a Stipulated Sum or Percentage of Construction Cost, payments for Basic Services shall be made as provided in Subparagraph 6.1.2, so that Basic Compensation for each Phase shall equal the following percentages of the total Basic Compensation payable:

(Include any additional Phases as appropriate.)

Schematic Design Phase:	percent (%)
Design Development Phase:	percent (%)
Construction Documents Phase:	percent (%)
Bidding or Negotiation Phase:	percent (%)
Construction Phase:	percent (%)

14.3 FOR PROJECT REPRESENTATION BEYOND BASIC SERVICES, as described in Paragraph 1.6, Compensation shall be computed separately in accordance with Subparagraph 1.6.2.

14.4 COMPENSATION FOR ADDITIONAL SERVICES

14.4.1 FOR ADDITIONAL SERVICES OF THE ARCHITECT, as described in Paragraph 1.7, and any other services included in Article 15 as part of Additional Services, but excluding Additional Services of consultants, Compensation shall be computed as follows:

(Here insert basis of compensation, including rates and/or multiples of Direct Personnel Expense for Principals and employees, and identify Principals and classify employees, if required. Identify specific services to which particular methods of compensation apply, if necessary.)

14.4.2 FOR ADDITIONAL SERVICES OF CONSULTANTS, including additional structural, mechanical and electrical engineering services and those provided under Subparagraph 1.7.21 or identified in Article 15 as part of Additional Services, a multiple of ( ) times the amounts billed to the Architect for such services.

(Identify specific types of consultants in Article 15, if required.)

14.5 FOR REIMBURSABLE EXPENSES, as described in Article 5, and any other items included in Article 15 as Reimbursable Expenses, a multiple of ( ) times the amounts expended by the Architect, the Architect's employees and consultants in the interest of the Project.

14.6 Payments due the Architect and unpaid under this Agreement shall bear interest from the date payment is due at the rate entered below, or in the absence thereof, at the legal rate prevailing at the principal place of business of the Architect.

(Here insert any rate of interest agreed upon.)

(Usury laws and requirements under the Federal Truth in Lending Act, similar state and local consumer credit laws and other regulations at the Owner's and Architect's principal places of business, the location of the Project and elsewhere may affect the validity of this provision. Specific legal advice should be obtained with respect to deletion, modification, or other requirements such as written disclosures or waivers.)

14.7 The Owner and the Architect agree in accordance with the Terms and Conditions of this Agreement that:

14.7.1 IF THE SCOPE of the Project or of the Architect's Services is changed materially, the amounts of compensation shall be equitably adjusted.

14.7.2 IF THE SERVICES covered by this Agreement have not been completed within

( ) months of the date hereof, through no fault of the Architect, the amounts of compensation, rates and multiples set forth herein shall be equitably adjusted.

ARTICLE 15

OTHER CONDITIONS OR SERVICES

This Agreement entered into as of the day and year first written above.

OWNER	ARCHITECT
______________________	______________________
______________________	______________________
______________________	______________________
BY______________________	BY______________________

THE AMERICAN INSTITUTE OF ARCHITECTS

AIA Document B801

Standard Form of Agreement Between Owner and Construction Manager

Recommended for use with the current editions of standard AIA agreement forms and documents.

THIS DOCUMENT HAS IMPORTANT LEGAL CONSEQUENCES; CONSULTATION WITH AN ATTORNEY IS ENCOURAGED WITH RESPECT TO ITS COMPLETION OR MODIFICATION

AGREEMENT

made this day of in the year of Nineteen Hundred and

BETWEEN the Owner:

and the Construction Manager:

For services in connection with the following described Project:
(Include detailed description of Project location and scope)

The Architect for the Project is:

The Owner and the Construction Manager agree as set forth below.

I. THE CONSTRUCTION MANAGER, as **agent of the Owner, shall provide services in accordance** with the Terms and Conditions of this Agreement.

The Construction Manager accepts the relationship of trust and confidence established between him and the Owner by this Agreement. He covenants with the Owner to furnish his professional skill and judgment and to cooperate with the Architect in furthering the interests of the Owner. He agrees to furnish efficient business administration and superintendence and to use his professional efforts at all times in an expeditious and economical manner consistent with the interests of the Owner.

II. THE OWNER shall compensate the Construction Manager in accordance with the Terms and Conditions of this Agreement.

A. *FOR BASIC SERVICES,* as described in Paragraph 1.1, compensation shall be computed as follows: **(select one)**

1. On the basis of a PROFESSIONAL FEE PLUS EXPENSES.

a. A Professional Fee of

dollars ($),

PLUS

b. An amount computed as follows:

Principals' time at the fixed rate of

dollars ($) per hour.

For the purposes of this Agreement, the Principals are:

Employees' time (other than Principals) at the following multiples of the employees' Direct Personnel Expense as defined in Article 4.

For those assigned to the Project in the office, a multiple of

( ).

For those assigned to the Project at the construction site, a multiple of

( ).

c. Services of professional consultants at a multiple of

( ) times

the amount billed to the Construction Manager for such services.

OR 2. On the basis of a MULTIPLE OF DIRECT PERSONNEL EXPENSE.

a. Principals' time at the fixed rate of dollars ($) per hour.

For the purposes of this Agreement, the Principals are:

b. Employees' time (other than Principals) at the following multiples of the employees' Direct Personnel Expense as defined in Article 4.

For those assigned to the Project in the office, a multiple of ( ).

For those assigned to the Project at the construction site, a multiple of ( ).

c. Services of professional consultants at a multiple of ( ) times the amount billed to the Construction Manager for such services.

OR 3. On the basis of a FIXED FEE of dollars ($).

OR 4. On the basis of a PERCENTAGE OF CONSTRUCTION COST, as defined in Article 3, of percent (%).

B. *AN INITIAL PAYMENT of* dollars ($). shall be made upon the execution of this Agreement and credited to the Owner's account.

C. *FOR ADDITIONAL SERVICES,* as described in Paragraph 1.2, compensation shall be computed as follows:

1. Principals' time at the fixed rate of dollars ($) per hour.

For the purposes of this Agreement, the Principals are:

2. Employees' time (other than Principals) at the following multiples of the employees' Direct Personnel Expense as defined in Article 4.

For those assigned to the Project in the office, a multiple of ( ).

For those assigned to the Project at the construction site, a multiple of ( ).

3. Services of professional consultants at a multiple of ( ) times the amount billed to the Construction Manager for such services.

D. *FOR REIMBURSABLE EXPENSES,* amounts expended as defined in Article 5.

III. THE PARTIES agree in accordance with the Terms and Conditions of this Agreement that:

A. *IF THE SCOPE* of the Project is changed materially, compensation for Basic Services shall be subject to renegotiation.

B. *IF THE SERVICES* covered by this Agreement have not been completed within ( ) months of the date hereof, the amounts of compensation, rates and multiples set forth in Paragraph II shall be subject to renegotiation.

TERMS AND CONDITIONS OF AGREEMENT BETWEEN OWNER AND CONSTRUCTION MANAGER

ARTICLE 1

CONSTRUCTION MANAGER'S SERVICES

1.1 BASIC SERVICES

The Construction Manager's Basic Services consist of the two phases described below and any other services included in Article 13 as Basic Services.

DESIGN PHASE

1.1.1 *Consultation During Project Development:* Review conceptual designs during development. Advise on site use and improvements, selection of materials, building systems and equipment. Provide recommendations on relative construction feasibility, availability of materials and labor, time requirements for installation and construction, and factors related to cost including costs of alternative designs or materials, preliminary budgets, and possible economies.

1.1.2 *Scheduling:* Provide and periodically update a Project time schedule that coordinates and integrates the Architect's services with construction schedules.

1.1.3 *Project Budget:* Prepare a Project budget for the Owner's approval as soon as major Project requirements have been identified and update periodically. Prepare an estimate of construction cost based on a quantity survey of Drawings and Specifications at the end of the Schematic Design Phase for approval by the Owner. Update and refine this estimate for Owner's approval as the development of the Drawings and Specifications proceeds, and advise the Owner and the Architect if it appears that the Project budget will not be met and make recommendations for corrective action.

1.1.4 *Coordination of Contract Documents:* Review the Drawings and Specifications as they are being prepared, recommending alternative solutions whenever design details affect construction feasibility or schedules.

1.1.4.1 Verify that the requirements and assignment of responsibilities for safety precautions and programs, temporary Project facilities and for equipment, materials and services for common use of Contractors are included in the proposed Contract Documents.

1.1.4.2 Advise on the method to be used for selecting Contractors and awarding contracts. If separate contracts are to be awarded, review the Drawings and Specifications to (1) ascertain if areas of jurisdiction overlap, (2) verify that all Work has been included, and (3) allow for phased construction.

1.1.4.3 Investigate and recommend a schedule for purchase by the Owner of all materials and equipment requiring long lead time procurement, and coordinate the schedule with the early preparation of Contract Documents by the Architect. Expedite and coordinate delivery of these purchases.

1.1.5 *Labor:* Provide an analysis of the types and quantity of labor required for the Project and review the availability of appropriate categories of labor required for critical phases.

1.1.5.1 Determine applicable requirements for equal employment opportunity programs for inclusion in the proposed Contract Documents.

1.1.6 *Bidding:* Prepare pre-qualification criteria for bidders and develop Contractor interest in the Project. Establish bidding schedules and conduct pre-bid conferences to familiarize bidders with the bidding documents and management techniques and with any special systems, materials or methods.

1.1.6.1 Receive bids, prepare bid analyses and make recommendations to the Owner for award of contracts or rejection of bids.

1.1.7 *Contract Awards:* Conduct pre-award conferences with successful bidders. Assist the Owner in preparing Construction Contracts and advise the Owner on the acceptability of Subcontractors and material suppliers proposed by Contractors.

CONSTRUCTION PHASE

The Construction Phase will commence with the award of the first Construction Contract or purchase order and will terminate 30 days after the final Certificate for Payment is issued by the Architect.

1.1.8 *Project Control:* Coordinate the Work of the Contractors with the activities and responsibilities of the Owner and Architect to complete the Project in accordance with the Owner's objectives on cost, time and quality. Provide sufficient personnel at the Project site with authority to achieve these objectives.

1.1.8.1 Schedule and conduct pre-construction and progress meetings at which Contractors, Owner, Architect and Construction Manager can discuss jointly such matters as procedures, progress, problems and scheduling.

1.1.8.2 Provide a detailed schedule for the operations of Contractors on the Project, including realistic activity sequences and durations, allocation of labor and materials, processing of shop drawings and samples, and delivery of products requiring long lead time procurement; include the Owner's occupancy requirements showing portions of the Project having occupancy priority.

1.1.8.3 Provide regular monitoring of the schedule as construction progresses. Identify potential variances between scheduled and probable completion dates. Review schedule for Work not started or incomplete and recommend to the Owner and Contractor adjustments in the schedule to meet the probable completion date. Provide summary reports of each monitoring, and document all changes in schedule.

1.1.8.4 Recommend courses of action to the Owner when requirements of a contract are not being fulfilled.

1.1.9 *Cost Control:* Revise and refine the approved estimate of construction cost, incorporate approved changes as they occur, and develop cash flow reports and forecasts as needed.

1.1.9.1 Provide regular monitoring of the approved estimate of construction cost, showing actual costs for activities in process and estimates for uncompleted tasks. Identify variances between actual and budgeted or estimated costs, and advise the Owner and Architect whenever projected costs exceed budgets or estimates.

1.1.9.2 Arrange for the maintenance of cost accounting records on authorized Work performed under unit costs, actual costs for labor and materials, or other bases requiring accounting records.

1.1.9.3 Develop and implement a system for review and processing of Change Orders.

1.1.9.4 Recommend necessary or desirable changes to the Owner and the Architect, review requests for changes, submit recommendations to the Owner and the Architect, and assist in negotiating Change Orders.

1.1.9.5 Develop and implement a procedure for the review and processing of applications by Contractors for progress and final payments. Make recommendations to the Architect for certification to the Owner for payment.

1.1.10 *Permits and Fees:* Assist in obtaining all building permits and special permits for permanent improvements, excluding permits for inspection or temporary facilities required to be obtained directly by the various Contractors. Verify that the Owner has paid all applicable fees and assessments for permanent facilities. Assist in obtaining approvals from all the authorities having jurisdiction.

1.1.11 *Owner's Consultants:* If required, assist the Owner in selecting and retaining professional services of a surveyor, special consultants and testing laboratories. Coordinate these services.

1.1.12 *Inspection:* Inspect the work of Contractors to assure that the Work is being performed in accordance with the requirements of the Contract Documents. Endeavor to guard the Owner against defects and deficiencies in the Work. Require any Contractor to stop Work or any portion thereof, and require special inspection or testing of any Work not in accordance with the provisions of the Contract Documents whether or not such Work be then fabricated, installed or completed. Reject Work which does not conform to the requirements of the Contract Documents.

1.1.12.1 The Construction Manager shall not be responsible for construction means, methods, techniques, sequences and procedures employed by Contractors in performance of their contract, and he shall not be responsible for the failure of any Contractors to carry out the Work in accordance with the Contract Documents.

1.1.13 *Contract Performance:* Consult with the Architect and the Owner if any Contractor requests interpretations of the meaning and intent of the Drawings and Specifications, and assist in the resolution of any questions which may arise.

1.1.14 *Shop Drawings and Samples:* In collaboration with the Architect, establish and implement procedures for expediting the processing and approval of shop drawings and samples.

1.1.15 *Reports and Records:* Record the progress of the Project. Submit written progress reports to the Owner and the Architect including information on the Contractors and Work, the percentage of completion and the number and amounts of Change Orders. Keep a daily log available to the Owner and the Architect.

1.1.15.1 Maintain at the Project site, on a current basis: records of all Contracts; shop drawings; samples; purchases; materials; equipment; applicable handbooks; federal, commercial and technical standards and specifications; maintenance and operating manuals and instructions; and any other related documents and revisions which arise out of the Contract or the Work. Obtain data from Contractors and maintain a current set of record drawings, specifications and operating manuals. At the completion of the Project, deliver all such records to the Owner.

1.1.16 *Owner-Purchased Items:* Accept delivery and arrange storage, protection and security for all Owner-purchased materials, systems and equipment which are a part of the Work until such items are turned over to the Contractors.

1.1.17 *Substantial Completion:* Upon the Contractors' determination of Substantial Completion of the Work or designated portions thereof, prepare for the Architect a list of incomplete or unsatisfactory items and a schedule for their completion. After the Architect certifies the Date of Substantial Completion, supervise the correction and completion of Work.

1.1.18 *Start-Up:* With the Owner's maintenance personnel, direct the checkout of utilities, operational systems and equipment for readiness and assist in their initial start-up and testing.

1.1.19 *Final Completion:* Determine final completion and provide written notice to the Owner and Architect that the Work is ready for final inspection. Secure and transmit to the Architect required guarantees, affidavits, releases, bonds and waivers. Turn over to the Owner all keys, manuals, record drawings and maintenance stocks.

1.2 ADDITIONAL SERVICES

The following Additional Services shall be performed upon authorization in writing from the Owner and shall be paid for as hereinbefore provided.

1.2.1 Services related to investigation, appraisals or valuations of existing conditions, facilities or equipment, or verifying the accuracy of existing drawings or other Owner-furnished information.

1.2.2 Services related to Owner-furnished equipment, furniture and furnishings which are not a part of the Work.

1.2.3 Services for tenant or rental spaces.

1.2.4 Services related to construction performed by the Owner.

1.2.5 Consultation on replacement of Work damaged by

fire or other cause during construction, and furnishing services for the replacement of such Work.

1.2.6 Services made necessary by the default of a Contractor.

1.2.7 Preparing to serve or serving as an expert witness in connection with any public hearing, arbitration proceeding, or legal proceeding.

1.2.8 Finding housing for construction labor, and defining requirements for establishment and maintenance of base camps.

1.2.9 Obtaining or training maintenance personnel or negotiating maintenance service contracts.

1.2.10 Services related to Work items required by the Conditions of the Contract and the Specifications which are not provided by Contractors.

1.2.11 Inspections of and services related to the Project after completion of the services under this Agreement.

1.2.12 Providing any other service not otherwise included in this Agreement.

ARTICLE 2

THE OWNER'S RESPONSIBILITIES

2.1 The Owner shall provide full information regarding his requirements for the Project.

2.2 The Owner shall designate a representative who shall be fully acquainted with the scope of the Work, and has authority to render decisions promptly and furnish information expeditiously.

2.3 The Owner shall retain an Architect for design and to prepare construction documents for the Project. The Architect's services, duties and responsibilities are described in the Agreement between the Owner and the Architect, pertinent parts of which will be furnished to the Construction Manager and will not be modified without written notification to him.

2.4 The Owner shall furnish such legal, accounting and insurance counselling services as may be necessary for the Project, and such auditing services as he may require to ascertain how or for what purposes the Contractors have used the moneys paid to them under the Construction Contracts.

2.5 The Owner shall furnish the Construction Manager with a sufficient quantity of construction documents.

2.6 If the Owner becomes aware of any fault or defect in the Project or nonconformance with the Contract Documents, he shall give prompt written notice thereof to the Construction Manager.

2.7 The services, information, surveys and reports required by Paragraphs 2.3 through 2.5 inclusive shall be furnished at the Owner's expense, and the Construction Manager shall be entitled to rely upon the accuracy and completeness thereof.

ARTICLE 3

CONSTRUCTION COST

3.1 If the Construction Cost is to be used as the basis for determining the Construction Manager's compensation for Basic Services, it shall be the cost of all Work, including work items in the Conditions of the Contract and the Specifications and shall be determined as follows:

3.1.1 For completed construction, the total construction cost of all such Work;

3.1.2 For Work not constructed, (1) the sum of the lowest bona fide bids received from qualified bidders for any or all of such Work or (2) if the Work is not bid, the sum of the bona fide negotiated proposals submitted for any or all of such Work; or

3.1.3 For Work for which no such bids or proposals are received, the Construction Cost contained in the Construction Manager's latest Project construction budget approved by the Owner.

3.2 Construction Cost shall not include the compensation of the Construction Manager (except for costs of Work items in the Conditions of the Contract and in the Specifications), the Architect and consultants, the cost of the land, rights-of-way, or other costs which are the responsibility of the Owner as provided in Paragraphs 2.3 through 2.5 inclusive.

3.3 The cost of labor, materials and equipment furnished by the Owner shall be included in the Construction Cost at current market rates, including a reasonable allowance for overhead and profit.

3.4 Cost estimates prepared by the Construction Manager represent his best judgment as a professional familiar with the construction industry. It is recognized, however, that neither the Construction Manager nor the Owner has any control over the cost of labor, materials or equipment, over Contractors' methods of determining bid prices, or other competitive bidding or market conditions.

3.5 No fixed limit of Construction Cost shall be deemed to have been established unless it is in writing and signed by the parties hereto. When a fixed limit of Construction Cost is established in writing as a condition of this Agreement, the Construction Manager shall advise what materials, equipment, component systems and types of construction should be included in the Contract Documents, and shall suggest reasonable adjustments in the scope of the Project to bring it within the fixed limit.

3.5.1. If responsive bids are not received as scheduled, any fixed limit of Construction Cost established as a condition of this Agreement shall be adjusted to reflect any change in the general level of prices occurring between the originally scheduled date and the date on which bids are received.

3.5.2 When the fixed limit of Construction Cost is exceeded, the Owner shall (1) give written approval of an increase in such fixed limit, (2) authorize rebidding within a reasonable time, or (3) cooperate in revising the scope and quality of the Work as required to reduce the Construction Cost. In the case of (3) the Construction Manager, without additional compensation, shall cooperate with the Architect as necessary to bring the Construction Cost within the fixed limit.

ARTICLE 4

DIRECT PERSONNEL EXPENSE

Direct Personnel Expense is defined as the salaries of professional, technical and clerical employees engaged on the Project by the Construction Manager, and the cost of their mandatory and customary benefits such as statutory employee benefits, insurance, sick leave, holidays, vacations, pensions and similar benefits.

ARTICLE 5

REIMBURSABLE EXPENSES

5.1 Reimbursable Expenses are in addition to the compensation for Basic and Additional Services and include actual expenditures made by the Construction Manager, his employees, or his professional consultants in the interest of the Project:

5.1.1 Long distance calls and telegrams, and fees paid for securing approval of authorities having jurisdiction over the Project.

5.1.2 Handling, shipping, mailing and reproduction of Project related materials.

5.1.3 Transportation and living when traveling in connection with the Project, relocation costs, and overtime work requiring higher than regular rates if authorized in advance by the Owner.

5.1.4 Electronic data processing service and rental of electronic data processing equipment when used in connection with Additional Services.

5.1.5 Premiums for insurance required in Article 10.

5.1.6 Providing construction support activities such as Work items included in the Conditions of the Contract and in the Specifications unless they are provided by the Contractors.

ARTICLE 6

PAYMENTS TO THE CONSTRUCTION MANAGER

6.1 Payments shall be made monthly upon presentation of the Construction Manager's statement of services as follows:

6.1.1 An initial payment as set forth in Paragraph IIB is the minimum payment under this Agreement.

6.1.2 When compensation is computed as described in Paragraphs IIA1a, IIA3, or IIA4, subsequent payments for Basic Services shall be made in proportion to services performed. The compensation at the completion of each Phase shall equal the following percentages of the total compensation for Basic Services:

Design Phase 20%
Construction Phase 100%

6.1.3 Payments for Reimbursable Expenses shall be made upon presentation of the Construction Manager's statement.

6.2 No deductions shall be made from the Construction Manager's compensation on account of penalty, liquidated damages, or other sums withheld from payments to Contractors.

6.3 If the Project is suspended for more than three months or abandoned in whole or in part, the Construction Manager shall be paid his compensation for services performed prior to receipt of written notice from the Owner of such suspension or abandonment, together with Reimbursable Expenses then due, and all termination expenses as defined in Paragraph 7.2 resulting from such suspension or abandonment. If the Project is resumed after being suspended for more than three months, the Construction Manager's compensation shall be subject to renegotiation.

6.4 If construction of the Project has started and is delayed by reason of strikes, or other circumstance not due to the fault of the Construction Manager, the Owner shall reimburse the Construction Manager for the costs of his Project-site staff as provided for by this Agreement. The Construction Manager shall reduce the size of his Project-site staff after a 60-day delay, or sooner if feasible, for the remainder of the delay period as directed by the Owner and, during the period, the Owner shall reimburse the Construction Manager for the direct personnel expense of such staff plus any relocation costs. Upon the termination of the delay, the Construction Manager shall restore his Project-site staff to its former size, subject to the approval of the Owner.

6.5 If the Project time schedule established in Subparagraph 1.1.2 is exceeded by more than thirty days through no fault of the Construction Manager, compensation for Basic Services performed by Principals, employees and professional consultants required beyond the thirtieth day to complete the services under this Agreement shall be as set forth in Paragraph IIC.

6.6 Payments due the Construction Manager which are unpaid for more than 60 days from date of billing shall bear interest at the legal rate of interest applicable at the construction site.

ARTICLE 7

TERMINATION OF AGREEMENT

7.1 This Agreement may be terminated by either party upon seven days' written notice should the other party fail substantially to perform in accordance with its terms through no fault of the party initiating termination. In the event of termination due to the fault of others than the Construction Manager, the Construction Manager shall be paid his compensation plus Reimbursable Expenses for services performed to termination date and all termination expenses.

7.2 Termination expenses are defined as Reimbursable Expenses directly attributable to termination, plus an amount computed as a percentage of the total compensation earned to the time of termination, as follows:

20 percent if termination occurs during the Design Phase; or

10 percent if termination occurs during the Construction Phase.

ARTICLE 8

SUCCESSORS AND ASSIGNS

The Owner and the Construction Manager each binds himself, his partners, successors, assigns and legal representatives to the other party to this Agreement and to the partners, successors, assigns and legal representatives of such other party with respect to all covenants of this Agreement. Neither the Owner nor the Construction Manager shall assign, sublet or transfer his interest in this Agreement without the written consent of the other.

ARTICLE 9

ARBITRATION

9.1 All claims, disputes and other matters in question between the parties to this Agreement, arising out of, or relating to this Agreement or the breach thereof, shall be decided by arbitration in accordance with the Construction Industry Arbitration Rules of the American Arbitration Association then obtaining unless the parties mutually agree otherwise. No arbitration, arising out of, or relating to this Agreement, shall include, by consolidation, joinder or in any other manner, any additional party not a party to this Agreement except by written consent containing a specific reference to this Agreement and signed by all the parties hereto. Any consent to arbitration involving an additional party or parties shall not constitute consent to arbitration of any dispute not described therein or with any party not named or described therein. This Agreement to arbitrate and any agreement to arbitrate with an additional party or parties duly consented to by the parties hereto shall be specifically enforceable under the prevailing arbitration law.

9.2 Notice of the demand for arbitration shall be filed in writing with the other party to this Agreement and with the American Arbitration Association. The demand shall be made within a reasonable time after the claim, dispute or other matter in question has arisen. In no event shall the demand for arbitration be made after the date when institution of legal or equitable proceedings based on such claim, dispute or other matter in question would be barred by the applicable statute of limitations.

9.3 The award rendered by the arbitrators shall be final, and judgment may be entered upon it in accordance with applicable law in any court having jurisdiction thereof.

ARTICLE 10

INSURANCE

The Construction Manager shall purchase and maintain insurance to protect himself from claims under workmen's compensation acts; claims for damages because of bodily injury including personal injury, sickness or disease, or death of any of his employees or of any person other than his employees; and from claims for damages because of injury to or destruction of tangible property including loss of use resulting therefrom; and from claims arising out of the performance of professional services caused by any errors, omissions or negligent acts for which he is legally liable.

ARTICLE 11

EXTENT OF AGREEMENT

11.1 This Agreement represents the entire and integrated agreement between the Owner and the Construction Manager and supersedes all prior negotiations, representations or agreements, either written or oral. This Agreement shall not be superseded by provisions of contracts for construction and may be amended only by written instrument signed by both Owner and Construction Manager.

11.2 Nothing contained herein shall be deemed to create any contractual relationship between the Construction Manager and the Architect or any of the Contractors, Subcontractors or material suppliers on the Project; nor shall anything contained herein be deemed to give any third party any claim or right of action against the Owner or the Construction Manager which does not otherwise exist without regard to this Agreement.

ARTICLE 12

GOVERNING LAW

Unless otherwise specified, this Agreement shall be governed by the law in effect at the location of the Project.

ARTICLE 13

OTHER CONDITIONS OR SERVICES

This Agreement executed the day and year first written above.

OWNER CONSTRUCTION MANAGER

THE AMERICAN INSTITUTE OF ARCHITECTS

COMPARISON OF THE 1970 AND 1976 EDITIONS, AIA DOCUMENT A201

General Conditions of the Contract for Construction

THIS DOCUMENT HAS IMPORTANT LEGAL CONSEQUENCES; CONSULTATION WITH AN ATTORNEY IS ENCOURAGED WITH RESPECT TO ITS MODIFICATION

SEPTEMBER 27, 1976 (REVISED TO CONFORM TO FINAL PRINTING OF A201)

This document has been published by The American Institute of Architects for the information of users of the AIA General Conditions. Additions to the 1970 Edition of A201 are indicated by underlined wording. Deletions from the 1970 Edition are indicated by wording which has been stricken.

This document IS NOT TO BE USED AS A CONTRACT DOCUMENT. The 1976 Edition has **been** published in a form suitable for use as a contract document, available **from** AIA and its distributors. Because of the numerous changes in the document, **it is** recommended that practitioners carefully coordinate their Supplementary **Conditions**, Specifications, and Instructions to Bidders with this document.

OWNER-ARCHITECT AGREEMENTS for projects which will be constructed under the terms of the 1976 Edition of A201 must also be coordinated with this document to insure that there are no variations between them. This can be accomplished through the ammendment of such Owner-Architect Agreements or through Supplementary Conditions to this document, or both.

Requests for additional information on, additional copies of, or interpretations of this document should be directed to the Department of Professional Practice at AIA Headquarters. Published commentary and cross-reference to the 1970 Edition and 1976 Edition are currently available. In addition, new editions of all AIA Documents relating to the General Conditions will be published during 1976, including Chapter 13 of the Architect's Handbook of Professional Practice, which describes the General Conditions in detail.

The AIA is indebted to the members of the Documents Board for their intensive efforts over more than two years, especially the A201 Task Force chaired by Dean F. Hilfinger, FAIA. The cooperation of, and the contributions made by, the following national organizations is greatly appreciated:

ACEC	**American Consulting Engineers Council**
ASA	American Subcontractors Association
ASC	Associated Specialty Contractors, Inc.
CSI	Construction Specifications Institute
NSPE	National Society of Professional Engineers

This 1976 Edition of A201 has been endorsed and approved by The Associated General Contractors of America (AGC), whose assistance is especially appreciated.

GENERAL CONDITIONS OF THE CONTRACT FOR CONSTRUCTION

1970 Edition

ARTICLE 1

CONTRACT DOCUMENTS

1.1 DEFINITIONS

1.1.1 THE CONTRACT DOCUMENTS

The Contract Documents consist of the Agreement, the Conditions of the Contract (General, Supplementary and other Conditions), the Drawings, the Specifications, all Addenda issued prior to execution of the Contract, and all Modifications thereto. A Modification is (1) a written amendment to the Contract signed by both parties, (2) a Change Order, (3) a written interpretation issued by the Architect pursuant to Subparagraph 1.2.5, or (4) a written order for a minor change in the Work issued by the Architect pursuant to Paragraph 12.3. A Modification may be made only after execution of the Contract.

New Language in 1.1.1 replaces "including the bidding documents" in former 1.1.2 →

1.1.2 THE CONTRACT

The Contract Documents form the Contract. The Contract represents the entire and integrated agreement between the parties hereto and supersedes all prior negotiations, representations, or agreements, either written or oral, including the bidding documents. The Contract may be amended or modified only by a Modification as defined in Subparagraph 1.1.1.

New language in 1.1.2 replaces former 2.1.2 and 5.1.3. →

1.1.3 THE WORK

The term Work includes all labor necessary to produce the construction required by the Contract Documents, and all materials and equipment incorporated or to be incorporated in such construction.

1.1.4 THE PROJECT

The Project is the total construction designed by the Architect of which the Work performed under the Contract Documents may be the whole or a part.

1976 Edition

ARTICLE I

CONTRACT DOCUMENTS

1.I DEFINITIONS

1.1.1 THE CONTRACT DOCUMENTS

The Contract Documents consist of the Owner-Contractor Agreement, the Conditions of the Contract (General, Supplementary and other Conditions), the Drawings, the Specifications, and all Addenda issued prior to and all Modifications issued after execution of the Contract . ~~and all Modifications thereto.~~ A Modification is (1) a written amendment to the Contract signed by both parties, (2) a Change Order, (3) a written interpretation issued by the Architect pursuant to Subparagraph ~~1.2.5~~ 2.2.8, or (4) a written order for a minor change in the Work issued by the Architect pursuant to Paragraph 12.3. ~~A Modification may be made only after execution of the Contract.~~ The Contract Documents do not include Bidding Documents such as the Advertisement or Invitation to Bid, the Instructions to Bidders, sample forms, the Contractor's Bid or portions of Addenda relating to any of these, or any other documents, unless specifically enumerated in the Owner-Contractor Agreement.

1.1.2 THE CONTRACT

The Contract Documents form the Contract for Construction. This Contract represents the entire and integrated agreement between the parties hereto and supersedes all prior negotiations, representations, or agreements, either written or oral . ~~including the bidding documents.~~ The Contract may be amended or modified only by a Modification as defined in Subparagraph 1.1.1. ~~Nothing contained in~~ The Contract Documents shall not be construed to create any contractual relationship of any kind between the Architect and the Contractor, but the Architect shall be entitled to performance of obligations intended for his benefit, and to enforcement thereof. Nothing contained in the Contract Documents shall create any contractual relationship between the Owner or the Architect and any Subcontractor or Sub-subcontractor.

1.1.3 THE WORK

The ~~term~~ Work comprises the completed construction required by the Contract Documents and includes all labor necessary to produce ~~the~~ such construction , ~~required by the Contract Documents,~~ and all materials and equipment incorporated or to be incorporated in such construction.

1.1.4 THE PROJECT

The Project is the total construction ~~designed by the Architect~~ of which the Work performed under the Contract Documents may be the whole or a part.

1970 Edition

1.2 EXECUTION, CORRELATION, INTENT AND INTERPRETATIONS

1.2.1 The Contract Documents shall be signed in not less than triplicate by the Owner and Contractor. If either the Owner or the Contractor or both do not sign the Conditions of the Contract, Drawings, Specifications, or any of the other Contract Documents, the Architect shall identify them.

1.2.2 By executing the Contract, the Contractor represents that he has visited the site, familiarized himself with the local conditions under which the Work is to be performed, and correlated his observations with the requirements of the Contract Documents.

1.2.3 The Contract Documents are complementary, and what is required by any one shall be as binding as if required by all. The intention of the Documents is to include all labor, materials, equipment and other items as provided in Subparagraph 4.4.1 necessary for the proper execution and completion of the Work. It is not intended that Work not covered under any heading, section, branch, class or trade of the Specifications shall be supplied unless it is required elsewhere in the Contract Documents or is reasonably inferable therefrom as being necessary to produce the intended results. Words which have well-known technical or trade meanings are used herein in accordance with such recognized meanings.

1.2.4 The organization of the Specifications into divisions, sections and articles, and the arrangement of Drawings shall not control the Contractor in dividing the Work among Subcontractors or in establishing the extent of Work to be performed by any trade.

1.2.5 Written interpretations necessary for the proper execution or progress of the Work, in the form of drawings or otherwise, will be issued with reasonable promptness by the Architect and in accordance with any schedule agreed upon. Either party to the Contract may make written request to the Architect for such interpretations. Such interpretations shall be consistent with and reasonably inferable from the Contract Documents, and may be effected by Field Order.

1.3 COPIES FURNISHED AND OWNERSHIP

1.3.1 Unless otherwise provided in the Contract Documents, the Contractor will be furnished, free of charge, all copies of Drawings and Specifications reasonably necessary for the execution of the Work.

1976 Edition

1.2 EXECUTION, CORRELATION AND INTENT ~~AND INTERPRETATIONS~~

1.2.1 The Contract Documents shall be signed in not less than triplicate by the Owner and Contractor. If either the Owner or the Contractor or both do not sign the Conditions of the Contract, Drawings, Specifications, or any of the other Contract Documents, the Architect shall identify ~~them~~ such Documents.

1.2.2 By executing the Contract, the Contractor represents that he has visited the site, familiarized himself with the local conditions under which the Work is to be performed, and correlated his observations with the requirements of the Contract Documents.

1.2.3 The ~~intention~~ intent of the Contract Documents is to include all ~~labor, materials, equipment and other items as provided in Subparagraph 4.4.1~~ items necessary for the proper execution and completion of the Work. The Contract Documents are complementary, and what is required by any one shall be as binding as if required by all. ~~It is not intended that~~ Work not covered ~~under any heading, section, branch, class or trade of the Specifications shall be supplied unless it is required elsewhere~~ in the Contract Documents will not be required unless it is consistent therewith and ~~or~~ is reasonably inferable therefrom as being necessary to produce the intended results. Words and abbreviations which have well-known technical or trade meanings are used ~~herein~~ in the Contract Documents in accordance with such recognized meanings.

1.2.4 The organization of the Specifications into divisions, sections and articles, and the arrangement of Drawings shall not control the Contractor in dividing the Work among Subcontractors or in establishing the extent of Work to be performed by any trade.

← Former 1.2.5 becomes new 2.2.8

1.3 ~~COPIES FURNISHED AND~~ OWNERSHIP AND USE OF DOCUMENTS

← Former 1.3.1 is new 3.2.5

1970 Edition

1.3.2 All Drawings, Specifications and copies thereof furnished by the Architect are and shall remain his property. They are not to be used on any other project, and, with the exception of one contract set for each party to the Contract, are to be returned to the Architect on request at the completion of the Work.

1976 Edition

1.3.1 All Drawings, Specifications and copies thereof furnished by the Architect are and shall remain his property. They are to be used only with respect to this Project and are not to be used on any other project . ~~, and,~~ With the exception of one contract set for each party to the Contract, such documents are to be returned or suitably accounted for to the Architect on request at the completion of the Work. Submission or distribution to meet official regulatory requirements or for other purposes in connection with the Project is not to be construed as publication in derogation of the Architect's common law copyright or other reserved rights.

1970 Edition

ARTICLE 2

ARCHITECT

2.1 DEFINITION

2.1.1 The Architect is the person or organization licensed to practice architecture and identified as such in the Agreement and is referred to throughout the Contract Documents as if singular in number and masculine in gender. The term Architect means the Architect or his authorized representative.

2.1.2 Nothing contained in the Contract Documents shall create any contractual relationship between the Architect and the Contractor.

2.2 ADMINISTRATION OF THE CONTRACT

2.2.1 The Architect will provide general Administration of the Construction Contract, including performance of the functions hereinafter described.

2.2.2 The Architect will be the Owner's representative during construction and until final payment. The Architect will have authority to act on behalf of the Owner to the extent provided in the Contract Documents, unless otherwise modified by written instrument which will be shown to the Contractor. The Architect will advise and consult with the Owner, and all of the Owner's instructions to the Contractor shall be issued through the Architect.

2.2.3 The Architect shall at all times have access to the Work wherever it is in preparation and progress. The Contractor shall provide facilities for such access so the Architect may perform his functions under the Contract Documents.

1976 Edition

ARTICLE 2

ARCHITECT

2.1 DEFINITION

2.1.1 The Architect is the person ~~or organization~~ lawfully licensed to practice architecture, ~~and~~ or an entity lawfully practicing architecture identified as such in the Owner-Contractor Agreement , and is referred to throughout the Contract Documents as if singular in number and masculine in gender. The term Architect means the Architect or his authorized representative.

Former 2.1.2 included in new 1.1.2

2.2 ADMINISTRATION OF THE CONTRACT

2.2.1 The Architect will provide ~~general~~ administration of the ~~construction~~ Contract ~~, including performance of the functions~~ as hereinafter described.

2.2.2 The Architect will be the Owner's representative during construction and until final payment is due. The Architect will advise and consult with the Owner. The Owner's instructions to the Contractor shall be forwarded through the Architect. The Architect will have authority to act on behalf of the Owner only to the extent provided in the Contract Documents, unless otherwise modified by written instrument ~~which will be shown to the Contractor~~ in accordance with Subparagraph 2.2.18. ~~The Architect will advise and consult with the Owner, and all of the Owner's instructions to the Contractor shall be issued through the Architect.~~

Former 2.2.3 is new 2.2.5

1970 Edition

2.2.4 The Architect will make periodic visits to the site to familiarize himself generally with the progress and quality of the Work and to determine in general if the Work is proceeding in accordance with the Contract Documents. On the basis of his on-site observations as an architect, he will keep the Owner informed of the progress of the Work, and will endeavor to guard the Owner against defects and deficiencies in the Work of the Contractor. The Architect will not be required to make exhaustive or continuous on-site inspections to check the quality or quantity of the Work. The Architect will not be responsible for construction means, methods, techniques, sequences or procedures, or for safety precautions and programs in connection with the Work, and he will not be responsible for the Contractor's failure to carry out the Work in accordance with the Contract Documents.

2.2.5 Based on such observations and the Contractor's Applications for Payment, the Architect will determine the amounts owing to the Contractor and will issue Certificates for Payment in such amounts, as provided in Paragraph 9.4.

2.2.6 The Architect will be, in the first instance, the interpreter of the requirements of the Contract Documents and the judge of the performance thereunder by both the Owner and Contractor. The Architect will, within a reasonable time, render such interpretations as he may deem necessary for the proper execution or progress of the Work.

1976 Edition

2.2.3 The Architect will ~~make periodic visits to~~ visit the site at intervals appropriate to the stage of construction to familiarize himself generally with the progress and quality of the Work and to determine in general if the Work is proceeding in accordance with the Contract Documents. However, the Architect will not be required to make exhaustive or continuous on-site inspections to check the quality or quantity of the Work. On the basis of his on-site observations as an architect, he will keep the Owner informed of the progress of the Work, and will endeavor to guard the Owner against defects and deficiencies in the Work of the Contractor. ~~The Architect will not be required to make exhaustive or continuous on-site inspections to check the quality or quantity of the Work.~~

New 2.2.4 replaces last sentence of former 2.2.4 and former 2.2.18 →

2.2.4 The Architect will not be responsible for and will not have control or charge of construction means, methods, techniques, sequences or procedures, or for safety precautions and programs in connection with the Work, and he will not be responsible for the Contractor's failure to carry out the Work in accordance with the Contract Documents. The Architect will not be responsible for or have control or charge over the acts or omissions of the Contractor, ~~any~~ Subcontractors, or any of their agents or employees, or any other persons performing any of the Work.

New 2.2.5 is former 2.2.3 →

2.2.5 The Architect shall at all times have access to the Work wherever it is in preparation and progress. The Contractor shall provide facilities for such access so the Architect may perform his functions under the Contract Documents.

2.2.6 Based on ~~such~~ the Architect's observations and an evaluation of the Contractor's Applications for Payment, the Architect will determine the amounts owing to the Contractor and will issue Certificates for Payment in such amounts, as provided in Paragraph 9.4.

2.2.7 The Architect will be ~~, in the first instance,~~ the interpreter of the requirements of the Contract Documents and the judge of the performance thereunder by both the Owner and Contractor. ~~The Architect will, within a reasonable time, render such interpretations as he may deem necessary for the proper execution or progress of the Work.~~

New 2.2.8 replaces former 1.2.5 →

2.2.8 ~~Written~~ The Architect will render interpretations necessary for the proper execution or progress of the Work, ~~in the form of drawings or otherwise, will be issued~~ with reasonable promptness ~~by the Architect~~ and in accordance with any ~~schedule~~ time limit agreed upon. Either party to the Contract may make written request to the Architect for such interpretations. ~~Such interpretations shall be consistent with and reasonably inferable from the Contract Documents, and may be effected by Field Order.~~

1970 Edition

2.2.7 Claims, disputes and other matters in question between the Contractor and the Owner relating to the execution or progress of the Work or the interpretation of the Contract Documents shall be referred initially to the Architect for decision which he will render in writing within a reasonable time.

2.2.8 All interpretations and decisions of the Architect shall be consistent with the intent of the Contract Documents. In his capacity as interpreter and judge, he will exercise his best efforts to insure faithful performance by both the Owner and the Contractor and will not show partiality to either.

2.2.9 The Architect's decisions in matters relating to artistic effect will be final if consistent with the intent of the Contract Documents.

2.2.10 Any claim, dispute or other matter that has been referred to the Architect, except those relating to artistic effect as provided in Subparagraph 2.2.9 and except any which have been waived by the making or acceptance of final payment as provided in Subparagraphs 9.7.5 and 9.7.6, shall be subject to arbitration upon the written demand of either party. However, no demand for arbitration of any such claim, dispute or other matter may be made until the earlier of:

2.2.10.1 The date on which the Architect has rendered his written decision, or

.2 the tenth day after the parties have presented their evidence to the Architect or have been given a reasonable opportunity to do so, if the Architect has not rendered his written decision by that date.

2.2.11 If a decision of the Architect is made in writing and states that it is final but subject to appeal, no demand for arbitration of a claim, dispute or other matter covered by such decision may be made later than thirty days after the date on which the party making the demand received the decision. The failure to demand arbitration within said thirty days' period will result in the Architect's decision becoming final and binding upon the Owner and the Contractor. If the Architect renders a decision after arbitration proceedings have been initiated, such decision may be entered as evidence but will not supersede any arbitration proceedings unless the decision is acceptable to the parties concerned.

1976 Edition

2.2.9 Claims, disputes and other matters in question between the Contractor and the Owner relating to the execution or progress of the Work or the interpretation of the Contract Documents shall be referred initially to the Architect for decision which he will render in writing within a reasonable time.

2.2.10 All interpretations and decisions of the Architect shall be consistent with the intent of and reasonably inferable from the Contract Documents and will be in writing or in the form of drawings. In his capacity as interpreter and judge, he will ~~exercise his best efforts to insure~~ endeavor to secure faithful performance by both the Owner and the Contractor , ~~and~~ will not show partiality to either , and will not be liable for the result of any interpretation or decision rendered in good faith in such capacity.

2.2.11 The Architect's decisions in matters relating to artistic effect will be final if consistent with the intent of the Contract Documents.

2.2.12 Any claim, dispute or other matter in question between the Contractor and the Owner ~~that has been~~ referred to the Architect, except those relating to artistic effect as provided in Subparagraph ~~2.2.9~~ 2.2.11 and except ~~any~~ those which have been waived by the making or acceptance of final payment as provided in Subparagraphs ~~9.7.5~~ 9.9.4 and ~~9.7.6~~ 9.9.5, shall be subject to arbitration upon the written demand of either party. However, no demand for arbitration of any such claim, dispute or other matter may be made until the earlier of (1) the date on which the Architect has rendered ~~his~~ a written decision, or (2) the tenth day after the parties have presented their evidence to the Architect or have been given a reasonable opportunity to do so, if the Architect has not rendered his written decision by that date. ~~If~~ When such a written decision of the Architect ~~is made in writing and~~ states (1) that ~~it~~ the decision is final but subject to appeal, ~~no,~~ and (2) that any demand for arbitration of a claim, dispute or other matter covered by such decision ~~may~~ must be made ~~later than~~ within thirty days after the date on which the party making the demand receives the written decision , ~~The~~ failure to demand arbitration within said thirty days' period will result in the Architect's decision becoming final and binding upon the Owner and the Contractor. If the Architect renders a decision after arbitration proceedings have been initiated, such decision may be entered as evidence but will not supersede any arbitration proceedings unless the decision is acceptable to ~~the~~ all parties concerned.

1970 Edition

2.2.12 The Architect will have authority to reject Work which does not conform to the Contract Documents. Whenever, in his reasonable opinion, he considers it necessary or advisable to insure the proper implementation of the intent of the Contract Documents, he will have authority to require special inspection or testing of the Work in accordance with Subparagraph 7.8.2 whether or not such Work be then fabricated, installed or completed. However, neither the Architect's authority to act under this Subparagraph 2.2.12, nor any decision made by him in good faith either to exercise or not to exercise such authority, shall give rise to any duty or responsibility of the Architect to the Contractor, any Subcontractor, any of their agents or employees, or any other person performing any of the Work.

2.2.13 The Architect will review Shop Drawings and Samples as provided in Subparagraphs 4.13.1 through 4.13.8 inclusive.

New 2.2.14 replaces former 2.2.13 and 4.13.5 →

2.2.14 The Architect will prepare Change Orders in accordance with Article 12, and will have authority to order minor changes in the Work as provided in Subparagraph 12.3.1.

2.2.15 The Architect will conduct inspections to determine the dates of Substantial Completion and final completion, will receive and review written guarantees and related documents required by the Contract and assembled by the Contractor and will issue a final Certificate for Payment.

2.2.16 If the Owner and Architect agree, the Architect will provide one or more Full-Time Project Representatives to assist the Architect in carrying out his responsibilities at the site. The duties, responsibilities and limitations of authority of any such Project Representative shall be as set forth in an exhibit to be incorporated in the Contract Documents.

2.2.17 The duties, responsibilities and limitations of authority of the Architect as the Owner's representative during construction as set forth in Articles 1 through 14 inclusive of these General Conditions will not be modified or extended without written consent of the Owner, the Contractor and the Architect.

1976 Edition

2.2.13 The Architect will have authority to reject Work which does not conform to the Contract Documents. Whenever, in his ~~reasonable~~ opinion, he considers it necessary or advisable ~~to insure~~ for the ~~proper~~ implementation of the intent of the Contract Documents, he will have authority to require special inspection or testing of the Work in accordance with Subparagraph ~~7.8.2~~ 7.7.2 whether or not such Work be then fabricated, installed or completed. However, neither the Architect's authority to act under this Subparagraph ~~2.2.14~~ 2.2.13, nor any decision made by him in good faith either to exercise or not to exercise such authority, shall give rise to any duty or responsibility of the Architect to the Contractor, any Subcontractor, any of their agents or employees, or any other person performing any of the Work.

2.2.14 The Architect will review and approve or take other appropriate action upon Contractor's submittals such as Shop Drawings, Product Data and Samples, ~~with reasonable promptness so as to cause no delay,~~ but only for conformance with the design concept of the ~~Project~~ Work and with the information given in the Contract Documents. Such action shall be taken with reasonable promptness so as to cause no delay. The Architect's approval of a ~~separate~~ specific item shall not indicate approval of an assembly ~~in~~ of which the item ~~functions.~~ is a component.

2.2.15 The Architect will prepare Change Orders in accordance with Article 12, and will have authority to order minor changes in the Work as provided in Subparagraph ~~12.3.1.~~ 12.4.1.

2.2.16 The Architect will conduct inspections to determine the dates of Substantial Completion and final completion, will receive and forward to the Owner for the Owner's review written ~~guarantees~~ warranties and related documents required by the Contract and assembled by the Contractor, and will issue a final Certificate for Payment upon compliance with Paragraph 9.9.

2.2.17 If the Owner and Architect agree, the Architect will provide one or more ~~Full-Time~~ Project Representatives to assist the Architect in carrying out his responsibilities at the site. The duties, responsibilities and limitations of authority of any such Project Representative shall be as set forth in an exhibit to be incorporated in the Contract Documents.

2.2.18 The duties, responsibilities and limitations of authority of the Architect as the Owner's representative during construction as set forth in ~~Articles 1 through 14 inclusive of these General Conditions~~ the Contract Documents will not be modified or extended without written consent of the Owner, the Contractor and the Architect.

1970 Edition

2.2.18 The Architect will not be responsible for the acts or omissions of the Contractor, any Subcontractors, or any of their agents or employees, or any other persons performing any of the Work.

2.2.19 In case of the termination of the employment of the Architect, the Owner shall appoint an architect against whom the Contractor makes no reasonable objection whose status under the Contract Documents shall be that of the former architect. Any dispute in connection with such appointment shall be subject to arbitration.

ARTICLE 3

OWNER

3.1 DEFINITION

3.1.1 The Owner is the person or organization identified as such in the Agreement and is referred to throughout the Contract Documents as if singular in number and masculine in gender. The term Owner means the Owner or his authorized representative.

3.2 INFORMATION AND SERVICES REQUIRED OF THE OWNER

New Subparagraph →

3.2.1 The Owner shall furnish all surveys describing the physical characteristics, legal limits and utility locations for the site of the Project.

3.2.2 The Owner shall secure and pay for easements for permanent structures or permanent changes in existing facilities.

3.2.3 Information or services under the Owner's control shall be furnished by the Owner with reasonable promptness to avoid delay in the orderly progress of the Work.

New 3.2.5 is former 1.3.1 →

3.2.4 The Owner shall issue all instructions to the Contractor through the Architect.

1976 Edition

← Former 2.2.18 included in new 2.2.4

2.2.19 In case of the termination of the employment of the Architect, the Owner shall appoint an architect against whom the Contractor makes no reasonable objection whose status under the Contract Documents shall be that of the former architect. Any dispute in connection with such appointment shall be subject to arbitration.

ARTICLE 3

OWNER

3.1 DEFINITION

3.1.1 The Owner is the person or ~~organization~~ entity identified as such in the Owner-Contractor Agreement and is referred to throughout the Contract Documents as if singular in number and masculine in gender. The term Owner means the Owner or his authorized representative.

3.2 INFORMATION AND SERVICES REQUIRED OF THE OWNER

3.2.1 The Owner shall, at the request of the Contractor, at the time of execution of the Owner-Contractor Agreement, furnish to the Contractor reasonable evidence that he has made financial arrangements to fulfill his obligations under the Contract. Unless such reasonable evidence is furnished, the Contractor is not required to execute the Owner-Contractor Agreement or to commence the Work.

3.2.2 The Owner shall furnish all surveys describing the physical characteristics, legal ~~limits~~ limitations and utility locations for the site of the Project, and a legal description of the site.

3.2.3 Except as provided in Subparagraph 4.7.1, the Owner shall secure and pay for necessary approvals, easements, assessments and charges required for the construction, use or occupancy of permanent structures or for permanent changes in existing facilities.

3.2.4 Information or services under the Owner's control shall be furnished by the Owner with reasonable promptness to avoid delay in the orderly progress of the Work.

3.2.5 Unless otherwise provided in the Contract Documents, the Contractor will be furnished, free of charge, all copies of Drawings and Specifications reasonably necessary for the execution of the Work.

3.2.6 The Owner shall ~~issue~~ forward all instructions to the Contractor through the Architect.

1970 Edition

3.2.5 The foregoing are in addition to other duties and responsibilities of the Owner enumerated herein and especially those in respect to Payment and Insurance in Articles 9 and 11 respectively.

3.3 OWNER'S RIGHT TO STOP THE WORK

3.3.1 If the Contractor fails to correct defective Work or persistently fails to supply materials or equipment in accordance with the Contract Documents, the Owner may order the Contractor to stop the Work, or any portion thereof, until the cause for such order has been eliminated.

3.4 OWNER'S RIGHT TO CARRY OUT THE WORK

3.4.1 If the Contractor defaults or neglects to carry out the Work in accordance with the Contract Documents or fails to perform any provision of the Contract, the Owner may, after seven days' written notice to the Contractor and without prejudice to any other remedy he may have, make good such deficiencies. In such case an appropriate Change Order shall be issued deducting from the payments then or thereafter due the Contractor the cost of correcting such deficiencies, including the cost of the Architect's additional services made necessary by such default, neglect or failure. The Architect must approve both such action and the amount charged to the Contractor. If the payments then or thereafter due the Contractor are not sufficient to cover such amount, the Contractor shall pay the difference to the Owner.

ARTICLE 4

CONTRACTOR

4.1 DEFINITION

4.1.1 The Contractor is the person or organization identified as such in the Agreement and is referred to throughout the Contract Documents as if singular in number and masculine in gender. The term Contractor means the Contractor or his authorized representative.

1976 Edition

3.2.7 The foregoing are in addition to other duties and responsibilities of the Owner enumerated herein and especially those in respect to Work by Owner or by Separate Contractors, Payments and Completion, and Insurance in Articles 6, 9 and 11 respectively.

3.3 OWNER'S RIGHT TO STOP THE WORK

3.3.1 If the Contractor fails to correct defective Work as required by Paragraph 13.2 or persistently fails to ~~supply materials or equipment~~ carry out the Work in accordance with the Contract Documents, the Owner, by a written order signed personally or by an agent specifically so empowered by the Owner in writing, may order the Contractor to stop the Work, or any portion thereof, until the cause for such order has been eliminated ; however, this right of the Owner to stop the Work shall not give rise to any duty on the part of the Owner to exercise this right for the benefit of the Contractor or any other person or entity, except to the extent required by Subparagraph 6.1.3.

3.4 OWNER'S RIGHT TO CARRY OUT THE WORK

3.4.1 If the Contractor defaults or neglects to carry out the Work in accordance with the Contract Documents ~~or~~ and fails within seven days after receipt of written notice from the Owner to ~~perform any provision of the Contract~~ commence and continue correction of such default or neglect with diligence and promptness, the Owner may, after seven days following receipt by the Contractor of an additional written notice ~~to the Contractor~~ and without prejudice to any other remedy he may have, make good such deficiencies. In such case an appropriate Change Order shall be issued deducting from the payments then or thereafter due the Contractor the cost of correcting such deficiencies, including ~~the cost of~~ compensation for the Architect's additional services made necessary by such default, neglect or failure. ~~The Architect must approve both~~ Such action by the Owner and the amount charged to the Contractor are both subject to the prior approval of the Architect. If the payments then or thereafter due the Contractor are not sufficient to cover such amount, the Contractor shall pay the difference to the Owner.

ARTICLE 4

CONTRACTOR

4.1 DEFINITION

4.1.1 The Contractor is the person or ~~organization~~ entity identified as such in the Owner-Contractor Agreement and is referred to throughout the Contract Documents as if singular in number and masculine in gender. The term Contractor means the Contractor or his authorized representative.

1970 Edition

4.2 REVIEW OF CONTRACT DOCUMENTS

4.2.1 The Contractor shall carefully study and compare the Contract Documents and shall at once report to the Architect any error, inconsistency or omission he may discover. The Contractor shall not be liable to the Owner or the Architect for any damage resulting from any such errors, inconsistencies or omissions in the Contract Documents. The Contractor shall do no Work without Drawings, Specifications or Modifications.

4.3 SUPERVISION AND CONSTRUCTION PROCEDURES

4.3.1 The Contractor shall supervise and direct the Work, using his best skill and attention. He shall be solely responsible for all construction means, methods, techniques, sequences and procedures and for coordinating all portions of the Work under the Contract.

New 4.3.2 replaces former 4.10.1 →

New 4.3.3 replaces former 7.8.5 →

4.4 LABOR AND MATERIALS

4.4.1 Unless otherwise specifically noted, the Contractor shall provide and pay for all labor, materials, equipment, tools, construction equipment and machinery, water, heat, utilities, transportation, and other facilities and services necessary for the proper execution and completion of the Work.

4.4.2 The Contractor shall at all times enforce strict discipline and good order among his employees and shall not employ on the Work any unfit person or anyone not skilled in the task assigned to him.

1976 Edition

4.2 REVIEW OF CONTRACT DOCUMENTS

4.2.1 The Contractor shall carefully study and compare the Contract Documents and shall at once report to the Architect any error, inconsistency or omission he may discover. The Contractor shall not be liable to the Owner or the Architect for any damage resulting from any such errors, inconsistencies or omissions in the Contract Documents. The Contractor shall ~~do~~ perform no portion of the Work at any time without ~~Drawings, Specifications or Modifications~~ Contract Documents or, where required, approved Shop Drawings, Product Data or Samples for such portion of the Work.

4.3 SUPERVISION AND CONSTRUCTION PROCEDURES

4.3.1 The Contractor shall supervise and direct the Work, using his best skill and attention. He shall be solely responsible for all construction means, methods, techniques, sequences and procedures and for coordinating all portions of the Work under the Contract.

4.3.2 The Contractor shall be responsible to the Owner for the acts and omissions of ~~all~~ his employees, ~~and all~~ Subcontractors and their agents and employees , and ~~all~~ other persons performing any of the Work under a contract with the Contractor.

4.3.3 ~~Neither the observations of the Architect in his Administration of the Construction Contract, nor inspections, tests or approvals by persons other than~~ The Contractor shall ~~relieve~~ not be relieved ~~the Contractor~~ from his obligations to perform the Work in accordance with the Contract Documents either by the activities or duties of the Architect in his administration of the Contract, or by inspections, tests or approvals required or performed under Paragraph 7.7 by persons other than the Contractor.

4.4 LABOR AND MATERIALS

4.4.1 Unless otherwise ~~specifically noted~~ provided in the Contract Documents, the Contractor shall provide and pay for all labor, materials, equipment, tools, construction equipment and machinery, water, heat, utilities, transportation, and other facilities and services necessary for the proper execution and completion of the Work , whether temporary or permanent and whether or not incorporated or to be incorporated in the Work.

4.4.2 The Contractor shall at all times enforce strict discipline and good order among his employees and shall not employ on the Work any unfit person or anyone not skilled in the task assigned to him.

1970 Edition

4.5 WARRANTY

4.5.1 The Contractor warrants to the Owner and the Architect that all materials and equipment furnished under this Contract will be new unless otherwise specified, and that all Work will be of good quality, free from faults and defects and in conformance with the Contract Documents. All Work not so conforming to these standards may be considered defective. If required by the Architect, the Contractor shall furnish satisfactory evidence as to the kind and quality of materials and equipment.

4.6 TAXES

4.6.1 The Contractor shall pay all sales, consumer, use and other similar taxes required by law.

4.7 PERMITS, FEES AND NOTICES

4.7.1 The Contractor shall secure and pay for all permits, governmental fees and licenses necessary for the proper execution and completion of the Work, which are applicable at the time the bids are received. It is not the responsibility of the Contractor to make certain that the Drawings and Specifications are in accordance with applicable laws, statutes, building codes and regulations.

4.7.2 The Contractor shall give all notices and comply with all laws, ordinances, rules, regulations and orders of any public authority bearing on the performance of the Work. If the Contractor observes that any of the Contract Documents are at variance therewith in any respect, he shall promptly notify the Architect in writing, and any necessary changes shall be adjusted by appropriate Modification. If the Contractor performs any Work knowing it to be contrary to such laws, ordinances, rules and regulations, and without such notice to the Architect, he shall assume full responsibility therefor and shall bear all costs attributable thereto.

New 4.7.3 replaces portions of former 4.7.1 and 4.7.2

1976 Edition

4.5 WARRANTY

4.5.1 The Contractor warrants to the Owner and the Architect that all materials and equipment furnished under this Contract will be new unless otherwise specified, and that all Work will be of good quality, free from faults and defects and in conformance with the Contract Documents. All Work not ~~so~~ conforming to these ~~standards~~ requirements, including substitutions not properly approved and authorized, may be considered defective. If required by the Architect, the Contractor shall furnish satisfactory evidence as to the kind and quality of materials and equipment. This warranty is not limited by the provisions of Paragraph 13.2.

4.6 TAXES

4.6.1 The Contractor shall pay all sales, consumer, use and other similar taxes ~~required by law.~~ for the Work or portions thereof provided by the Contractor which are legally enacted at the time bids are received, whether or not yet effective.

4.7 PERMITS, FEES AND NOTICES

4.7.1 Unless otherwise provided in the Contract Documents, the Contractor shall secure and pay for the building permit and for all other permits and governmental fees , ~~and~~ licenses and inspections necessary for the proper execution and completion of the Work which are customarily secured after execution of the Contract and which are ~~applicable~~ legally required at the time the bids are received.

Last sentence of former 4.7.1 included in new 4.7.3

4.7.2 The Contractor shall give all notices and comply with all laws, ordinances, rules, regulations and lawful orders of any public authority bearing on the performance of the Work.

Second sentence of former 4.7.2 included in new 4.7.3

Last sentence of former 4.7.2 is new 4.7.4

4.7.3 It is not the responsibility of the Contractor to make certain that the ~~Drawings and Specifications~~ Contract Documents are in accordance with applicable laws, statutes, building codes and regulations. If the Contractor observes that any of the Contract Documents are at variance therewith in any respect, he shall promptly notify the Architect in writing, and any necessary changes shall be ~~adjusted~~ accomplished by appropriate Modification.

1970 Edition

New 4.7.4 is last sentence of former 4.7.2 →

1976 Edition

4.7.4 If the Contractor performs any Work knowing it to be contrary to such laws, ordinances, rules and regulations, and without such notice to the Architect, he shall assume full responsiblility therefor and shall bear all costs attributable thereto.

1970 Edition

4.8 CASH ALLOWANCES

4.8.1 The Contractor shall include in the Contract Sum all allowances stated in the Contract Documents. These allowances shall cover the net cost of the materials and equipment delivered and unloaded at the site, and all applicable taxes. The Contractor's handling costs on the site, labor, installation costs, overhead, profit and other expenses contemplated for the original allowance shall be included in the Contract Sum and not in the allowance. The Contractor shall cause the Work covered by these allowances to be performed for such amounts and by such persons as the Architect may direct, but he will not be required to employ persons against whom he makes a reasonable objection. If the cost, when determined, is more than or less than the allowance, the Contract Sum shall be adjusted accordingly by Change Order which will include additional handling costs on the site, labor, installation costs, overhead, profit and other expenses resulting to the Contractor from any increase over the original allowance.

New 4.8.2 replaces portions of former 4.8.1 →

1976 Edition

4.8 ~~CASH~~ ALLOWANCES

4.8.1 The Contractor shall include in the Contract Sum all allowances stated in the Contract Documents. Items covered by these allowances shall be supplied for such amounts and by such persons as the Owner may direct, but the Contractor will not be required to employ persons against whom he makes a reasonable objection.

4.8.2 Unless otherwise provided in the Contract Documents:

.1 these allowances shall cover the ~~net~~ cost to the Contractor, less any applicable trade discount, of the materials and equipment required by the allowance delivered ~~and unloaded~~ at the site, and all applicable taxes;

.2 the Contractor's ~~handling~~ costs for unloading and handling on the site, labor, installation costs, overhead, profit and other expenses contemplated for the original allowance shall be included in the Contract Sum and not in the allowance ; ~~The Contractor shall cause the Work covered by these allowances to be performed for such amounts and by such persons as the Architect may direct, but he will not be required to employ persons against whom he makes a reasonable objection.~~

.3 ~~If~~ whenever the cost ~~, when determined,~~ is more than or less than the allowance, the Contract Sum shall be adjusted accordingly by Change Order , the amount of which will ~~include additional~~ recognize changes, if any, in handling costs on the site, labor, installation costs, overhead, profit and other expenses . ~~resulting to the Contractor from any increase over the original allowance.~~

1970 Edition

4.9 SUPERINTENDENT

4.9.1 The Contractor shall employ a competent superintendent and necessary assistants who shall be in attendance at the Project site during the progress of the Work. The superintendent shall be satisfactory to the Architect, and shall not be changed except with the consent of the Architect, unless the superintendent proves to be unsatisfactory to the Contractor and ceases to be in his employ. The superintendent shall represent the Contractor and all communications given to the superintendent shall be as binding as if given to the Contractor. Important communications will be confirmed in writing. Other communications will be so confirmed on written request in each case.

1976 Edition

4.9 SUPERINTENDENT

4.9.1 The Contractor shall employ a competent superintendent and necessary assistants who shall be in attendance at the Project site during the progress of the Work. ~~The superintendent shall be satisfactory to the Architect, and shall not be changed except with the consent of the Architect, unless the superintendent proves to be unsatisfactory to the Contractor and ceases to be in his employ.~~ The superintendent shall represent the Contractor and all communications given to the superintendent shall be as binding as if given to the Contractor. Important communications ~~will~~ shall be confirmed in writing. Other communications ~~will~~ shall be so confirmed on written request in each case.

1970 Edition

4.10 RESPONSIBILITY FOR THOSE PERFORMING THE WORK

4.10.1 The Contractor shall be responsible to the Owner for the acts and omissions of all his employees and all Subcontractors, their agents and employees, and all other persons performing any of the Work under a contract with the Contractor.

4.11 PROGRESS SCHEDULE

4.11.1 The Contractor, immediately after being awarded the Contract, shall prepare and submit for the Architect's approval an estimated progress schedule for the Work. The progress schedule shall be related to the entire Project to the extent required by the Contract Documents. This schedule shall indicate the dates for the starting and completion of the various stages of construction and shall be revised as required by the conditions of the Work, subject to the Architect's approval.

4.12 DRAWINGS AND SPECIFICATIONS AT THE SITE

4.12.1 The Contractor shall maintain at the site for the Owner one copy of all Drawings, Specifications, Addenda, approved Shop Drawings, Change Orders and other Modifications, in good order and marked to record all changes made during construction. These shall be available to the Architect. The Drawings, marked to record all changes made during construction, shall be delivered to him for the Owner upon completion of the Work.

4.13 SHOP DRAWINGS AND SAMPLES

4.13.1 Shop Drawings are drawings, diagrams, illustrations, schedules, performance charts, brochures and other data which are prepared by the Contractor or any Subcontractor, manufacturer, supplier or distributor, and which illustrate some portion of the Work.

New Subparagraph →

4.13.2 Samples are physical examples furnished by the Contractor to illustrate materials, equipment or workmanship, and to establish standards by which the Work will be judged.

1976 Edition

← Former 4.10.1 becomes new 4.3.2

4.10 PROGRESS SCHEDULE

4.10.1 The Contractor, immediately after being awarded the Contract, shall prepare and submit for the Owner's and Architect's ~~approval~~ information an estimated progress schedule for the Work. The progress schedule shall be related to the entire Project to the extent required by the Contract Documents , and shall provide for expeditious and practicable execution of the Work. ~~This schedule shall indicate the dates for the starting and completion of the various stages of construction and shall be revised as required by the conditions of the Work, subject to the Architect's approval.~~

4.11 ~~DRAWINGS AND SPECIFICATIONS~~ DOCUMENTS AND SAMPLES AT THE SITE

4.11.1 The Contractor shall maintain at the site for the Owner one record copy of all Drawings, Specifications, Addenda, ~~approved Shop Drawings,~~ Change Orders and other Modifications, in good order and marked currently to record all changes made during construction , and approved Shop Drawings, Product Data and Samples. These shall be available to the Architect ~~. The Drawings, marked to record all changes made during construction~~ and shall be delivered to him for the Owner upon completion of the Work.

4.12 SHOP DRAWINGS , PRODUCT DATA AND SAMPLES

4.12.1 Shop Drawings are drawings, diagrams, ~~illustrations,~~ schedules ~~, performance charts, brochures~~ and other data ~~which are~~ specially prepared for the Work by the Contractor or any Subcontractor, manufacturer, supplier or distributor ~~, and which~~ to illustrate some portion of the Work.

4.12.2 Product Data are illustrations, standard schedules, performance charts, instructions, brochures, diagrams and other information furnished by the Contractor to illustrate a material, product or system for some portion of the Work.

4.12.3 Samples are physical examples ~~furnished by the Contractor to~~ which illustrate materials, equipment or workmanship . and ~~to~~ establish standards by which the Work will be judged.

1970 Edition

4.13.3 The Contractor shall review, stamp with his approval and submit, with reasonable promptness and in orderly sequence so as to cause no delay in the Work or in the work of any other contractor, all Shop Drawings and Samples required by the Contract Documents or subsequently by the Architect as covered by Modifications. Shop Drawings and Samples shall be properly identified as specified, or as the Architect may require. At the time of submission the Contractor shall inform the Architect in writing of any deviation in the Shop Drawings or Samples from the requirements of the Contract Documents.

1976 Edition

4.12.4 The Contractor shall review, ~~stamp with his approval~~ approve and submit, with reasonable promptness and in ~~orderly~~ such sequence ~~so~~ as to cause no delay in the Work or in the work of the Owner or any ~~other~~ separate contractor, all Shop Drawings , Product Data and Samples required by the Contract Documents . ~~or subsequently by the Architect as covered by Modifications. Shop Drawings and Samples shall be properly identified as specified, or as the Architect may require. At the time of submission the Contractor shall inform the Architect in writing of any deviation in the Shop Drawings or Samples from the requirements of the Contract Documents.~~

4.13.4 By approving and submitting Shop Drawings and Samples, the Contractor thereby represents that he has determined and verified all field measurements, field construction criteria, materials, catalog numbers and similar data, or will do so, and that he has checked and coordinated each Shop Drawing and Sample with the requirements of the Work and of the Contract Documents.

4.12.5 By approving and submitting Shop Drawings , Product Data and Samples, the Contractor ~~thereby~~ represents that he has determined and verified all materials, field measurements, and field construction criteria ~~, materials, catalog numbers and similar data,~~ related thereto, or will do so, and that he has checked and coordinated ~~each Shop Drawing and Sample~~ the information contained within such submittals with the requirements of the Work and of the Contract Documents.

4.13.5 The Architect will review and approve Shop Drawings and Samples with reasonable promptness so as to cause no delay, but only for conformance with the design concept of the Project and with the information given in the Contract Documents. The Architect's approval of a separate item shall not indicate approval of an assembly in which the item functions.

← Former 4.13.5 included in new 2.2.14

New 4.12.6 replaces former 4.13.7 →

4.12.6 ~~The Architect's approval of Shop Drawings and Samples~~ The Contractor shall not ~~relieve~~ be relieved ~~the Contractor~~ of responsibility for any deviation from the requirements of the Contract Documents by the Architect's approval of Shop Drawings, Product Data or Samples under Subparagraph 2.2.14 unless the Contractor has specifically informed the Architect in writing of such deviation at the time of submission and the Architect has given written approval to the specific deviation . ~~nor shall~~ The ~~Architect's approval relieve the~~ Contractor shall not be relieved from responsibility for errors or omissions in the Shop Drawings, Product Data or Samples by the Architect's approval thereof.

4.13.6 The Contractor shall make any corrections required by the Architect and shall resubmit the required number of corrected copies of Shop Drawings or new Samples until approved. The Contractor shall direct specific attention in writing or on resubmitted Shop Drawings to revisions other than the corrections requested by the Architect on previous submissions.

4.12.7 ~~The Contractor shall make any corrections required by the Architect and shall resubmit the required number of corrected copies of Shop Drawings or new Samples until approved.~~ The Contractor shall direct specific attention, in writing or on resubmitted Shop Drawings , Product Data or Samples, to revisions other than ~~the corrections~~ those requested by the Architect on previous ~~submissions.~~ submittals.

1970 Edition

4.13.7 The Architect's approval of Shop Drawings or Samples shall not relieve the Contractor of responsibility for any deviation from the requirements of the Contract Documents unless the Contractor has informed the Architect in writing of such deviation at the time of submission and the Architect has given written approval to the specific deviation, nor shall the Architect's approval relieve the Contractor from responsibility for errors or omissions in the Shop Drawings or Samples.

4.13.8 No portion of the Work requiring a Shop Drawing or Sample submission shall be commenced until the submission has been approved by the Architect. All such portions of the Work shall be in accordance with approved Shop Drawings and Samples.

4.14 USE OF SITE

4.14.1 The Contractor shall confine operations at the site to areas permitted by law, ordinances, permits and the Contract Documents and shall not unreasonably encumber the site with any materials or equipment.

4.15 CUTTING AND PATCHING OF WORK

4.15.1 The Contractor shall do all cutting, fitting or patching of his Work that may be required to make its several parts fit together properly, and shall not endanger any Work by cutting, excavating or otherwise altering the Work or any part of it.

New 4.14.2 replaces former 6.3.1 →

4.16 CLEANING UP

4.16.1 The Contractor at all times shall keep the premises free from accumulation of waste materials or rubbish caused by his operations. At the completion of the Work he shall remove all his waste materials and rubbish from and about the Project as well as all his tools, construction equipment, machinery and surplus materials, and shall clean all glass surfaces and leave the Work "broom-clean" or its equivalent, except as otherwise specified.

1976 Edition

← Former 4.13.7 becomes new 4.12.6

4.12.8 No portion of the Work requiring submission of a Shop Drawing , Product Data or Sample ~~submission~~ shall be commenced until the ~~submission~~ submittal has been approved by the Architect as provided in Subparagraph 2.2.14. All such portions of the Work shall be in accordance with approved ~~Shop Drawings and Samples.~~ submittals.

4.13 USE OF SITE

4.13.1 The Contractor shall confine operations at the site to areas permitted by law, ordinances, permits and the Contract Documents and shall not unreasonably encumber the site with any materials or equipment.

4.14 CUTTING AND PATCHING OF WORK

4.14.1 The Contractor shall ~~do~~ be responsible for all cutting, fitting or patching ~~of his Work~~ that may be required to complete the Work or to make its several parts fit together properly . ~~and shall not endanger any Work by cutting, excavating or otherwise altering the Work or any part of it.~~

4.14.2 ~~The Contractor shall be responsible for any cutting, fitting and patching that may be required to complete his Work except as otherwise specifically provided in the Contract Documents.~~ The Contractor shall not damage or endanger any portion of the Work or the work of the Owner or any ~~other~~ separate contractors by cutting, ~~excavating~~ patching or otherwise altering any work , or by excavation. ~~and~~ The Contractor shall not cut or otherwise alter the work of the Owner or any ~~other~~ separate contractor except with the written consent of the ~~Architect~~ Owner and of such separate contractor. The Contractor shall not unreasonably withhold from the Owner or any separate contractor his consent to cutting or otherwise altering the Work.

4.15 CLEANING UP

4.15.1 The Contractor at all times shall keep the premises free from accumulation of waste materials or rubbish caused by his operations. At the completion of the Work he shall remove all his waste materials and rubbish from and about the Project as well as all his tools, construction equipment, machinery and surplus materials . ~~and shall clean all glass surfaces and leave the Work "broom-clean" or its equivalent, except as otherwise specified.~~

1970 Edition

4.16.2 If the Contractor fails to clean up, the Owner may do so and the cost thereof shall be charged to the Contractor as provided in Paragraph 3.4.

4.17 COMMUNICATIONS

4.17.1 The Contractor shall forward all communications to the Owner through the Architect.

New 4.17.1 is former 7.7.1 →

4.18 INDEMNIFICATION

4.18.1 The Contractor shall indemnify and hold harmless the Owner and the Architect and their agents and employees from and against all claims, damages, losses and expenses including attorneys' fees arising out of or resulting from the performance of the Work, provided that any such claim, damage, loss or expense (1) is attributable to bodily injury, sickness, disease or death, or to injury to or destruction of tangible property (other than the Work itself) including the loss of use resulting therefrom, and (2) is caused in whole or in part by any negligent act or omission of the Contractor, any Subcontractor, anyone directly or indirectly employed by any of them or anyone for whose acts any of them may be liable, regardless of whether or not it is caused in part by a party indemnified hereunder.

4.18.2 In any and all claims against the Owner or the Architect or any of their agents or employees by any employee of the Contractor, any Subcontractor, anyone directly or indirectly employed by any of them or anyone for whose acts any of them may be liable, the indemnification obligation under this Paragraph 4.18 shall not be limited in any way by any limitation on the amount or type of damages, compensation or benefits payable by or for the Contractor or any Subcontractor under workmen's compensation acts, disability benefit acts or other employee benefit acts.

1976 Edition

4.15.2 If the Contractor fails to clean up at the completion of the Work, the Owner may do so as provided in Paragraph 3.4 and the cost thereof shall be charged to the Contractor . ~~as provided in Paragraph 3.4.~~

4.16 COMMUNICATIONS

4.16.1 The Contractor shall forward all communications to the Owner through the Architect.

4.17 ROYALTIES AND PATENTS

4.17.1 The Contractor shall pay all royalties and license fees. He shall defend all suits or claims for infringement of any patent rights and shall save the Owner harmless from loss on account thereof, except that the Owner shall be responsible for all such loss when a particular design, process or the product of a particular manufacturer or manufacturers is specified, but if the Contractor has reason to believe that the design, process or product specified is an infringement of a patent, he shall be responsible for such loss unless he promptly gives such information to the Architect.

4.18 INDEMNIFICATION

4.18.1 To the fullest extent permitted by law, the Contractor shall indemnify and hold harmless the Owner and the Architect and their agents and employees from and against all claims, damages, losses and expenses, including but not limited to attorneys' fees, arising out of or resulting from the performance of the Work, provided that any such claim, damage, loss or expense (1) is attributable to bodily injury, sickness, disease or death, or to injury to or destruction of tangible property (other than the Work itself) including the loss of use resulting therefrom, and (2) is caused in whole or in part by any negligent act or omission of the Contractor, any Subcontractor, anyone directly or indirectly employed by any of them or anyone for whose acts any of them may be liable, regardless of whether or not it is caused in part by a party indemnified hereunder. Such obligation shall not be construed to negate, abridge, or otherwise reduce any other right or obligation of indemnity which would otherwise exist as to any party or person described in this Paragraph 4.18.

4.18.2 In any and all claims against the Owner or the Architect or any of their agents or employees by any employee of the Contractor, any Subcontractor, anyone directly or indirectly employed by any of them or anyone for whose acts any of them may be liable, the indemnification obligation under this Paragraph 4.18 shall not be limited in any way by any limitation on the amount or type of damages, compensation or benefits payable by or for the Contractor or any Subcontractor under workers' or workmen's compensation acts, disability benefit acts or other employee benefit acts.

1970 Edition

4.18.3 The obligations of the Contractor under this Paragraph 4.18 shall not extend to the liability of the Architect, his agents or employees arising out of (1) the preparation or approval of maps, drawings, opinions, reports, surveys, Change Orders, designs or specifications, or (2) the giving of or the failure to give directions or instructions by the Architect, his agents or employees provided such giving or failure to give is the primary cause of the injury or damage.

ARTICLE 5

SUBCONTRACTORS

5.1 DEFINITION

5.1.1 A Subcontractor is a person or organization who has a direct contract with the Contractor to perform any of the Work at the site. The term Subcontractor is referred to throughout the Contract Documents as if singular in number and masculine in gender and means a Subcontractor or his authorized representative.

5.1.2 A Sub-subcontractor is a person or organization who has a direct or indirect contract with a Subcontractor to perform any of the Work at the site. The term Sub-subcontractor is referred to throughout the Contract Documents as if singular in number and masculine in gender and means a Sub-subcontractor or an authorized representative thereof.

5.1.3 Nothing contained in the Contract Documents shall create any contractual relation between the Owner or the Architect and any Subcontractor or Sub-subcontractor.

5.2 AWARD OF SUBCONTRACTS AND OTHER CONTRACTS FOR PORTIONS OF THE WORK

5.2.1 Unless otherwise specified in the Contract Documents or in the Instructions to Bidders, the Contractor, as soon as practicable after the award of the Contract, shall furnish to the Architect in writing for acceptance by the Owner and the Architect a list of the names of the Subcontractors proposed for the principal portions of the Work. The Architect shall promptly notify the Contractor in writing if either the Owner or the Architect, after due investigation, has reasonable objection to any Subcontractor on such list and does not accept him. Failure of the Owner or Architect to make objection promptly to any Subcontractor on the list shall constitute acceptance of such Subcontractor.

1976 Edition

4.18.3 The obligations of the Contractor under this Paragraph 4.18 shall not extend to the liability of the Architect, his agents or employees, arising out of (1) the preparation or approval of maps, drawings, opinions, reports, surveys, change orders, designs or specifications, or (2) the giving of or the failure to give directions or instructions by the Architect, his agents or employees providing such giving or failure to give is the primary cause of the injury or damage.

ARTICLE 5

SUBCONTRACTORS

5.1 DEFINITION

5.1.1 A Subcontractor is a person or ~~organization~~ entity who has a direct contract with the Contractor to perform any of the Work at the site. The term Subcontractor is referred to throughout the Contract Documents as if singular in number and masculine in gender and means a Subcontractor or his authorized representative. The term Subcontractor does not include any separate contractor or his subcontractors.

5.1.2 A Sub-subcontractor is a person or ~~organization~~ entity who has a direct or indirect contract with a Subcontractor to perform any of the Work at the site. The term Sub-subcontractor is referred to throughout the Contract Documents as if singular in number and masculine in gender and means a Sub-subcontractor or an authorized representative thereof.

← Former 5.1.3 included in new 1.1.2

5.2 AWARD OF SUBCONTRACTS AND OTHER CONTRACTS FOR PORTIONS OF THE WORK

5.2.1 Unless otherwise ~~specified in~~ required by the Contract Documents or ~~in~~ the ~~Instructions to Bidders~~ Bidding Documents, the Contractor, as soon as practicable after the award of the Contract, shall furnish to the Owner and the Architect in writing ~~for acceptance by the Owner and the Architect a list of~~ the names of the ~~Subcontractors~~ persons or entities (including those who are to furnish materials or equipment fabricated to a special design) proposed for each of the principal portions of the Work. The Architect ~~shall~~ will promptly ~~notify~~ reply to the Contractor in writing stating whether or not ~~if either~~ the Owner or the Architect, after due investigation, has reasonable objection to any ~~Subcontractor~~ such proposed person or entity. ~~on such list and does not accept him.~~ Failure of the Owner or Architect to ~~make objection promptly to any Subcontractor on the list~~ reply promptly shall constitute ~~acceptance of such Subcontractor~~ notice of no reasonable objection.

1970 Edition

5.2.2 The Contractor shall not contract with any Subcontractor or any person or organization (including those who are to furnish materials or equipment fabricated to a special design) proposed for portions of the Work designated in the Contract Documents or in the Instructions to Bidders or, if none is so designated, with any Subcontractor proposed for the principal portions of the Work who has been rejected by the Owner and the Architect. The Contractor will not be required to contract with any Subcontractor or person or organization against whom he has a reasonable objection.

5.2.3 If the Owner or Architect refuses to accept any Subcontractor or person or organization on a list submitted by the Contractor in response to the requirements of the Contract Documents or the Instructions to Bidders, the Contractor shall submit an acceptable substitute and the Contract Sum shall be increased or decreased by the difference in cost occasioned by such substitution and an appropriate Change Order shall be issued; however, no increase in the Contract Sum shall be allowed for any such substitution unless the Contractor has acted promptly and responsively in submitting for acceptance any list or lists of names as required by the Contract Documents or the Instructions to Bidders.

5.2.4 If the Owner or the Architect requires a change of any proposed Subcontractor or person or organization previously accepted by them, the Contract Sum shall be increased or decreased by the difference in cost occasioned by such change and an appropriate Change Order shall be issued.

5.2.5 The Contractor shall not make any substitution for any Subcontractor or person or organization who has been accepted by the Owner and the Architect, unless the substitution is acceptable to the Owner and the Architect.

5.3 SUBCONTRACTUAL RELATIONS

5.3.1 All work performed for the Contractor by a Subcontractor shall be pursuant to an appropriate agreement between the Contractor and the Subcontractor (and where appropriate between Subcontractors and Sub-subcontractors) which shall contain provisions that:

- **.1** preserve and protect the rights of the Owner and the Architect under the Contract with respect to the Work to be performed under the subcontract so that the subcontracting thereof will not prejudice such rights;
- **.2** require that such Work be performed in accordance with the requirements of the Contract Documents;

1976 Edition

5.2.2 The Contractor shall not contract with any ~~Subcontractor or any~~ such proposed person or entity ~~organization (including those who are to furnish materials or equipment fabricated to a special design) proposed for portions of the Work designated in the Contract Documents or in the Instructions to Bidders or, if none is so designated with any Subcontractor proposed for the principal portions of the Work who has been rejected by~~ to whom the Owner ~~and~~ or the Architect has made reasonable objection under the provisions of Subparagraph 5.2.1. The Contractor ~~will~~ shall not be required to contract with ~~any Subcontractor or person or organization against~~ anyone to whom he has a reasonable objection.

5.2.3 If the Owner or the Architect ~~refuses to accept~~ has reasonable objection to any such proposed ~~Subcontractor or~~ person or entity, ~~organization on a list submitted by the Contractor in response to the requirements of the Contract Documents or the Instructions to Bidders,~~ the Contractor shall submit ~~an acceptable~~ a substitute to whom the Owner or the Architect has no reasonable objection, and the Contract Sum shall be increased or decreased by the difference in cost occasioned by such substitution and an appropriate Change Order shall be issued; however, no increase in the Contract Sum shall be allowed for any such substitution unless the Contractor has acted promptly and responsively in submitting ~~for acceptance any list or lists of~~ names as required by ~~the Contract Documents or the Instructions to Bidders~~ Subparagraph 5.2.1.

Former 5.2.4 Deleted

5.2.4 The Contractor shall ~~not~~ make ~~any~~ no substitution for any Subcontractor ~~or~~ , person or entity previously selected if ~~organization who has been accepted by~~ the Owner ~~and the~~ or Architect ~~unless the~~ makes reasonable objection to such substitution . ~~is acceptable to the Owner and the Architect.~~

5.3 SUBCONTRACTUAL RELATIONS

5.3.1 By an appropriate agreement, written where legally required for validity, the Contractor shall require each Subcontractor, to the extent of the Work to be performed by the Subcontractor, to be bound to the Contractor by the terms of the Contract Documents, and to assume toward the Contractor all the obligations and responsibilitites which the Contractor, by these Documents, assumes toward the Owner and the Architect. Said agreement shall preserve and protect the rights of the Owner and the Architect under the Contract Documents with respect to the Work to be performed ~~under the Subcontract~~ by the Subcontractor so that the subcontracting thereof will

1970 Edition	1976 Edition
	not prejudice such rights, and shall allow to the Subcontractor, unless specifically provided otherwise in the Contractor-Subcontractor agreement, the benefit of all rights, remedies and redress against the Contractor that the Contractor, by these Documents, has against the Owner. Where appropriate, the Contractor shall require each Subcontractor to enter into similar agreements with his Sub-subcontractors. The Contractor shall make available to each proposed Subcontractor, prior to the execution of the Subcontract, copies of the Contract Documents to which the Subcontractor will be bound by this Paragraph 5.3, and identify to the Subcontractor any terms and conditions of the proposed Subcontract which may be at variance with the Contract Documents. Each Subcontractor shall similarly make copies of such Documents available to his Sub-subcontractors.
.3 require submission to the Contractor of applications for payment under each subcontract to which the Contractor is a party, in reasonable time to enable the Contractor to apply for payment in accordance with Article 9; ←	Former 5.3.1.3 and 5.3.1.4 Deleted
.4 require that all claims for additional costs, extensions of time, damages for delays or otherwise with respect to subcontracted portions of the Work shall be submitted to the Contractor (via any Subcontractor or Sub-subcontractor where appropriate) in sufficient time so that the Contractor may comply in the manner provided in the Contract Documents for like claims by the Contractor upon the Owner; ←	
.5 waive all rights the contracting parties may have against one another for damages caused by fire or other perils covered by the property insurance described in Paragraph 11.3, except such rights as they may have to the proceeds of such insurance held by the Owner as trustee under Paragraph 11.3; and ←	Former 5.3.1.5 included in new 11.3.6
.6 obligate each Subcontractor specifically to consent to the provisions of this Paragraph 5.3. ←	Former 5.3.1.6 Deleted
5.4 PAYMENTS TO SUBCONTRACTORS	
5.4.1 The Contractor shall pay each Subcontractor, upon receipt of payment from the Owner, an amount equal to the percentage of completion allowed to the Contractor on account of such Subcontractor's Work, less the percentage retained from payments to the Contractor. The Contractor shall also require each Subcontractor to make similar payments to his subcontractors. ←	Former 5.4.1 becomes new 9.5.2
5.4.2 If the Architect fails to issue a Certificate for Payment for any cause which is the fault of the Contractor and not the fault of a particular Subcontractor, the Contractor shall pay that Subcontractor on demand, made at any time after the Certificate for Payment should otherwise have been issued, for his Work to the extent completed, less the retained percentage. ←	Former 5.4.2 Deleted
5.4.3 The Contractor shall pay each Subcontractor a just share of any insurance moneys received by the Contractor under Article 11, and he shall require each Subcontractor to make similar payments to his subcontractors. ←	Former 5.4.3 included in new 11.3.3

1970 Edition

5.4.4 The Architect may, on request and at his discretion, furnish to any Subcontractor, if practicable, information regarding percentages of completion certified to the Contractor on account of Work done by such Subcontractors.

5.4.5 Neither the Owner nor the Architect shall have any obligation to pay or to see to the payment of any moneys to any Subcontractor except as may otherwise be required by law.

ARTICLE 6

SEPARATE CONTRACTS

6.1 OWNER'S RIGHT TO AWARD SEPARATE CONTRACTS

6.1.1 The Owner reserves the right to award other contracts in connection with other portions of the Project under these or similar Conditions of the Contract.

6.1.2 When separate contracts are awarded for different portions of the Project, "the Contractor" in the contract documents in each case shall be the contractor who signs each separate contract.

New Subparagraph →

6.2 MUTUAL RESPONSIBILITY OF CONTRACTORS

6.2.1 The Contractor shall afford other contractors reasonable opportunity for the introduction and storage of their materials and equipment and the execution of their work, and shall properly connect and coordinate his Work with theirs.

6.2.2 If any part of the Contractor's Work depends for proper execution or results upon the work of any other separate contractor, the Contractor shall inspect and promptly report to the Architect any apparent discrepancies or defects in such work that render it unsuitable for such proper execution and results. Failure of the Contractor so to inspect and report shall constitute an acceptance of the other contractor's work as fit and proper to receive his Work, except as to defects which may develop in the other separate contractor's work after the execution of the Contractor's Work.

1976 Edition

← Former 5.4.4 becomes new 9.5.3

← Former 5.4.5 is new 9.5.4

ARTICLE 6

WORK BY OWNER OR BY SEPARATE ~~CONTRACTS~~ CONTRACTORS

6.1 OWNER'S RIGHT TO PERFORM WORK AND TO AWARD SEPARATE CONTRACTS

6.1.1 The Owner reserves the right to perform work related to the Project with his own forces, and to award ~~other~~ separate contracts in connection with other portions of the Project or other work on the site under these or similar Conditions of the Contract. If the Contractor claims that delay or additional cost is involved because of such action by the Owner, he shall make such claim as provided elsewhere in the Contract Documents.

6.1.2 When separate contracts are awarded for different portions of the Project or other work on the site, the term Contractor in the Contract Documents in each case shall ~~be~~ mean the Contractor who ~~signs~~ executes each separate ~~contract~~ Owner-Contractor Agreement.

6.1.3 The Owner will provide for the coordination of the work of his own forces and of each separate contractor with the Work of the Contractor, who shall cooperate therewith as provided in Paragraph 6.2.

6.2 MUTUAL RESPONSIBILITY ~~OF CONTRACTORS~~

6.2.1 The Contractor shall afford ~~other~~ the Owner and separate contractors reasonable opportunity for the introduction and storage of their materials and equipment and the execution of their work, and shall ~~properly~~ connect and coordinate his Work with theirs as required by the Contract Documents.

6.2.2 If any part of the Contractor's Work depends for proper execution or results upon the work of the Owner or any ~~other~~ separate contractor, the Contractor shall, ~~inspect and~~ prior to proceeding with the Work, promptly report to the Architect any apparent discrepancies or defects in such other work that render it unsuitable for such proper execution and results. Failure of the Contractor so to ~~inspect and~~ report shall constitute an acceptance of the ~~other~~ Owner's or separate contractors' work as fit and proper to receive his Work, except as to defects which may ~~develop~~ subsequently become apparent in ~~the other separate contractor's~~ such work by others. ~~after the execution of the Contractor's Work.~~

1970 Edition

New 6.2.3 is former 6.3.2

New Subparagraph

6.2.3 Should the Contractor cause damage to the work or property of any separate contractor on the Project, the Contractor shall, upon due notice, settle with such other contractor by agreement or arbitration, if he will so settle. If such separate contractor sues the Owner or initiates an arbitration proceeding on account of any damage alleged to have been so sustained, the Owner shall notify the Contractor who shall defend such proceedings at the Owner's expense, and if any judgment or award against the Owner arises therefrom the Contractor shall pay or satisfy it and shall reimburse the Owner for all attorneys' fees and court or arbitration costs which the Owner has incurred.

6.3 CUTTING AND PATCHING UNDER SEPARATE CONTRACTS

6.3.1 The Contractor shall be responsible for any cutting, fitting and patching that may be required to complete his Work except as otherwise specifically provided in the Contract Documents. The Contractor shall not endanger any work of any other contractors by cutting, excavating or otherwise altering any work and shall not cut or alter the work of any other contractor except with the written consent of the Architect.

6.3.2 Any costs caused by defective or ill-timed work shall be borne by the party responsible therefor.

6.4 OWNER'S RIGHT TO CLEAN UP

6.4.1 If a dispute arises between the separate contractors as to their responsibility for cleaning up as required by Paragraph 4.16, the Owner may clean up and charge the cost thereof to the several contractors as the Architect shall determine to be just.

ARTICLE 7

MISCELLANEOUS PROVISIONS

7.1 GOVERNING LAW

7.1.1 The Contract shall be governed by the law of the place where the Project is located.

1976 Edition

6.2.3 Any costs caused by defective or ill-timed work shall be borne by the party responsible therefor.

6.2.4 Should the Contractor wrongfully cause damage to the work or property of the Owner, or to other work on the site, the Contractor shall promptly remedy such damage as provided in Subparagraph 10.2.5.

6.2.5 Should the Contractor wrongfully cause damage to the work or property of any separate contractor ~~on the Project~~, the Contractor shall upon due notice promptly attempt to settle with such other contractor by agreement, or otherwise to resolve the dispute. ~~arbitration, if he will so settle.~~ If such separate contractor sues ~~the Owner~~ or initiates an arbitration proceeding against the Owner on account of any damage alleged to have been ~~so sustained,~~ caused by the Contractor, the Owner shall notify the Contractor who shall defend such proceedings at the Owner's expense, and if any judgment or award against the Owner arises therefrom the Contractor shall pay or satisfy it and shall reimburse the Owner for all attorneys' fees and court or arbitration costs which the Owner has incurred.

Former 6.3.1 becomes new 4.14.2

Former 6.3.2 is new 6.2.3

6.3 OWNER'S RIGHT TO CLEAN UP

6.3.1 If a dispute arises between the Contractor and separate contractors as to their responsibility for cleaning up as required by Paragraph ~~4.16~~ 4.15, the Owner may clean up and charge the cost thereof to the ~~several~~ contractors responsible therefor as the Architect shall determine to be just.

ARTICLE 7

MISCELLANEOUS PROVISIONS

7.1 GOVERNING LAW

7.1.1 The Contract shall be governed by the law of the place where the Project is located.

1970 Edition

7.2 SUCCESSORS AND ASSIGNS

7.2.1 The Owner and the Contractor each binds himself, his partners, successors, assigns and legal representatives to the other party hereto and to the partners, successors, assigns and legal representatives of such other party in respect to all covenants, agreements and obligations contained in the Contract Documents. Neither party to the Contract shall assign the Contract or sublet it as a whole without the written consent of the other, nor shall the Contractor assign any moneys due or to become due to him hereunder, without the previous written consent of the Owner.

7.3 WRITTEN NOTICE

7.3.1 Written notice shall be deemed to have been duly served if delivered in person to the individual or member of the firm or to an officer of the corporation for whom it was intended, or if delivered at or sent by registered or certified mail to the last business address known to him who gives the notice.

7.4 CLAIMS FOR DAMAGES

7.4.1 Should either party to the Contract suffer injury or damage to person or property because of any act or omission of the other party or of any of his employees, agents or others for whose acts he is legally liable, claim shall be made in writing to such other party within a reasonable time after the first observance of such injury or damage.

7.5 PERFORMANCE BOND AND LABOR AND MATERIAL PAYMENT BOND

7.5.1 The Owner shall have the right to require the Contractor to furnish bonds covering the faithful performance of the Contract and the payment of all obligations arising thereunder if and as required in the Instructions to Bidders or elsewhere in the Contract Documents.

7.6 RIGHTS AND REMEDIES

7.6.1 The duties and obligations imposed by the Contract Documents and the rights and remedies available thereunder shall be in addition to and not a limitation of any duties, obligations, rights and remedies otherwise imposed or available by law.

New Subparagraph

1976 Edition

7.2 SUCCESSORS AND ASSIGNS

7.2.1 The Owner and the Contractor each binds himself, his partners, successors, assigns and legal representatives to the other party hereto and to the partners, successors, assigns and legal representatives of such other party in respect to all covenants, agreements and obligations contained in the Contract Documents. Neither party to the Contract shall assign the Contract or sublet it as a whole without the written consent of the other, nor shall the Contractor assign any moneys due or to become due to him hereunder, without the previous written consent of the Owner.

7.3 WRITTEN NOTICE

7.3.1 Written notice shall be deemed to have been duly served if delivered in person to the individual or member of the firm or entity or to an officer of the corporation for whom it was intended, or if delivered at or sent by registered or certified mail to the last business address known to him who gives the notice.

7.4 CLAIMS FOR DAMAGES

7.4.1 Should either party to the Contract suffer injury or damage to person or property because of any act or omission of the other party or of any of his employees, agents or others for whose acts he is legally liable, claim shall be made in writing to such other party within a reasonable time after the first observance of such injury or damage.

7.5 PERFORMANCE BOND AND LABOR AND MATERIAL PAYMENT BOND

7.5.1 The Owner shall have the right to require the Contractor to furnish bonds covering the faithful performance of the Contract and the payment of all obligations arising thereunder if and as required in the ~~Instructions to Bidders~~ Bidding Documents or ~~elsewhere~~ in the Contract Documents.

7.6 RIGHTS AND REMEDIES

7.6.1 The duties and obligations imposed by the Contract Documents and the rights and remedies available thereunder shall be in addition to and not a limitation of any duties, obligations, rights and remedies otherwise imposed or available by law.

7.6.2 No action or failure to act by the Owner, Architect or Contractor shall constitute a waiver of any right or duty afforded any of them under the Contract, nor shall any such action or failure to act constitute an approval of or acquiescence in any breach thereunder, except as may be specifically agreed in writing.

1970 Edition | **1976 Edition**

7.7 ROYALTIES AND PATENTS

7.7.1 The Contractor shall pay all royalties and license fees. He shall defend all suits or claims for infringement of any patent rights and shall save the Owner harmless from loss on account thereof, except that the Owner shall be responsible for all such loss when a particular design, process or the product of a particular manufacturer or manufacturers is specified, but if the Contractor has reason to believe that the design, process or product specified is an infringement of a patent, he shall be responsible for such loss unless he promptly gives such information to the Architect.

← Former 7.7.1 is new 4.17.1

7.8 TESTS

7.8.1 If the Contract Documents, laws, ordinances, rules, regulations or orders of any public authority having jurisdiction require any Work to be inspected, tested or approved, the Contractor shall give the Architect timely notice of its readiness and of the date arranged so the Architect may observe such inspection, testing or approval. The Contractor shall bear all costs of such inspections, tests and approvals unless otherwise provided.

7.7 TESTS

7.7.1 If the Contract Documents, laws, ordinances, rules, regulations or orders of any public authority having jurisdiction require any portion of the Work to be inspected, tested or approved, the Contractor shall give the Architect timely notice of its readiness ~~and of the date arranged~~ so the Architect may observe such inspection, testing or approval. The Contractor shall bear all costs of such inspections, tests ~~and~~ or approvals conducted by public authorities. Unless otherwise provided, the Owner shall bear all costs of other inspections, tests or approvals.

7.8.2 If after the commencement of the Work the Architect determines that any Work requires special inspection, testing, or approval which Subparagraph 7.8.1 does not include, he will, upon written authorization from the Owner, instruct the Contractor to order such special inspection, testing or approval, and the Contractor shall give notice as in Subparagraph 7.8.1. If such special inspection or testing reveals a failure of the Work to comply (1) with the requirements of the Contract Documents or (2), with respect to the performance of the Work, with laws, ordinances, rules, regulations or orders of any public authority having jurisdiction, the Contractor shall bear all costs thereof, including the Architect's additional services made necessary by such failure; otherwise the Owner shall bear such costs, and an appropriate Change Order shall be issued.

7.7.2 If ~~after the commencement of the Work~~ the Architect determines that any Work requires special inspection, testing, or approval which Subparagraph ~~7.8.1~~ 7.7.1 does not include, he will, upon written authorization from the Owner, instruct the Contractor to order such special inspection, testing or approval, and the Contractor shall give notice as provided in Subparagraph ~~7.8.1~~ 7.7.1. If such special inspection or testing reveals a failure of the Work to comply ~~(1)~~ with the requirements of the Contract Documents, ~~or (2) with respect to the performance of the Work, with laws, ordinances, rules, regulations or orders of any public authority having jurisdiction,~~ the Contractor shall bear all costs thereof, including compensation for the Architect's additional services made necessary by such failure; otherwise the Owner shall bear such costs, and an appropriate Change Order shall be issued.

7.8.3 Required certificates of inspection, testing or approval shall be secured by the Contractor and promptly delivered by him to the Architect.

7.7.3 Required certificates of inspection, testing or approval shall be secured by the Contractor and promptly delivered by him to the Architect.

7.8.4 If the Architect wishes to observe the inspections, tests or approvals required by this Paragraph 7.8, he will do so promptly and, where practicable, at the source of supply.

7.7.4 If the Architect ~~wishes~~ is to observe the inspections, tests or approvals required by ~~this Paragraph 7.8~~ the Contract Documents, he will do so promptly and, where practicable, at the source of supply.

7.8.5 Neither the observations of the Architect in his Administration of the Construction Contract, nor inspections, tests or approvals by persons other than the Contractor shall relieve the Contractor from his obligations to perform the Work in accordance with the Contract Documents.

← Former 7.8.5 becomes new 4.3.3

1970 Edition

7.9 INTEREST

7.9.1 Any moneys not paid when due to either party under this Contract shall bear interest at the legal rate in force at the place of the Project.

7.10 ARBITRATION

7.10.1 All claims, disputes and other matters in question arising out of, or relating to, this Contract or the breach thereof, except as set forth in Subparagraph 2.2.9 with respect to the Architect's decisions on matters relating to artistic effect, and except for claims which have been waived by the making or acceptance of final payment as provided by Subparagraphs 9.7.5 and 9.7.6, shall be decided by arbitration in accordance with the Construction Industry Arbitration Rules of the American Arbitration Association then obtaining unless the parties mutually agree otherwise. This agreement to arbitrate shall be specifically enforceable under the prevailing arbitration law. The award rendered by the arbitrators shall be final, and judgment may be entered upon it in accordance with applicable law in any court having jurisdiction thereof.

1976 Edition

7.8 INTEREST

7.8.1 ~~Any moneys not paid when due to either party~~ Payments due and unpaid under ~~this~~ the Contract Documents shall bear interest from the date payment is due at such rate as the parties may agree upon in writing or, in the absence thereof, at the legal rate ~~in force~~ prevailing at the place of the Project.

7.9 ARBITRATION

7.9.1 All claims, disputes and other matters in question between the Contractor and the Owner arising out of, or relating to, ~~this~~ the Contract Documents or the breach thereof, except as ~~set forth~~ provided in Subparagraph ~~2.2.9~~ 2.2.11 with respect to the Architect's decisions on matters relating to artistic effect, and except for claims which have been waived by the making or acceptance of final payment as provided by Subparagraphs ~~9.7.4~~ 9.9.4 and ~~9.7.5,~~ 9.9.5, shall be decided by arbitration in accordance with the Construction Industry Arbitration Rules of the American Arbitration Association then obtaining unless the parties mutually agree otherwise. No arbitration arising out of or relating to the Contract Documents shall include, by consolidation, joinder or in any other manner, the Architect, his employees or consultants except by written consent containing a specific reference to the Owner-Contractor Agreement and signed by the Architect, the Owner, the Contractor and any other person sought to be joined. No arbitration shall include by consolidation, joinder or in any other manner, parties other than the Owner, the Contractor and any other persons substantially involved in a common question of fact or law, whose presence is required if complete relief is to be accorded in the arbitration. No person other than the Owner or Contractor shall be included as an original third party or additional third party to an arbitration whose interest or responsibility is insubstantial. Any consent to arbitration involving an additional person or persons shall not constitute consent to arbitration of any dispute not described therein or with any person not named or described therein. ~~This~~ The foregoing agreement to arbitrate and any other agreement to arbitrate with an additional person or perons duly consented to by the parties to the Owner-Contractor Agreement shall be specifically enforceable under the prevailing arbitration law. The award rendered by the arbitrators shall be final, and judgment may be entered upon it in accordance with applicable law in any court having jurisdiction thereof.

1970 Edition

7.10.2 Notice of the demand for arbitration shall be filed in writing with the other party to the Contract and with the American Arbitration Association, and a copy shall be filed with the Architect. The demand for arbitration shall be made within the time limits specified in Subparagraphs 2.2.10 and 2.2.11 where applicable, and in all other cases within a reasonable time after the claim, dispute or other matter in question has arisen, and in no event shall it be made after the date when institution of legal or equitable proceedings based on such claim, dispute or other matter in question would be barred by the applicable statute of limitations.

7.10.3 The Contractor shall carry on the Work and maintain the progress schedule during any arbitration proceedings, unless otherwise agreed by him and the Owner in writing.

ARTICLE 8

TIME

8.1 DEFINITIONS

8.1.1 The Contract Time is the period of time alloted in the Contract Documents for completion of the Work.

8.1.2 The date of commencement of the Work is the date established in a notice to proceed. If there is no notice to proceed, it shall be the date of the Agreement or such other date as may be established therein.

8.1.3 The Date of Substantial Completion of the Work or designated portion thereof is the Date certified by the Architect when construction is sufficiently complete, in accordance with the Contract Documents, so the Owner may occupy the Work or designated portion thereof for the use for which it is intended.

8.1.4 The term day as used in the Contract Documents shall mean calendar day.

8.2 PROGRESS AND COMPLETION

8.2.1 All time limits stated in the Contract Documents are of the essence of the Contract

8.2.2 The Contractor shall begin the Work on the date of commencement as defined in Subparagraph 8.1.2. He shall carry the Work forward expeditiously with adequate forces and shall complete it within the Contract Time.

1976 Edition

7.9.2 Notice of the demand for arbitration shall be filed in writing with the other party to the ~~Contract~~ Owner-Contractor Agreement and with the American Arbitration Association, and a copy shall be filed with the Architect. The demand for arbitration shall be made within the time limits specified in ~~Subparagraphs 2.2.10 and 2.2.11~~ Subparagraph 2.2.12 where applicable, and in all other cases within a reasonable time after the claim, dispute or other matter in question has arisen, and in no event shall it be made after the date when institution of legal or equitable proceedings based on such claim, dispute or other matter in question would be barred by the applicable statute of limitations.

7.9.3 Unless otherwise agreed in writing, the Contractor shall carry on the Work and maintain ~~the~~ its progress ~~schedule~~ during any arbitration proceedings, ~~unless otherwise agreed by him and the Owner in writing.~~ and the Owner shall continue to make payments to the Contractor in accordance with the Contract Documents.

ARTICLE 8

TIME

8.1 DEFINITIONS

8.1.1 Unless otherwise provided, the Contract Time is the period of time allotted in the Contract Documents for Substantial Completion of the Work as defined in Subparagraph 8.1.3, including authorized ~~extensions~~ adjustments thereto.

8.1.2 The date of commencement of the Work is the date established in a notice to proceed. If there is no notice to proceed, it shall be the date of the Owner-Contractor Agreement or such other date as may be established therein.

8.1.3 The Date of Substantial Completion of the Work or designated portion thereof is the Date certified by the Architect when construction is sufficiently complete, in accordance with the Contract Documents, so the Owner ~~may~~ can occupy or utilize the Work or designated portion thereof for the use for which it is intended.

8.1.4 The term day as used in the Contract Documents shall mean calendar day unless otherwise specifically designated.

8.2 PROGRESS AND COMPLETION

8.2.1 All time limits stated in the Contract Documents are of the essence of the Contract.

8.2.2 The Contractor shall begin the Work on the date of commencement as defined in Subparagraph 8.1.2. He shall carry the Work forward expeditiously with adequate forces and shall ~~complete it~~ achieve Substantial Completion within the Contract Time.

1970 Edition

8.2.3 If a date or time of completion is included in the Contract, it shall be the Date of Substantial Completion as defined in Subparagraph 8.1.3, including authorized extensions thereto, unless otherwise provided.

8.3 DELAYS AND EXTENSIONS OF TIME

8.3.1 If the Contractor is delayed at any time in the progress of the Work by any act or neglect of the Owner or the Architect, or by any employee of either, or by any separate contractor employed by the Owner, or by changes ordered in the Work, or by labor disputes, fire, unusual delay in transportation, unavoidable casualties or any causes beyond the Contractor's control, or by delay authorized by the Owner pending arbitration, or by any cause which the Architect determines may justify the delay, then the Contract Time shall be extended by Change Order for such reasonable time as the Architect may determine.

8.3.2 All claims for extension of time shall be made in writing to the Architect no more than twenty days after the occurrence of the delay; otherwise they shall be waived. In the case of a continuing cause of delay only one claim is necessary.

8.3.3 If no schedule or agreement is made stating the dates upon which written interpretations as set forth in Subparagraph 1.2.5 shall be furnished, then no claim for delay shall be allowed on account of failure to furnish such interpretations until fifteen days after demand is made for them, and not then unless such claim is reasonable.

8.3.4 This Paragraph 8.3 does not exclude the recovery of damages for delay by either party under other provisions of the Contract Documents.

ARTICLE 9

PAYMENTS AND COMPLETION

9.1 CONTRACT SUM

9.1.1 The Contract Sum is stated in the Agreement and is the total amount payable by the Owner to the Contractor for the performance of the Work under the Contract Documents.

1976 Edition

← Former 8.2.3 included in new 8.1.1

8.3 DELAYS AND EXTENSIONS OF TIME

8.3.1 If the Contractor is delayed at any time in the progress of the Work by any act or neglect of the Owner or the Architect, or by any employee of either, or by any separate contractor employed by the Owner, or by changes ordered in the Work, or by labor disputes, fire, unusual delay in transportation, adverse weather conditions not reasonably anticipatable, unavoidable casualties, or any causes beyond the Contractor's control, or by delay authorized by the Owner pending arbitration, or by any other cause which the Architect determines may justify the delay, then the Contract Time shall be extended by Change Order for such reasonable time as the Architect may determine.

8.3.2 ~~All claims~~ Any claim for extension of time shall be made in writing to the Architect not more than twenty days after the ~~occurrence~~ commencement of the delay; otherwise ~~they~~ it shall be waived. In the case of a continuing delay only one claim is necessary. The Contractor shall provide an estimate of the probable effect of such delay on the progress of the Work.

8.3.3 If no ~~schedule or~~ agreement is made stating the dates upon which ~~written~~ interpretations as ~~set forth~~ provided in Subparagraph ~~1.2.5~~ 2.2.8 shall be furnished, then no claim for delay shall be allowed on account of failure to furnish such interpretations until fifteen days after ~~demand~~ written request is made for them, and not then unless such claim is reasonable.

8.3.4 This Paragraph 8.3 does not exclude the recovery of damages for delay by either party under other provisions of the Contract Documents.

ARTICLE 9

PAYMENTS AND COMPLETION

9.1 CONTRACT SUM

9.1.1 The Contract Sum is stated in the Owner-Contractor Agreement and, including authorized adjustments thereto, is the total amount payable by the Owner to the Contractor for the performance of the Work under the Contract Documents.

1970 Edition

9.2 SCHEDULE OF VALUES

9.2.1 Before the first Application for Payment, the Contractor shall submit to the Architect a schedule of values of the various portions of the Work, including quantities if required by the Architect, aggregating the total Contract Sum, divided so as to facilitate payments to Subcontractors in accordance with Paragraph 5.4, prepared in such form as specified or as the Architect and the Contractor may agree upon, and supported by such data to substantiate its correctness as the Architect may require. Each item in the schedule of values shall include its proper share of overhead and profit. This schedule, when approved by the Architect, shall be used only as a basis for the Contractor's Applications for Payment.

9.3 PROGRESS PAYMENTS

9.3.1 At least ten days before each progress payment falls due, the Contractor shall submit to the Architect an itemized Application for Payment, supported by such data substantiating the Contractor's right to payment as the Owner or the Architect may require.

9.3.2 If payments are to be made on account of materials or equipment not incorporated in the Work but delivered and suitably stored at the site, or at some other location agreed upon in writing, such payments shall be conditioned upon submission by the Contractor of bills of sale or such other procedures satisfactory to the Owner to establish the Owner's title to such materials or equipment or otherwise protect the Owner's interest including applicable insurance and transportation to the site.

9.3.3 The Contractor warrants and guarantees that title to all Work, materials and equipment covered by an Application for Payment, whether incorporated in the Project or not, will pass to the Owner upon the receipt of such payment by the Contractor, free and clear of all liens, claims, security interests or encumbrances, hereinafter referred to in this Article 9 as "liens"; and that no Work, materials or equipment covered by an Application for Payment will have been acquired by the Contractor; or by any other person performing the Work at the site or furnishing materials and equipment for the Project, subject to an agreement under which an interest therein or an encumbrance thereon is retained by the seller or otherwise imposed by the Contractor or such other person.

1976 Edition

9.2 SCHEDULE OF VALUES

9.2.1 Before the first Application for Payment, the Contractor shall submit to the Architect a schedule of values ~~of~~ allocated to the various portions of the Work, ~~including quantities if required by the Architect, aggregating the total Contract Sum, divided so as to facilitate payments to Subcontractors in accordance with Paragraph 5.4,~~ prepared in such form ~~as specified or as the Architect and the Contractor may agree upon,~~ and supported by such data to substantiate its ~~correctness~~ accuracy as the Architect may require. ~~Each item in the schedule of values shall include its proper share of overhead and profit.~~ This schedule, ~~when approved~~ unless objected to by the Architect, shall be used only as a basis for the Contractor's Applications for Payment.

9.3 ~~PROGRESS PAYMENTS~~ APPLICATIONS FOR PAYMENT

9.3.1 At least ten days before ~~each progress payment falls due~~ the date for each progress payment established in the Owner-Contractor Agreement, the Contractor shall submit to the Architect an itemized Application for Payment, notarized if required, supported by such data substantiating the Contractor's right to payment as the Owner or the Architect may require , and reflecting retainage, if any, as provided elsewhere in the Contract Documents.

9.3.2 ~~If~~ Unless otherwise provided in the Contract Documents, payments ~~are to~~ will be made on account of materials or equipment not incorporated in the Work but delivered and suitably stored at the site ~~or~~ and, if approved in advance by the Owner, payments may similarly be made for materials or equipment suitably stored at some other location agreed upon in writing . ~~, such~~ Payments for materials or equipment stored on or off the site shall be conditioned upon submission by the Contractor of bills of sale or such other procedures satisfactory to the Owner to establish the Owner's title to such materials or equipment or otherwise protect the Owner's interest, including applicable insurance and transportation to the site for those materials and equipment stored off the site.

9.3.3 The Contractor warrants ~~and guarantees~~ that title to all Work, materials and equipment covered by an Application for Payment ~~, whether incorporated in the Project or not,~~ will pass to the Owner either by incorporation in the construction or upon the receipt of ~~such~~ payment by the Contractor, whichever occurs first, free and clear of all liens, claims, security interests or encumbrances, hereinafter referred to in this Article 9 as "liens"; and that no Work, materials or equipment covered by an Application for Payment will have been acquired by the Contractor, or by any other person performing ~~the~~ Work at the site or furnishing materials and equipment for the Project, subject to an agreement under which an interest therein or an encumbrance thereon is retained by the seller or otherwise imposed by the Contractor or such other person.

1970 Edition

9.4 CERTIFICATES FOR PAYMENT

9.4.1 If the Contractor has made Application for Payment as above, the Architect will, with reasonable promptness but not more than seven days after the receipt of the Application, issue a Certificate for Payment to the Owner, with a copy to the Contractor, for such amount as he determines to be properly due, or state in writing his reasons for withholding a Certificate as provided in Subparagraph 9.5.1.

9.4.2 The issuance of a Certificate for Payment will constitute a representation by the Architect to the Owner, based on his observations at the site as provided in Subparagraph 2.2.4 and the data comprising the Application for Payment, that the Work has progressed to the point indicated; that, to the best of his knowledge, information and belief, the quality of the Work is in accordance with the Contract Documents (subject to an evaluation of the Work for conformance with the Contract Documents upon Substantial Completion, to the results of any subsequent tests required by the Contract Documents, to minor deviations from the Contract Documents correctable prior to completion, and to any specific qualifications stated in his Certificate); and that the Contractor is entitled to payment in the amount certified. In addition, the Architect's final Certificate for Payment will constitute a further representation that the conditions precedent to the Contractor's being entitled to final payment as set forth in Subparagraph 9.7.2 have been fulfilled. However, by issuing a Certificate for Payment, the Architect shall not thereby be deemed to represent that he has made exhaustive or continuous on-site inspections to check the quality or quantity of the Work or that he has reviewed the construction means, methods, techniques, sequences or procedures, or that he has made any examination to ascertain how or for what purpose the Contractor has used the moneys previously paid on account of the Contract Sum.

9.4.3 After the Architect has issued a Certificate for Payment, the Owner shall make payment in the manner provided in the Agreement.

New 9.5.2 replaces former 5.4.1 ⟶

1976 Edition

9.4 CERTIFICATES FOR PAYMENT

9.4.1 ~~If the Contractor has made Application for Payment as above,~~ The Architect will, ~~with reasonable promptness but not more than~~ within seven days after the receipt of the Contractor's Application for Payment, either issue a Certificate for Payment to the Owner, with a copy to the Contractor, for such amount as ~~he~~ the Architect determines ~~to be~~ is properly due, or ~~state~~ notify the Contractor in writing his reasons for withholding a Certificate as provided in Subparagraph ~~9.5.1~~ 9.6.1.

9.4.2 The issuance of a Certificate for Payment will constitute a representation by the Architect to the Owner, based on his observations at the site as provided in Subparagraph ~~2.2.4~~ 2.2.3 and the data comprising the Application for Payment, that the Work has progressed to the point indicated; that, to the best of his knowledge, information and belief, the quality of the Work is in accordance with the Contract Documents (subject to an evaluation of the Work for conformance with the Contract Documents upon Substantial Completion, to the results of any subsequent tests required by or performed under the Contract Documents, to minor deviations from the Contract Documents correctable prior to completion, and to any specific qualifications stated in his Certificate); and that the Contractor is entitled to payment in the amount certified. ~~In addition, the Architect's final Certificate for Payment will constitute a further representation that the conditions precedent to the Contractor's being entitled to final payment as set forth in Subparagraph 9.7.2 have been fulfilled.~~ However, by issuing a Certificate for Payment, the Architect shall not thereby be deemed to represent that he has made exhaustive or continuous on-site inspections to check the quality or quantity of the Work or that he has reviewed the construction means, methods, techniques, sequences or procedures, or that he has made any examination to ascertain how or for what purpose the Contractor has used the moneys previously paid on account of the Contract Sum.

9.5 PROGRESS PAYMENTS

9.5.1 After the Architect has issued a Certificate for Payment, the Owner shall make payment in the manner and within the time provided in the ~~Agreement~~ Contract Documents.

9.5.2 The Contractor shall promptly pay each Subcontractor, upon receipt of payment from the Owner, ~~an amount equal to the percentage of completion allowed~~ out of the amount paid to the Contractor on account of such Subcontractor's Work, the amount to which said Subcontractor is entitled, ~~less~~ reflecting the percentage actually retained , if any, from payments to the Contractor on account of such Subcontractor's Work. The Contractor shall, by an appropriate agreement with each Subcontractor, ~~also~~ require each Subcontractor to make ~~similar~~ payments to his Sub-subcontractors in similar manner.

1970 Edition	1976 Edition
New 9.5.3 replaces former 5.4.4 →	9.5.3 The Architect may, on request and at his discretion, furnish to any Subcontractor, if practicable, information regarding the percentages of completion ~~certified to~~ or the amounts applied for by the Contractor and the action taken thereon by the Architect on account of Work done by such ~~Subcontractors~~ Subcontractor.
New 9.5.4 is former 5.4.5 →	9.5.4 Neither the Owner nor the Architect shall have any obligation to pay or to see to the payment of any moneys to any Subcontractor except as may otherwise be required by law.
9.4.4 No certificate for a progress payment, nor any progress payment, nor any partial or entire use or occupancy of the Project by the Owner, shall constitute an acceptance of any Work not in accordance with the Contract Documents.	9.5.5 No Certificate for a progress payment, nor any progress payment, nor any partial or entire use or occupancy of the Project by the Owner, shall constitute an acceptance of any Work not in accordance with the Contract Documents.
9.5 PAYMENTS WITHHELD	9.6 PAYMENTS WITHHELD
9.5.1 The Architect may decline to approve an Application for Payment and may withhold his Certificate in whole or in part, to the extent necessary reasonably to protect the Owner, if in his opinion he is unable to make representations to the Owner as provided in Subparagraph 9.4.2. The Architect may also decline to approve any Applications for Payment or, because of subsequently discovered evidence or subsequent inspections, he may nullify the whole or any part of any Certificate for Payment previously issued, to such extent as may be necessary in his opinion to protect the Owner from loss because of:	9.6.1 The Architect may decline to ~~approve an Application for Payment~~ certify payment and may withhold his Certificate in whole or in part, to the extent necessary reasonably to protect the Owner, if in his opinion he is unable to make representations to the Owner as provided in Subparagraph 9.4.2. If the Architect is unable to make representations to the Owner as provided in Subparagraph 9.4.2 and to certify payment in the amount of the Appplication, he will notify the Contractor as provided in Subparagraph 9.4.1. If the Contractor and the Architect cannot agree on a revised amount, the Architect will promptly issue a Certificate for Payment for the amount for which he is able to make such representations to the Owner. The Architect may also decline to ~~approve any Applications for Payment~~ certify payment or, because of subsequently discovered evidence or subsequent ~~inspections~~ observations, he may nullify the whole or any part of any Certificate for Payment previously issued, to such extent as may be necessary in his opinion to protect the Owner from loss because of:
.1 defective work not remedied,	.1 defective work not remedied,
.2 third party claims filed or reasonable evidence indicating probable filing of such claims,	.2 third party claims filed or reasonable evidence indicating probable filing of such claims,
.3 failure of the Contractor to make payments properly to Subcontractors or for labor, materials or equipment,	.3 failure of the Contractor to make payments properly to Subcontractors or for labor, materials or equipment,
.4 reasonable doubt that the Work can be completed for the unpaid balance of the Contract Sum,	.4 reasonable ~~doubt~~ evidence that the Work ~~can~~ cannot be completed for the unpaid balance of the Contract Sum,
.5 damage to another contractor,	.5 damage to the Owner or another contractor,
.6 reasonable indication that the Work will not be completed within the Contract Time, or	.6 reasonable ~~indication~~ evidence that the Work will not be completed within the Contract Time, or
.7 unsatisfactory prosecution of the Work by the Contractor.	.7 ~~unsatisfactory prosecution of~~ persistent failure to carry out the Work ~~by the Contractor.~~ in accordance with the Contract Documents.

1970 Edition

9.5.2 When the above grounds in Subparagraph 9.5.1 are removed, payment shall be made for amounts withheld because of them.

9.6 FAILURE OF PAYMENT

9.6.1 If the Architect should fail to issue any Certificate for Payment, through no fault of the Contractor, within seven days after receipt of the Contractor's Application for Payment, or if the Owner should fail to pay the Contractor within seven days after the date of payment established in the Agreement any amount certified by the Architect or awarded by arbitration, then the Contractor may, upon seven additional days' written notice to the Owner and the Architect, stop the Work until payment of the amount owing has been received.

9.7 SUBSTANTIAL COMPLETION AND FINAL PAYMENT

9.7.1 When the Contractor determines that the Work or a designated portion thereof acceptable to the Owner is substantially complete, the Contractor shall prepare for submission to the Architect a list of items to be completed or corrected. The failure to include any items on such list does not alter the responsibility of the Contractor to complete all Work in accordance with the Contract Documents. When the Architect on the basis of an inspection determines that the Work is substantially complete, he will then prepare a Certificate of Substantial Completion which shall establish the Date of Substantial Completion, shall state the responsibilities of the Owner and the Contractor for maintenance, heat, utilities, and insurance, and shall fix the time within which the Contractor shall complete the items listed therein. The Certificate of Substantial Completion shall be submitted to the Owner and the Contractor for their written acceptance of the responsibilities assigned to them in such Certificate.

New Subparagraph →

1976 Edition

9.6.2 When the above grounds in Subparagraph ~~9.5.1~~ 9.6.1 are removed, payment shall be made for amounts withheld because of them.

9.7 FAILURE OF PAYMENT

9.7.1 If the Architect ~~should fail to~~ does not issue a Certificate for Payment, through no fault of the Contractor, within seven days after receipt of the Contractor's Application for Payment, or if the Owner ~~should fail to~~ does not pay the Contractor within seven days after the date ~~of payment~~ established in the ~~Agreement~~ Contract Documents any amount certified by the Architect or awarded by arbitration, then the Contractor may, upon seven additional days' written notice to the Owner and the Architect, stop the Work until payment of the amount owing has been received. The Contract Sum shall be increased by the amount of the Contractor's reasonable costs of shut-down, delay and start-up, which shall be effected by appropriate Change Order in accordance with Paragraph 12.3.

9.8 SUBSTANTIAL COMPLETION ~~AND FINAL PAYMENT~~

9.8.1 When the Contractor ~~determines~~ considers that the Work, or a designated portion thereof which is acceptable to the Owner, is substantially complete as defined in Subparagraph 8.1.3, the Contractor shall prepare for submission to the Architect a list of items to be completed or corrected. The failure to include any items on such list does not alter the responsibility of the Contractor to complete all Work in accordance with the Contract Documents. When the Architect on the basis of an inspection determines that the Work or designated portion thereof is substantially complete, he will then prepare a Certificate of Substantial Completion which shall establish the Date of Substantial Completion, shall state the responsibilities of the Owner and the Contractor for security, maintenance, heat, utilities, damage to the Work, and insurance, and shall fix the time within which the Contractor shall complete the items listed therein. Warranties required by the Contract Documents shall commence on the Date of Substantial Completion of the Work or designated portion thereof unless otherwise provided in the Certificate of Substantial Completion. The Certificate of Substantial Completion shall be submitted to the Owner and the Contractor for their written acceptance of the responsibilities assigned to them in such Certificate.

9.8.2 Upon Substantial Completion of the Work or designated portion thereof and upon application by the Contractor and certification by the Architect, the Owner shall make payment, reflecting adjustment in retainage, if any, for such Work or portion thereof, as provided in the Contract Documents.

1970 Edition

9.7.2 Upon receipt of written notice that the Work is ready for final inspection and acceptance and upon receipt of a final Application for Payment, the Architect will promptly make such inspection and, when he finds the Work acceptable under the Contract Documents and the Contract fully performed, he will promptly issue a final Certificate for Payment stating that to the best of his knowledge, information and belief, and on the basis of his observations and inspections, the Work has been completed in accordance with the terms and conditions of the Contract Documents and that the entire balance found to be due the Contractor, and noted in said final Certificate, is due and payable.

9.7.3 Neither the final payment nor the remaining retained percentage shall become due until the Contractor submits to the Architect (1) an Affidavit that all payrolls, bills for materials and equipment, and other indebtedness connected with the Work for which the Owner or his property might in any way be responsible, have been paid or otherwise satisfied, (2) consent of surety, if any, to final payment and (3), if required by the Owner, other data establishing payment or satisfaction of all such obligations, such as receipts, releases and waivers of liens arising out of the Contract, to the extent and in such form as may be designated by the Owner. If any Subcontractor refuses to furnish a release or waiver required by the Owner, the Contractor may furnish a bond satisfactory to the Owner to indemnify him against any such lien. If any such lien remains unsatisfied after all payments are made, the Contractor shall refund to the Owner all moneys that the latter may be compelled to pay in discharging such lien, including all costs and reasonable attorneys' fees.

9.7.4 If after Substantial Completion of the Work final completion thereof is materially delayed through no fault of the Contractor, and the Architect so confirms, the Owner shall, upon certification by the Architect, and without terminating the Contract, make payment of the balance due for that portion of the Work fully completed and accepted. If the remaining balance for Work not fully completed or corrected is less than the retainage stipulated in the Agreement, and if bonds have been furnished as required in Subparagraph 7.5.1, the written consent of the surety to the payment of the balance due for that portion of the Work fully completed and accepted shall be submitted by the Contractor to the Architect prior to certification of such payment. Such payment shall be made under the terms and conditions governing final payment, except that it shall not constitute a waiver of claims.

1976 Edition

9.9 FINAL COMPLETION AND FINAL PAYMENT

9.9.1 Upon receipt of written notice that the Work is ready for final inspection and acceptance and upon receipt of a final Application for Payment, the Architect will promptly make such inspection and, when he finds the Work acceptable under the Contract Documents and the Contract fully performed, he will promptly issue a final Certificate for Payment stating that to the best of his knowledge, information and belief, and on the basis of his observations and inspections, the Work has been completed in accordance with the terms and conditions of the Contract Documents and that the entire balance found to be due the Contractor, and noted in said final Certificate, is due and payable. The Architect's final Certificate for Payment will constitute a further representation that the conditions precedent to the Contractor's being entitled to final payment as set forth in Subparagraph ~~9.7.2~~ 9.9.2 have been fulfilled.

9.9.2 Neither the final payment nor the remaining retained percentage shall become due until the Contractor submits to the Architect (1) an affidavit that all payrolls, bills for materials and equipment, and other indebtedness connected with the Work for which the Owner or his property might in any way be responsible, have been paid or otherwise satisfied, (2) consent of surety, if any, to final payment and (3), if required by the Owner, other data establishing payment or satisfaction of all such obligations, such as receipts, releases and waivers of liens arising out of the Contract, to the extent and in such form as may be designated by the Owner. If any Subcontractor refuses to furnish a release or waiver required by the Owner, the Contractor may furnish a bond satisfactory to the Owner to indemnify him against any such lien. If any such lien remains unsatisfied after all payments are made, the Contractor shall refund to the Owner all moneys that the latter may be compelled to pay in discharging such lien, including all costs and reasonable attorneys' fees.

9.9.3 If, after Substantial Completion of the Work, final completion thereof is materially delayed through no fault of the Contractor or by the issuance of Change Orders affecting final completion, and the Architect so confirms, the Owner shall, upon application by the Contractor and certification by the Architect, and without terminating the Contract, make payment of the balance due for that portion of the Work fully completed and accepted. If the remaining balance for Work not fully completed or corrected is less than the retainage stipulated in the ~~Agreement~~ Contract Documents, and if bonds have been furnished as ~~required~~ provided in ~~Subparagraph 7.5.1,~~ Paragraph 7.5, the written consent of the surety to the payment of the balance due for that portion of the Work fully completed and accepted shall be submitted by the Contractor to the Architect prior to certification of such payment. Such payment shall be made under the terms and conditions governing final payment, except that it shall not constitute a waiver of claims.

1970 Edition

9.7.5 The making of final payment shall constitute a waiver of all claims by the Owner except those arising from:

- **.1** unsettled liens,
- **.2** faulty or defective Work appearing after Substantial Completion,
- **.3** failure of the Work to comply with the requirements of the Contract Documents, or
- **.4** terms of any special guarantees required by the Contract Documents.

9.7.6 The acceptance of final payment shall constitute a waiver of all claims by the Contractor except those previously made in writing and still unsettled.

ARTICLE 10

PROTECTION OF PERSONS AND PROPERTY

10.1 SAFETY PRECAUTIONS AND PROGRAMS

10.1.1 The Contractor shall be responsible for initiating, maintaining and supervising all safety precautions and programs in connection with the Work.

10.2 SAFETY OF PERSONS AND PROPERTY

10.2.1 The Contractor shall take all reasonable precautions for the safety of, and shall provide all reasonable protection to prevent damage, injury or loss to:

- **.1** all employees on the Work and all other persons who may be affected thereby;
- **.2** all the Work and all materials and equipment to be incorporated therein, whether in storage on or off the site, under the care, custody or control of the Contractor or any of his Subcontractors or Sub-subcontractors; and
- **.3** other property at the site or adjacent thereto, including trees, shrubs, lawns, walks, pavements, roadways, structures and utilities not designated for removal, relocation or replacement in the course of construction.

10.2.2 The Contractor shall comply with all applicable laws, ordinances, rules, regulations and lawful orders of any public authority having jurisdiction for the safety of persons or property or to protect them from damage, injury or loss. He shall erect and maintain, as required by existing conditions and progress of the Work, all reasonable safeguards for safety and protection, including posting danger signs and other warnings against hazards, promulgating safety regulations and notifying owners and users of adjacent utilities.

New 10.2.3 replaces last sentence of former 10.2.2 →

1976 Edition

9.9.4 The making of final payment shall constitute a waiver of all claims by the Owner except those arising from:

- .1 unsettled liens,
- .2 faulty or defective Work appearing after Substantial Completion,
- .3 failure of the Work to comply with the requirements of the Contract Documents, or
- .4 terms of any special ~~guarantees~~ warranties required by the Contract Documents.

9.9.5 The acceptance of final payment shall constitute a waiver of all claims by the Contractor except those previously made in writing and ~~still~~ identified by the Contractor as unsettled at the time of the final Application for Payment.

ARTICLE 10

PROTECTION OF PERSONS AND PROPERTY

10.1 SAFETY PRECAUTIONS AND PROGRAMS

10.1.1 The Contractor shall be responsible for initiating, maintaining and supervising all safety precautions and programs in connection with the Work.

10.2 SAFETY OF PERSONS AND PROPERTY

10.2.1 The Contractor shall take all reasonable precautions for the safety of, and shall provide all reasonable protection to prevent damage, injury or loss to:

- .1 all employees on the Work and all other persons who may be affected thereby;
- .2 all the Work and all materials and equipment to be incorporated therein, whether in storage on or off the site, under the care, custody or control of the Contractor or any of his Subcontractors or Subsubcontractors; and
- .3 other property at the site or adjacent thereto, including trees, shrubs, lawns, walks, pavements, roadways, structures and utilities not designated for removal, relocation or replacement in the course of construction.

10.2.2 The Contractor shall give all notices and comply with all applicable laws, ordinances, rules, regulations and lawful orders of any public authority ~~having jurisdiction for~~ bearing on the safety of persons or property or ~~to protect them~~ their protection from damage, injury or loss.

10.2.3 ~~He~~ The Contractor shall erect and maintain, as required by existing conditions and progress of the Work, all reasonable safeguards for safety and protection, including posting danger signs and other warnings against hazards, promulgating safety regulations and notifying owners and users of adjacent utilities.

1970 Edition

10.2.3 When the use or storage of explosives or other hazardous materials or equipment is necessary for the execution of the Work, the Contractor shall exercise the utmost care and shall carry on such activities under the supervision of properly qualified personnel.

10.2.4 All damage or loss to any property referred to in Clauses 10.2.1.2 and 10.2.1.3 caused in whole or in part by the Contractor, any Subcontractor, any Sub-subcontractor, or anyone directly or indirectly employed by any of them, or by anyone for whose acts any of them may be liable, shall be remedied by the Contractor, except damage or loss attributable to faulty Drawings or Specifications or to the acts or omissions of the Owner or Architect or anyone employed by either of them or for whose acts either of them may be liable, and not attributable to the fault or negligence of the Contractor.

10.2.5 The Contractor shall designate a responsible member of his organization at the site whose duty shall be the prevention of accidents. This person shall be the Contractor's superintendent unless otherwise designated in writing by the Contractor to the Owner and the Architect.

10.2.6 The Contractor shall not load or permit any part of the Work to be loaded so as to endanger its safety.

10.3 EMERGENCIES

10.3.1 In any emergency affecting the safety of persons or property, the Contractor shall act, at his discretion, to prevent threatened damage, injury or loss. Any additional compensation or extension of time claimed by the Contractor on account of emergency work shall be determined as provided in Article 12 for Changes in the Work.

ARTICLE 11

INSURANCE

11.1 CONTRACTOR'S LIABILITY INSURANCE

11.1.1 The Contractor shall purchase and maintain such insurance as will protect him from claims set forth below which may arise out of or result from the Contractor's operations under the Contract, whether such operations be by himself or by any Subcontractor or by anyone directly or indirectly employed by any of them, or by anyone for whose acts any of them may be liable:

1976 Edition

10.2.4 When the use or storage of explosives or other hazardous materials or equipment is necessary for the execution of the Work, the Contractor shall exercise the utmost care and shall carry on such activities under the supervision of properly qualified personnel.

10.2.5 The Contractor shall promptly remedy all damage or loss (other than damage or loss insured under Paragraph 11.3) to any property referred to in Clauses 10.2.1.2 and 10.2.1.3 caused in whole or in part by the Contractor, any Subcontractor, any Sub-subcontractor, or anyone directly or indirectly employed by any of them, or by anyone for whose acts any of them may be liable ~~shall be remedied by the Contractor~~ and for which the Contractor is responsible under Clauses 10.2.1.2 and 10.2.1.3, except damage or loss attributable to ~~faulty Drawings or Specifications or~~ the acts or omissions of the Owner or Architect or anyone directly or indirectly employed by either of them, or by anyone for whose acts either of them may be liable, and not attributable to the fault or negligence of the Contractor. The foregoing obligations of the Contractor are in addition to his obligations under Paragraph 4.18.

10.2.6 The Contractor shall designate a responsible member of his organization at the site whose duty shall be the prevention of accidents. This person shall be the Contractor's superintendent unless otherwise designated ~~in writing~~ by the Contractor in writing to the Owner and the Architect.

10.2.7 The Contractor shall not load or permit any part of the Work to be loaded so as to endanger its safety.

10.3 EMERGENCIES

10.3.1 In any emergency affecting the safety of persons or property, the Contractor shall act, at his discretion, to prevent threatened damage, injury or loss. Any additional compensation or extension of time claimed by the Contractor on account of emergency work shall be determined as provided in Article 12 for Changes in the Work.

ARTICLE 11

INSURANCE

11.1 CONTRACTOR'S LIABILITY INSURANCE

11.1.1 The Contractor shall purchase and maintain such insurance as will protect him from claims set forth below which may arise out of or result from the Contractor's operations under the Contract, whether such operations be by himself or by any Subcontractor or by anyone directly or indirectly employed by any of them, or by anyone for whose acts any of them may be liable:

1970 Edition

.1 claims under workmen's compensation, disability benefit and other similar employee benefit acts;
.2 claims for damages because of bodily injury, occupational sickness or disease, or death of his employees;
.3 claims for damages because of bodily injury, sickness or disease, or death of any person other than his employees;
.4 claims for damages insured by usual personal injury liability coverage which are sustained (1) by any person as a result of an offense directly or indirectly related to the employment of such person by the Contractor, or (2) by any other person; and
.5 claims for damages because of injury to or destruction of tangible property, including loss of use resulting therefrom.

New Clause →

11.1.2 The insurance required by Subparagraph 11.1.1 shall be written for not less than any limits of liability specified in the Contract Documents, or required by law, whichever is greater, and shall include contractual liability insurance as applicable to the Contractor's obligations under Paragraph 4.18.

New 11.1.3 replaces last portion of former 11.1.2 →

11.1.3 Certificates of Insurance acceptable to the Owner shall be filed with the Owner prior to commencement of the Work. These Certificates shall contain a provision that coverages afforded under the policies will not be cancelled until at least fifteen days' prior written notice has been given to the Owner.

11.2 OWNER'S LIABILITY INSURANCE

11.2.1 The Owner shall be responsible for purchasing and maintaining his own liability insurance and, at his option, may purchase and maintain such insurance as will protect him against claims which may arise from operations under the Contract.

11.3 PROPERTY INSURANCE

11.3.1 Unless otherwise provided, the Owner shall purchase and maintain property insurance upon the entire Work at the site to the full insurable value thereof. This insurance shall include the interests of the Owner, the Contractor, Subcontractors and Sub-subcontractors in the Work and shall insure against the perils of Fire, Extended Coverage, Vandalism and Malicious Mischief.

1976 Edition

.1 claims under workers' or workmen's compensation, disability benefit and other similar employee benefit acts;
.2 claims for damages because of bodily injury, occupational sickness or disease, or death of his employees;
.3 claims for damages because of bodily injury, sickness or disease, or death of any person other than his employees;
.4 claims for damages insured by usual personal injury liability coverage which are sustained (1) by any person as a result of an offense directly or indirectly related to the employment of such person by the Contractor, or (2) by any other person;
.5 claims for damages, other than to the Work itself, because of injury to or destruction of tangible property, including loss of use resulting therefrom; and
.6 claims for damages because of bodily injury or death of any person or property damage arising out of the ownership, maintenance or use of any motor vehicle.

11.1.2 The insurance required by Subparagraph 11.1.1 shall be written for not less than any limits of liability specified in the Contract Documents, or required by law, whichever is greater. ~~, and~~

11.1.3 The insurance required by Subparagraph 11.1.1 shall include contractual liability insurance ~~as~~ applicable to the Contractor's obligations under Paragraph 4.18.

11.1.4 Certificates of Insurance acceptable to the Owner shall be filed with the Owner prior to commencement of the Work. These Certificates shall contain a provision that coverages afforded under the policies will not be cancelled until at least ~~fifteen~~ thirty days' prior written notice has been given to the Owner.

11.2 OWNER'S LIABILITY INSURANCE

11.2.1 The Owner shall be responsible for purchasing and maintaining his own liability insurance and, at his option, may purchase and maintain such insurance as will protect him against claims which may arise from operations under the Contract.

11.3 PROPERTY INSURANCE

11.3.1 Unless otherwise provided, the Owner shall purchase and maintain property insurance upon the entire Work at the site to the full insurable value thereof. This insurance shall include the interests of the Owner, the Contractor, Subcontractors and Sub-subcontractors in the Work and shall insure against the perils of fire and extended coverage and shall include "all risk" insurance for physical loss or damage including, without duplication of coverage, theft, vandalism, and malicious mischief. If the

1970 Edition

New 11.3.1 replaces last portion of former 11.3.4 →

11.3.2 The Owner shall purchase and maintain such steam boiler and machinery insurance as may be required by the Contract Documents or by law. This insurance shall include the interests of the Owner, the Contractor, Subcontractors and Sub-subcontractors in the Work.

11.3.3 Any insured loss is to be adjusted with the Owner and made payable to the Owner as trustee for the insureds, as their interests may appear, subject to the requirements of any applicable mortgagee clause and of Subparagraph 11.3.8.

New 11.3.3 includes former 5.4.3 →

11.3.4 The Owner shall file a copy of all policies witn the Contractor before an exposure to loss may occur. If the Owner does not intend to purchase such insurance, he shall inform the Contractor in writing prior to commencement of the Work. The Contractor may then effect insurance which will protect the interests of himself, his Subcontractors and the Sub-subcontractors in the Work, and by appropriate Change Order the cost thereof shall be charged to the Owner. If the Contractor is damaged by failure of the Owner to purchase or maintain such insurance and so to notify the Contractor, then the Owner shall bear all reasonable costs properly attributable thereto.

11.3.5 If the Contractor requests in writing that insurance for special hazards be included in the property insurance policy, the Owner shall, if possible, include such insurance, and the cost thereof shall be charged to the Contractor by appropriate Change Order.

1976 Edition

Owner does not intend to purchase such insurance for the full insurable value of the entire Work, he shall inform the Contractor in writing prior to commencement of the Work. The Contractor may then effect insurance which will protect the interests of himself, his Subcontractors and the Sub-subcontractors in the Work, and by appropriate Change Order the cost thereof shall be charged to the Owner. If the Contractor is damaged by failure of the Owner to purchase or maintain such insurance and to so notify the Contractor, then the Owner shall bear all reasonable costs properly attributable thereto. If not covered under the all risk insurance or otherwise provided in the Contract Documents, the Contractor shall effect and maintain similar property insurance on portions of the Work stored off the site or in transit when such portions of the Work are to be included in an Application for Payment under Subparagraph 9.3.2.

11.3.2 The Owner shall purchase and maintain such ~~steam~~ boiler and machinery insurance as may be required by the Contract Documents or by law. This insurance shall include the interests of the Owner, the Contractor, Subcontractors and Sub-subcontractors in the Work.

11.3.3 Any ~~insured~~ loss insured under Subparagraph 11.3.1 is to be adjusted with the Owner and made payable to the Owner as trustee for the insureds, as their interests may appear, subject to the requirements of any applicable mortagagee clause and of Subparagraph 11.3.8. The Contractor shall pay each Subcontractor a just share of any insurance moneys received by the Contractor, ~~under Article 11~~ and by appropriate agreement, written where legally required for validity, ~~he~~ shall require each Subcontractor to make ~~similar~~ payments to his Sub-subcontractors in similar manner.

11.3.4 The Owner shall file a copy of all policies with the Contractor before an exposure to loss may occur.

← Last portion of former 11.3.4 included in new 11.3.1

11.3.5 If the Contractor requests in writing that insurance for risks other than those described in Subparagraphs 11.3.1 and 11.3.2 or other special hazards be included in the property insurance policy, the Owner shall, if possible, include such insurance, and the cost thereof shall be charged to the Contractor by appropriate Change Order.

1970 Edition	1976 Edition
11.3.6 The Owner and Contractor waive all rights against each other for damages caused by fire or other perils to the extent covered by insurance provided under this Paragraph 11.3, except such rights as they may have to the proceeds of such insurance held by the Owner as trustee. The Contractor shall require similar waivers by Subcontractors and Sub-subcontractors in accordance with Clause 5.3.1.5.	11.3.6 The Owner and Contractor waive all rights against 1) each other and the subcontractors, sub-subcontractors, agents and employees each of the other, and 2) the Architect and Separate Contractors, if any, and their subcontractors, sub-subcontractors, agents and employees, for damages caused by fire or other perils to the extent covered by insurance obtained pursuant to this Paragraph 11.3 or any other property insurance applicable to the Work, except such rights as they may have to the proceeds of such insurance held by the Owner as trustee. The foregoing waiver afforded the Architect, his agents and employees shall not extend to the liability imposed by Subparagraph 4.18.3. The Owner or the Contractor, as appropriate, shall require of the Architect, Separate Contractors, subcontractors and sub-subcontractors by appropriate agreements, written where legally required for validity, similar waivers each in favor of all other parties enumerated in this Subparagraph 11.3.6.
11.3.7 If required in writing by any party in interest, the Owner as trustee shall, upon the occurrence of an insured loss, give bond for the proper performance of his duties. He shall deposit in a separate account any money so received, and he shall distribute it in accordance with such agreement as the parties in interest may reach, or in accordance with an award by arbitration in which case the procedure shall be as provided in Paragraph 7.10. If after such loss no other special agreement is made, replacement of damaged work shall be covered by an appropriate Change Order.	11.3.7 If required in writing by any party in interest, the Owner as trustee shall, upon the occurrence of an insured loss, give bond for the proper performance of his duties. He shall deposit in a separate account any money so received, and he shall distribute it in accordance with such agreement as the parties in interest may reach, or in accordance with an award by arbitration in which case the procedure shall be as provided in Paragraph ~~7.10~~ 7.9. If after such loss no other special agreement is made, replacement of damaged work shall be covered by an appropriate Change Order.
11.3.8 The Owner as trustee shall have power to adjust and settle any loss with the insurers unless one of the parties in interest shall object in writing within five days after the occurrence of loss to the Owner's exercise of this power, and if such objection be made, arbitrators shall be chosen as provided in Paragraph 7.10. The Owner as trustee shall, in that case, make settlement with the insurers in accordance with the directions of such arbitrators. If distribution of the insurance proceeds by arbitration is required, the arbitrators will direct such distribution.	11.3.8 The Owner as trustee shall have power to adjust and settle any loss with the insurers unless one of the parties in interest shall object in writing within five days after the occurrence of loss to the Owner's exercise of this power, and if such objection be made, arbitrators shall be chosen as provided in Paragraph ~~7.10~~ 7.9. The Owner as trustee shall, in that case, make settlement with the insurers in accordance with the directions of such arbitrators. If distribution of the insurance proceeds by arbitration is required, the arbitrators will direct such distribution.
New Subparagraph →	11.3.9 If the Owner finds it necessary to occupy or use a portion or portions of the Work prior to Substantial Completion thereof, such occupancy shall not commence prior to a time mutually agreed to by the Owner and Contractor and to which the insurance company or companies providing the property insur-

1970 Edition

11.4 LOSS OF USE INSURANCE

11.4.1 The Owner, at his option, may purchase and maintain such insurance as will insure him against loss of use of his property due to fire or other hazards, however caused.

ARTICLE 12

CHANGES IN THE WORK

12.1 CHANGE ORDERS

New 12.1.1 replaces former 12.1.2 →

12.1.1 The Owner, without invalidating the Contract, may order Changes in the Work within the general scope of the Contract consisting of additions, deletions or other revisions, the Contract Sum and the Contract Time being adjusted accordingly. All such Changes in the Work shall be authorized by Change Order, and shall be executed under the applicable conditions of the Contract Documents.

12.1.2 A Change Order is a written order to the Contractor signed by the Owner and the Architect, issued after the execution of the Contract, authorizing a Change in the Work or an adjustment in the Contract Sum or the Contract Time. Alternatively, the Change Order may be signed by the Architect alone, provided he has written authority from the Owner for such procedure and that a copy of such written authority is furnished to the Contractor upon request. A Change Order may also be signed by the Contractor if he agrees to the adjustment in the Contract Sum or the Contract Time. The Contract Sum and the Contract Time may be changed only by Change Order.

1976 Edition

ance have consented by endorsement to the policy or policies. This insurance shall not be cancelled or lapsed on account of such partial occupancy. Consent of the Contractor and of the insurance company or companies to such occupancy or use shall not be unreasonably withheld.

11.4 LOSS OF USE INSURANCE

11.4.1 The Owner, at his option, may purchase and maintain such insurance as will insure him against loss of use of his property due to fire or other hazards, however caused. The Owner waives all rights of action against the Contractor for loss of use of his property, including consequential losses due to fire or other hazards however caused, to the extent covered by insurance under this Paragraph 11.4.

ARTICLE 12

CHANGES IN THE WORK

12.1 CHANGE ORDERS

12.1.1 A Change Order is a written order to the Contractor signed by the Owner and the Architect, issued after execution of the Contract, authorizing a change in the Work or an adjustment in the Contract Sum or the Contract Time. ~~Alternatively, the Change Order may be signed by the Architect alone, provided he has written authority from the Owner for such procedure and that a copy of such written authority is furnished to the Contractor upon request.~~ The Contract Sum and the Contract Time may be changed only by Change Order. A Change Order ~~may also be~~ signed by the Contractor ~~if he agrees to~~ indicates his agreement therewith, including the adjustment in the Contract Sum or the Contract Time. ~~The Contract Sum and the Contract Time may be changed only by Change Order.~~

12.1.2 The Owner, without invalidating the Contract, may order changes in the Work within the general scope of the Contract consisting of additions, deletions or other revisions, the Contract Sum and the Contract Time being adjusted accordingly. All such changes in the Work shall be authorized by Change Order, and shall be ~~executed~~ performed under the applicable conditions of the Contract Documents.

← Former 12.1.2 becomes new 12.1.1

1970 Edition

12.1.3 The cost or credit to the Owner resulting from a Change in the Work shall be determined in one or more of the following ways:

- **.1** by mutual acceptance of a lump sum properly itemized;
- **.2** by unit prices stated in the Contract Documents or subsequently agreed upon; or
- **.3** by cost and a mutually acceptable fixed or percentage fee.

New Clause →

12.1.4 If none of the methods set forth in Subparagraph 12.1.3 is agreed upon, the Contractor, provided he receives a Change Order, shall promptly proceed with the Work involved. The cost of such Work shall then be determined by the Architect on the basis of the Contractor's reasonable expenditures and savings, including, in the case of an increase in the Contract Sum, a reasonable allowance for overhead and profit. In such case, and also under Clause 12.1.3.3 above, the Contractor shall keep and present, in such form as the Architect may prescribe, an itemized accounting together with appropriate supporting data. Pending final determination of cost to the Owner, payments on account shall be made on the Architect's Certificate for Payment. The amount of credit to be allowed by the Contractor to the Owner for any deletion or change which results in a net decrease in cost will be the amount of the actual net decrease as confirmed by the Architect. When both additions and credits are involved in any one change, the allowance for overhead and profit shall be figured on the basis of net increase, if any.

12.1.5 If unit prices are stated in the Contract Documents or subsequently agreed upon, and if the quantities originally contemplated are so changed in a proposed Change Order that application of the agreed unit prices to the quantities of Work proposed will create a hardship on the Owner or the Contractor, the applicable unit prices shall be equitably adjusted to prevent such hardship.

1976 Edition

12.1.3 The cost or credit to the Owner resulting from a change in the Work shall be determined in one or more of the following ways:

- .1 by mutual acceptance of a lump sum properly itemized and supported by sufficient substantiating data to permit evaluation;
- .2 by unit prices stated in the Contract Documents or subsequently agreed upon;
- .3 by cost to be determined in a manner agreed upon by the parties and a mutually acceptable fixed or percentage fee ; or
- .4 by the method provided in Subparagraph 12.1.4.

12.1.4 If none of the methods set forth in ~~Subparagraph 12.1.3~~ Clauses 12.1.3.1, 12.1.3.2 or 12.1.3.3 is agreed upon, the Contractor, provided he receives a ~~Change Order~~ written order signed by the Owner, shall promptly proceed with the Work involved. The cost of such Work shall then be determined by the Architect on the basis of the ~~Contractor's~~ reasonable expenditures and savings of those performing the Work attributable to the change, including, in the case of an increase in the Contract Sum, a reasonable allowance for overhead and profit. In such case, and also under ~~Clause~~ Clauses 12.1.3.3 and 12.1.3.4 above, the Contractor shall keep and present, in such form as the Architect may prescribe, an itemized accounting together with appropriate supporting data for inclusion in a Change Order. Unless otherwise provided in the Contract Documents, cost shall be limited to the following: cost of materials, including sales tax and cost of delivery; cost of labor, including social security, old age and unemployment insurance, and fringe benefits required by agreement or custom; workers' or workmen's compensation insurance; bond premiums; rental value of equipment and machinery; and the additional costs of supervision and field office personnel directly attributable to the change. Pending final determination of cost to the Owner, payments on account shall be made on the Architect's Certificate for Payment. The amount of credit to be allowed by the Contractor to the Owner for any deletion or change which results in a net decrease in ~~cost~~ the Contract Sum will be the amount of the actual net ~~decrease~~ cost as confirmed by the Architect. When both additions and credits covering related Work or substitutions are involved in any one change , the allowance for overhead and profit shall be figured on the basis of the net increase, if any , with respect to that change.

12.1.5 If unit prices are stated in the Contract Documents or subsequently agreed upon, and if the quantities originally contemplated are so changed in a proposed Change Order that application of the agreed unit prices to the quantities of Work proposed will ~~create a hardship on~~ cause substantial inequity to the Owner or the Contractor, the applicable unit prices shall be equitably adjusted. ~~to prevent such hardship.~~

1970 Edition

12.1.6 Should concealed conditions encountered in the performance of the Work below the surface of the ground be at variance with the conditions indicated by the Contract Documents or should unknown physical conditions below the surface of the ground of an unusual nature, differing materially from those ordinarily encountered and generally recognized as inherent in work of the character provided for in this Contract, be encountered, the Contract Sum shall be equitably adjusted by Change Order upon claim by either party made within twenty days after the first observance of the conditions.

12.1.7 If the Contractor claims that additional cost is involved because of (1) any written interpretation issued pursuant to Subparagraph 1.2.5, (2) any order by the Owner to stop the Work pursuant to Paragraph 3.3 where the Contractor was not at fault, or (3) any written order for a minor change in the Work issued pursuant to Paragraph 12.3, the Contractor shall make such claim as provided in Paragraph 12.2.

12.2 CLAIMS FOR ADDITIONAL COST

12.2.1 If the Contractor wishes to make a claim for an increase in the Contract Sum, he shall give the Architect written notice thereof within twenty days after the occurrence of the event giving rise to such claim. This notice shall be given by the Contractor before proceeding to execute the Work, except in an emergency endangering life or property in which case the Contractor shall proceed in accordance with Subparagraph 10.3.1. No such claim shall be valid unless so made. If the Owner and the Contractor cannot agree on the amount of the adjustment in the Contract Sum, it shall be determined by the Architect. Any change in the Contract Sum resulting from such claim shall be authorized by Change Order.

New 12.3.2 replaces former 12.1.7 →

1976 Edition

12.2 CONCEALED CONDITIONS

12.2.1 Should concealed conditions encountered in the performance of the Work below the surface of the ground or should concealed or unknown conditions in an existing structure be at variance with the conditions indicated by the Contract Documents, or should unknown physical conditions below the surface of the ground or should concealed or unknown conditions in an existing structure of an unusual nature, differing materially from those ordinarily encountered and generally recognized as inherent in work of the character provided for in this Contract, be encountered, the Contract Sum shall be equitably adjusted by Change Order upon claim by either party made within twenty days after the first observance of the conditions.

← Former 12.1.7 becomes new 12.3.2

12.3 CLAIMS FOR ADDITIONAL COST

12.3.1 If the Contractor wishes to make a claim for an increase in the Contract Sum, he shall give the Architect written notice thereof within twenty days after the occurrence of the event giving rise to such claim. This notice shall be given by the Contractor before proceeding to execute the Work, except in an emergency endangering life or property in which case the Contractor shall proceed in accordance with ~~Subparagraph 10.3.1~~ Paragraph 10.3. No such claim shall be valid unless so made. If the Owner and the Contractor cannot agree on the amount of the adjustment in the Contract Sum, it shall be determined by the Architect. Any change in the Contract Sum resulting from such claim shall be authorized by Change Order.

12.3.2 If the Contractor claims that additional cost is involved because of , but not limited to, (1) any written interpretation pursuant to Subparagraph ~~1.2.5~~ 2.2.8, (2) any order by the Owner to stop the Work pursuant to Paragraph 3.3 where the Contractor was not at fault, ~~or~~ (3) any written order for a minor change in the Work issued pursuant to Paragraph ~~12.3~~ 12.4, or (4) failure of payment by the Owner pursuant to Paragraph 9.7, the Contractor shall make such claim as provided in ~~Paragraph 12.2~~ Subparagraph 12.3.1.

1970 Edition

12.3 MINOR CHANGES IN THE WORK

12.3.1 The Architect shall have authority to order minor changes in the Work not involving an adjustment in the Contract Sum or an extension of the Contract Time and not inconsistent with the intent of the Contract Documents. Such changes may be effected by Field Order or by other written order. Such changes shall be binding on the Owner and the Contractor.

12.4 FIELD ORDERS

12.4.1 The Architect may issue written Field Orders which interpret the Contract Documents in accordance with Subparagraph 1.2.5 or which order minor changes in the Work in accordance with Paragraph 12.3 without change in Contract Sum or Contract Time. The Contractor shall carry out such Field Orders promptly.

ARTICLE 13

UNCOVERING AND CORRECTION OF WORK

13.1 UNCOVERING OF WORK

13.1.1 If any Work should be covered contrary to the request of the Architect, it must, if required by the Architect, be uncovered for his observation and replaced, at the Contractor's expense.

13.1.2 If any other Work has been covered which the Architect has not specifically requested to observe prior to being covered, the Architect may request to see such Work and it shall be uncovered by the Contractor. If such Work be found in accordance with the Contract Documents, the cost of uncovering and replacement shall, by appropriate Change Order, be charged to the Owner. If such Work be found not in accordance with the Contract Documents, the Contractor shall pay such costs unless it be found that this condition was caused by a separate contractor employed as provided in Article 6, and in that event the Owner shall be responsible for the payment of such costs.

13.2 CORRECTION OF WORK

13.2.1 The Contractor shall promptly correct all Work rejected by the Architect as defective or as failing to conform to the Contract Documents whether observed before or after Substantial Completion and whether or not fabricated, installed or completed. The Contractor shall bear all cost of correcting such rejected Work, including the cost of the Architect's additional services thereby made necessary.

1976 Edition

12.4 MINOR CHANGES IN THE WORK

12.4.1 The Architect ~~shall~~ will have authority to order minor changes in the Work not involving an adjustment in the Contract Sum or an extension of the Contract Time and not inconsistent with the intent of the Contract Documents. Such changes ~~may~~ shall be effected by ~~Field Order or by other~~ written order , and ~~Such changes~~ shall be binding on the Owner and the Contractor. The Contractor shall carry out such written orders promptly.

←

Former 12.4.1 Deleted

ARTICLE 13

UNCOVERING AND CORRECTION OF WORK

13.1 UNCOVERING OF WORK

13.1.1 If any portion of the Work should be covered contrary to the request of the Architect or to requirements specifically expressed in the Contract Documents, it must, if required in writing by the Architect, be uncovered for his observation and shall be replaced at the Contractor's expense.

13.1.2 If any other portion of the Work has been covered which the Architect has not specifically requested to observe prior to being covered, the Architect may request to see such Work and it shall be uncovered by the Contractor. If such Work be found in accordance with the Contract Documents, the cost of uncovering and replacement shall, by appropriate Change Order, be charged to the Owner. If such Work be found not in accordance with the Contract Documents, the Contractor shall pay such costs unless it be found that this condition was caused by the Owner or a separate contractor ~~employed~~ as provided in Article 6, ~~and~~ in ~~that~~ which event the Owner shall be responsible for the payment of such costs.

13.2 CORRECTION OF WORK

13.2.1 The Contractor shall promptly correct all Work rejected by the Architect as defective or as failing to conform to the Contract Documents whether observed before or after Substantial Completion and whether or not fabricated, installed or completed. The Contractor shall bear all ~~cost~~ costs of correcting such rejected Work, including ~~the cost of~~ compensation for the Architect's additional services ~~thereby~~ made necessary thereby.

1970 Edition

13.2.2 If, within one year after the Date of Substantial Completion or within such longer period of time as may be prescribed by law or by the terms of any applicable special guarantee required by the Contract Documents, any of the Work is found to be defective or not in accordance with the Contract Documents, the Contractor shall correct it promptly after receipt of a written notice from the Owner to do so unless the Owner has previously given the Contractor a written acceptance of such condition. The Owner shall give such notice promptly after discovery of the condition.

13.2.3 All such defective or non-conforming Work under Subparagraphs 13.2.1 and 13.2.2 shall be removed from the site if necessary, and the Work shall be corrected to comply with the Contract Documents without cost to the Owner.

13.2.4 The Contractor shall bear the cost of making good all work of separate contractors destroyed or damaged by such removal or correction.

New 13.2.4 replaces former 13.2.6

13.2.5 If the Contractor does not remove such defective or non-conforming Work within a reasonable time fixed by written notice from the Architect, the Owner may remove it and may store the materials or equipment at the expense of the Contractor. If the Contractor does not pay the cost of such removal and storage within ten days thereafter, the Owner may upon ten additional days' written notice sell such Work at auction or at private sale and shall account for the net proceeds thereof, after deducting all the costs that should have been borne by the Contractor including compensation for additional architectural services. If such proceeds of sale do not cover all costs which the Contractor should have borne, the difference shall be charged to the Contractor and an appropriate Change Order shall be issued. If the payments then or thereafter due the Contractor are not sufficient to cover such amount, the Contractor shall pay the difference to the Owner.

13.2.6 If the Contractor fails to correct such defective or non-conforming Work, the Owner may correct it in accordance with Paragraph 3.4.

1976 Edition

13.2.2 If, within one year after the Date of Substantial Completion of the Work or designated portion thereof or within one year after acceptance by the Owner of designated equipment or within such longer period of time as may be prescribed by law or by the terms of any applicable special ~~guarantee~~ warranty required by the Contract Documents, any of the Work is found to be defective or not in accordance with the Contract Documents, the Contractor shall correct it promptly after receipt of a written notice from the Owner to do so unless the Owner has previously given the Contractor a written acceptance of such condition. This obligation shall survive termination of the Contract. The Owner shall give such notice promptly after discovery of the condition.

13.2.3 The Contractor shall remove from the site all portions of the ~~such defective or non-conforming~~ Work which are defective or non-conforming and which have not been corrected under Subparagraphs 4.5.1, 13.2.1 and 13.2.2, unless removal is waived by the Owner. ~~shall be removed from the site if necessary, and the Work shall be corrected to comply with the Contract Documents without cost to the Owner.~~

Former 13.2.4 becomes new 13.2.6

13.2.4 If the Contractor fails to correct ~~such~~ defective or non-conforming Work as provided in Subparagraphs 4.5.1, 13.2.1 and 13.2.2, the Owner may correct it in accordance with Paragraph 3.4.

13.2.5 If the Contractor does not ~~remove~~ proceed with the correction of such defective or non-conforming Work within a reasonable time fixed by written notice from the Architect, the Owner may remove it and may store the materials or equipment at the expense of the Contractor. If the Contractor does not pay the cost of such removal and storage within ten days thereafter, the Owner may upon ten additional days' written notice sell such Work at auction or at private sale and shall account for the net proceeds thereof, after deducting all the costs that should have been borne by the Contractor, including compensation for the Architect's additional ~~architectural~~ services made necessary thereby. If such proceeds of sale do not cover all costs which the Contractor should have borne, the difference shall be charged to the Contractor and an appropriate Change Order shall be issued. If the payments then or thereafter due the Contractor are not sufficient to cover such amount, the Contractor shall pay the difference to the Owner.

Former 13.2.6 becomes new 13.2.4

1970 Edition

New 13.2.6 replaces former 13.214 →

1976 Edition

13.2.6 The Contractor shall bear the cost of making good all work of the Owner or separate contractors destroyed or damaged by such ~~removal or~~ correction or removal.

New Subparagraph →

13.2.7 Nothing contained in this Paragraph 13.2 shall be construed to establish a period of limitation with respect to any other obligation which the Contractor might have under the Contract Documents, including Paragraph 4.5 hereof. The establishment of the time period of one year after the Date of Substantial Completion or such longer period of time as may be prescribed by law or by the terms of any warranty required by the Contract Documents relates only to the specific obligation of the Contractor to correct the Work, and has no relationship to the time within which his obligation to comply with the Contract Documents may be sought to be enforced, nor to the time within which proceedings may be commenced to establish the Contractor's liability with respect to his obligations other than specifically to correct the Work.

1970 Edition:

13.3 ACCEPTANCE OF DEFECTIVE OR NON-CONFORMING WORK

13.3.1 If the Owner prefers to accept defective or non-conforming Work, he may do so instead of requiring its removal and correction, in which case a Change Order will be issued to reflect an appropriate reduction in the Contract Sum, or, if the amount is determined after final payment, it shall be paid by the Contractor.

1976 Edition:

13.3 ACCEPTANCE OF DEFECTIVE OR NON-CONFORMING WORK

13.3.1 If the Owner prefers to accept defective or non-conforming Work, he may do so instead of requiring its removal and correction, in which case a Change Order will be issued to reflect ~~an appropriate~~ a reduction in the Contract Sum where appropriate and equitable. ~~or if the amount is determined after final payment, it shall be paid by the Contractor.~~ Such adjustment shall be effected whether or not final payment has been made.

1970 Edition:

ARTICLE 14

TERMINATION OF THE CONTRACT

14.1 TERMINATION BY THE CONTRACTOR

14.1.1 If the Work is stopped for a period of thirty days under an order of any court or other public authority having jurisdiction, or as a result of an act of government, such as a declaration of a national emergency making materials unavailable, through no act or fault of the Contractor or a Subcontractor or their agents or employees or any other persons performing any of the Work under a contract with the Contractor, or if the Work should be stopped for a period of thirty days by the Contractor for the Architect's failure to issue a Certificate for Payment as provided in Paragraph 9.6 or for the Owner's failure to make payment thereon as provided in Paragraph 9.6, then the Contractor may, upon seven days' written notice to the Owner and the Architect, terminate the Contract and recover from the Owner payment for all Work executed and for any proven loss sustained upon any materials, equipment, tools, construction equipment and machinery, including reasonable profit and damages.

1976 Edition:

ARTICLE 14

TERMINATION OF THE CONTRACT

14.1 TERMINATION BY THE CONTRACTOR

14.1.1 If the Work is stopped for a period of thirty days under an order of any court or other public authority having jurisdiction, or as a result of an act of government, such as a declaration of a national emergency making materials unavailable, through no act or fault of the Contractor or a Subcontractor or their agents or employees or any other persons performing any of the Work under a contract with the Contractor, or if the Work should be stopped for a period of thirty days by the Contractor ~~for~~ because the ~~Architect's failure to issue~~ Architect has not issued a Certificate for Payment as provided in Paragraph ~~9.6~~ 9.7 or ~~for~~ because the ~~Owner's failure to make~~ Owner has not made payment thereon as provided in Paragraph ~~9.6;~~ 9.7, then the Contractor may, upon seven additional days' written notice to the Owner and the Architect, terminate the Contract and recover from the Owner payment for all Work executed and for any proven loss sustained upon any materials, equipment, tools, construction equipment and machinery, including reasonable profit and damages.

1970 Edition

14.2 TERMINATION BY THE OWNER

14.2.1 If the Contractor is adjudged a bankrupt, or if he makes a general assignment for the benefit of his creditors, or if a receiver is appointed on account of his insolvency, or if he persistently or repeatedly refuses or fails, except in cases for which extension of time is provided, to supply enough properly skilled workmen or proper materials, or if he fails to make prompt payment to Subcontractors or for materials or labor, or persistently disregards laws, ordinances, rules, regulations or orders of any public authority having jurisdiction, or otherwise is guilty of a substantial violation of a provision of the Contract Documents, then the Owner, upon certification by the Architect that sufficient cause exists to justify such action, may, without prejudice to any right or remedy and after giving the Contractor and his surety, if any, seven days' written notice, terminate the employment of the Contractor and take possession of the site and of all materials, equipment, tools, construction equipment and machinery thereon owned by the Contractor and may finish the Work by whatever method he may deem expedient. In such case the Contractor shall not be entitled to receive any further payment until the Work is finished.

14.2.2 If the unpaid balance of the Contract Sum exceeds the costs of finishing the Work, including compensation for the Architect's additional services, such excess shall be paid to the Contractor. If such costs exceed such unpaid balance, the Contractor shall pay the difference to the Owner. The costs incurred by the Owner as herein provided shall be certified by the Architect.

1976 Edition

14.2 TERMINATION BY THE OWNER

14.2.1 If the Contractor is adjudged a bankrupt, or if he makes a general assignment for the benefit of his creditors, or if a receiver is appointed on account of his insolvency, or if he persistently or repeatedly refuses or fails, except in cases for which extension of time is provided, to supply enough properly skilled workmen or proper materials, or if he fails to make prompt payment to Subcontractors or for materials or labor, or persistently disregards laws, ordinances, rules, regulations or orders of any public authority having jurisdiction, or otherwise is guilty of a substantial violation of a provision of the Contract Documents, then the Owner, upon certification by the Architect that sufficient cause exists to justify such action, may, without prejudice to any right or remedy and after giving the Contractor and his surety, if any, seven days' written notice, terminate the employment of the Contractor and take possession of the site and of all materials, equipment, tools, construction equipment and machinery thereon owned by the Contractor and may finish the Work by whatever method he may deem expedient. In such case the Contractor shall not be entitled to receive any further payment until the Work is finished.

14.2.2 If the unpaid balance of the Contract Sum exceeds the costs of finishing the Work, including compensation for the Architect's additional services made necessary thereby, such excess shall be paid to the Contractor. If such costs exceed the unpaid balance, the Contractor shall pay the difference to the Owner. ~~The costs incurred by~~ The amount to be paid to the Contractor or to the Owner, as ~~herein provided~~ the case may be, shall be certified by the Architect, upon application, in the manner provided in Paragraph 9.4, and this obligation for payment shall survive the termination of the Contract.

THE AMERICAN INSTITUTE OF ARCHITECTS

AIA Document C141

Standard Form of Agreement Between Architect and Engineer

1979 EDITION

Recommended for use with the current editions of standard AIA agreement forms and documents.

THIS DOCUMENT HAS IMPORTANT LEGAL CONSEQUENCES; CONSULTATION WITH AN ATTORNEY IS ENCOURAGED WITH RESPECT TO ITS COMPLETION OR MODIFICATION

AGREEMENT

made as of the day of in the year of Nineteen Hundred and

BETWEEN the Architect:

and the Engineer:

The Architect has made an agreement dated with the Owner:

For the following Project:

which agreement is hereinafter referred to as the Prime Agreement and which provides for furnishing professional services in connection with the Project described therein. A copy of the Prime Agreement (from which compensation amounts may be deleted) is attached, made a part hereof and marked Exhibit A.

The Architect and the Engineer agree as set forth below.

I. THE ENGINEER shall provide for the Architect, in accordance with the Terms and Conditions of this Agreement, the following professional services which the Architect is required to provide for the Owner under the attached copy of the Prime Agreement:

(Describe the type of engineering services.)

The part of the Project for which the Engineer is to provide such services is hereinafter called This Part of the Project.

The Engineer is the Architect's independent professional consultant for This Part of the Project, responsible for methods and means used in performing the Engineer's services under this Agreement, and is not a joint venturer with the Architect.

II. THE ARCHITECT shall be the general administrator of the professional services for the Project, and shall facilitate the exchange of information among the engineers and consultants retained by the Architect for the Project as necessary for the coordination of their work.

Except as authorized by the Architect, all communication between the Engineer and the Owner or the Contractor shall be forwarded through the Architect.

TERMS AND CONDITIONS OF AGREEMENT BETWEEN ARCHITECT AND ENGINEER

ARTICLE 1
ENGINEER'S SERVICES AND RESPONSIBILITIES

1.1 GENERAL

1.1.1 The Engineer shall perform the services undertaken according to this Agreement with the Architect in the same manner and to the same extent that the Architect is bound by the attached Prime Agreement to perform such services for the Owner. Except as set forth herein, the Engineer shall not have any duties or responsibilities for any other part of the Project.

1.1.2 The Engineer shall designate, when necessary, a representative authorized to act in the Engineer's behalf with respect to This Part of the Project.

1.1.3 The Engineer's services shall be performed in character, sequence and timing so that they will be coordinated with those of the Architect and all other consultants for the Project.

1.1.4 The Engineer shall recommend to the Architect the obtaining of such investigations, surveys, tests, analyses and reports as may be necessary for the proper execution of the Engineer's services.

1.1.5 The Engineer shall provide progress copies of drawings, reports, specifications and other necessary information to the Architect and other consultants. All aspects of the Work designed by the Engineer shall be coordinated by the Engineer, and the Engineer shall also become familiar with the Work designed by the Architect and other consultants as necessary for the proper coordination of the Project.

1.1.6 The Engineer shall cooperate with the Architect in determining the proper share of the construction budget to be allocated to This Part of the Project.

1.1.7 The Engineer shall not be responsible for the acts or omissions of the Architect, the Architect's other consultants, the Contractor, any Subcontractors, any of their agents or employees, or any other persons performing any of the Work.

1.2 BASIC SERVICES

The Engineer's Basic Services consist of the five Phases described below and any other services included in Article 17 as Basic Services.

SCHEMATIC DESIGN PHASE

1.2.1 The Engineer shall ascertain the requirements for This Part of the Project and shall confirm such requirements to the Architect.

1.2.2 The Engineer shall review alternative systems with the Architect, attend necessary conferences, prepare necessary analyses, drawings and other documents, be available for general consultation, and make recommendations regarding basic systems for This Part of the Project. When necessary, the Engineer shall consult with public agencies and other organizations concerning utility services and requirements.

1.2.3 The Engineer shall prepare and submit to the Architect a Statement of Probable Construction Cost of This Part of the Project based on current area, volume or other unit costs, as directed by the Architect.

DESIGN DEVELOPMENT PHASE

1.2.4. When authorized by the Architect, the Engineer shall prepare from the Schematic Design Studies approved by the Owner and confirmed by the Architect the Design Development Documents. These shall consist of drawings and other documents to fix and describe This Part of the Project, including materials, equipment, component systems and types of construction as may be appropriate, all of which are to be approved by the Owner and the Architect.

1.2.5 The Engineer shall submit to the Architect a further Statement of Probable Construction Cost of This Part of the Project, as directed by the Architect.

CONSTRUCTION DOCUMENTS PHASE

1.2.6 When authorized by the Architect, the Engineer shall prepare from the Design Development Documents approved by the Owner and confirmed by the Architect Drawings and Specifications setting forth in detail the requirements for the construction of This Part of the Project, all of which are to be approved by the Owner and the Architect. The Engineer shall prepare the Drawings and Specifications in such format as the Architect may reasonably require.

1.2.7 The Engineer shall advise the Architect of any adjustments to previous Statements of Probable Construction Cost of This Part of the Project indicated by changes in requirements or general market conditions.

1.2.8 The Engineer shall assist the Architect as necessary in connection with the Owner's responsibility for filing the documents concerning This Part of the Project required for the approval of governmental authorities having jurisdiction over the Project.

BIDDING OR NEGOTIATION PHASE

1.2.9 If required by the Architect, the Engineer shall assist the Architect and the Owner in obtaining and evaluating bids or negotiated proposals, and in awarding and preparing contracts for construction.

CONSTRUCTION PHASE—ADMINISTRATION OF THE CONSTRUCTION CONTRACT

1.2.10 The Construction Phase will commence with the award of the Contract for Construction and, together with the Engineer's obligation to provide Basic Services under this Agreement, will terminate when final payment to the Contractor is due or, in the absence of a final Certificate for Payment or of such due date, sixty days after the date of Substantial Completion of the Work, whichever occurs first.

1.2.11 Unless otherwise provided in this Agreement and incorporated in the Contract Documents, the Engineer shall assist the Architect in the Administration of the Contract for Construction with respect to This Part of the Project, as set forth below and in the edition of AIA Document A201, General Conditions of the Contract for Construction, identified in the Prime Agreement.

1.2.12 The Engineer shall visit the site at intervals appropriate to the stage of construction for This Part of the

Project, or as otherwise agreed with the Architect in writing, to become generally familiar with the progress and quality of the Work for This Part of the Project and to determine in general if such Work is proceeding in accordance with the Contract Documents. However, the Engineer shall not be required to make exhaustive or continuous on-site inspections to check the quality or quantity of the Work for This Part of the Project. On the basis of such on-site observations as an engineer, the Engineer shall keep the Architect informed of the progress and quality of the Work for This Part of the Project and shall endeavor to guard the Owner against defects and deficiencies in such Work of the Contractor.

1.2.13 The Engineer shall not have control or charge of, and shall not be responsible for, construction means, methods, techniques, sequences or procedures, for safety precautions and programs in connection with the Work, for the acts or omissions of the Contractor, Subcontractors or any other persons performing any of the Work, or for the failure of any of them to carry out the Work in accordance with the Contract Documents.

1.2.14 The Engineer shall at all times have access to the Work for This Part of the Project wherever it is in preparation or progress.

1.2.15 The Engineer, based on observations at the site and on evaluations of the Contractor's Applications for Payment, shall assist the Architect in determining the amounts owing to the Contractor for This Part of the Project and shall certify such amounts to the Architect. Such certification shall be in writing if requested.

1.2.16 Certification by the Engineer to the Architect of an amount owing to the Contractor shall constitute a representation by the Engineer to the Architect that, based on the Engineer's observations at the site as provided in Subparagraph 1.2.12 and the data comprising the Contractor's Application for Payment, the Work for This Part of the Project has progressed to the point indicated; that to the best of the Engineer's knowledge, information and belief, the quality of such Work is in accordance with the Contract Documents (subject to an evaluation of such Work for conformance with the Contract Documents upon Substantial Completion, to the results of any subsequent tests required by or performed under the Contract Documents, to minor deviations from the Contract Documents correctable prior to completion, and to any specific qualifications stated by the Engineer); and that the Contractor is entitled to payment in the amount certified.

1.2.17 Upon written request of the Architect, the Engineer shall furnish to the Architect, with reasonable promptness, written interpretations of the Contract Documents prepared by the Engineer, if, in the opinion of the Architect, such interpretations are necessary for the proper execution or progress of the Work.

1.2.18 The Engineer shall render written recommendations, within a reasonable time, on all claims, disputes and other matters in question between the Owner and the Contractor relating to the execution or progress of This Part of the Project or the interpretation of the Contract Documents.

1.2.19 The Engineer shall assist the Architect in determining whether the Architect shall reject Work for This Part of the Project which does not conform to the Contact Documents or whether special inspection or testing is required.

1.2.20 The Engineer shall review and approve, or take other appropriate action upon, and forward to the Architect for final disposition the Contractor's submittals such as Shop Drawings, Product Data and Samples with respect to This Part of the Project; but only for conformance with the design concept of the Work and with the information given in the Contract Documents. Such action shall be taken with reasonable promptness so as to cause no delay. The Engineer's approval of a specific item shall not indicate approval of an assembly of which the item is a component.

1.2.21 The Engineer shall assist the Architect in preparing Change Orders for This Part of the Project for the Owner's approval and execution in accordance with the Contract Documents. The Engineer shall recommend to the Architect minor changes in the Work not involving an adjustment in the Contract Sum or an extension of the Contract Time which are not inconsistent with the intent of the Contract Documents.

1.2.22 The Engineer shall assist the Architect in conducting inspections, with respect to This Part of the Project, to determine the dates of Substantial Completion and final completion, and shall review and approve, or take other appropriate action on, the Contractor's list of items to be completed or corrected and shall forward the list to the Architect for final disposition. The Engineer shall assist the Architect in receiving and forwarding to the Owner for the Owner's review written warranties and related documents required by the Contract Documents and assembled by the Contractor with respect to This Part of the Project. If requested, the Engineer shall issue to the Architect a final certificate in writing with respect to final payment for This Part of the Project.

1.3 PROJECT REPRESENTATION BEYOND BASIC SERVICES

1.3.1 If more extensive representation at the site than is described under Subparagraphs 1.2.10 through 1.2.22, inclusive, is required for This Part of the Project, the Engineer shall, if requested by the Architect, provide one or more Project Representatives to assist the Engineer in carrying out such responsibilities at the site.

1.3.2 Such Project Representatives shall be selected, employed and directed by the Engineer, and the Engineer shall be compensated therefor as set forth in an exhibit appended hereto or an amendment to this Agreement. Such exhibit or amendment shall describe the duties, responsibilities and limitations of authority of such Project Representatives.

1.3.3 Through the observations of such Project Representatives, the Engineer shall endeavor to provide further protection for the Owner against defects and deficiencies in the Work for This Part of the Project, but the furnishing of such Project representation shall not modify the rights, responsibilities or obligations of the Engineer as described in Subparagraphs 1.2.10 through 1.2.22, inclusive.

1.4 ADDITIONAL SERVICES

The following Services are not included in Basic Services unless so identified in Article 17. They shall be provided if authorized or confirmed in writing by the Architect, and they shall be paid for by the Architect as provided in this Agreement, in addition to the compensation for Basic Services.

1.4.1. Providing special analyses of the Owner's needs, and programming the requirements for This Part of the Project.

1.4.2 Providing financial feasibility or other special studies.

1.4.3 Providing planning surveys, site evaluations, environmental studies or comparative studies of prospective sites, and preparing special surveys, studies and submissions required for approvals of governmental authorities or others having jurisdiction over the Project.

1.4.4 Providing services relative to future facilities, systems and equipment which are not intended to be constructed during the Construction Phase.

1.4.5 Providing services to investigate existing conditions or facilities, or to make measured drawings thereof, or to verify the accuracy of drawings or other information related thereto.

1.4.6 Preparing documents for alternate, separate or sequential bids, or providing extra services in connection with bidding, negotiation or construction prior to the completion of the Construction Documents Phase, when requested by the Architect.

1.4.7 Providing coordination of work performed by separate contractors or by the Owner's own forces.

1.4.8 Providing services in connection with the work of a construction manager or separate consultants retained by the Owner.

1.4.9. Providing Detailed Estimates of Construction Cost, analyses of owning and operating costs, or detailed quantity surveys or inventories of material, equipment and labor.

1.4.10 Providing engineering services or special consultants related to interior design services and other similar services required for, or in connection with, the selection, procurement or installation of furniture, furnishings and related equipment.

1.4.11 Providing services for planning tenant or rental spaces.

1.4.12 Making revisions in Drawings, Specifications or other documents when such revisions are inconsistent with written approvals or instructions previously given, are required by the enactment or revision of codes, laws or regulations subsequent to the preparation of such documents, or are due to other causes not solely within the control of the Engineer.

1.4.13 Preparing Drawings, Specifications and supporting data, and providing other services in connection with Change Orders to the extent that the adjustment in the Basic Compensation resulting from the adjusted Construction Cost is not commensurate with the services required of the Engineer, provided such Change Orders are required by causes not solely within the control of the Engineer.

1.4.14 Making investigations, surveys, valuations, inventories or detailed appraisals of existing facilities, and providing services required in connection with construction performed by the Owner.

1.4.15 Providing consultation concerning replacement of any Work damaged by fire or other cause during construction, and furnishing services as may be required in connection with the replacement of such Work.

1.4.16 Providing services made necessary by the default of the Contractor, or by major defects or deficiencies in the Work of the Contractor, or by failure of performance of either the Owner or the Contractor under the Contract for Construction.

1.4.17 Preparing a set of reproducible record drawings showing significant changes in the Work made during construction, based on marked-up prints, drawings and other data furnished by the Contractor to the Architect.

1.4.18 Providing extensive assistance in the utilization of any equipment or system, such as initial start-up or testing, adjusting and balancing, preparation of operation and maintenance manuals, training personnel for operation and maintenance, and consultation during operation.

1.4.19 Providing services after issuance to the Owner of the final Certificate for Payment, or in the absence of a final Certificate for Payment, more than sixty days after the Date of Substantial Completion of the Work.

1.4.20 Preparing to serve or serving as an expert witness in connection with any public hearing, arbitration proceeding or legal proceeding.

1.4.21 Providing services of consultants for other than the normal engineering services for This Part of the Project.

1.4.22 Providing any other services not otherwise included in this Agreement or not customarily furnished in accordance with generally accepted engineering practice.

1.5 TIME

1.5.1 The Engineer shall perform Basic and Additional Services as expeditiously as is consistent with professional skill and care and the orderly progress of the Work. Upon request of the Architect, the Engineer shall submit, for the Architect's approval, a schedule for the performance of the Engineer's services which shall be adjusted as required as the Project proceeds, and which shall include allowances for periods of time required for the Owner's and the Architect's review and approval of submissions and for approvals of authorities having jurisdiction over the Project. This schedule, when approved by the Architect, shall not, except for reasonable cause, be exceeded by the Engineer.

ARTICLE 2
THE ARCHITECT'S RESPONSIBILITIES

2.1 The Architect shall, with reasonable promptness, provide all available information regarding the requirements for This Part of the Project.

2.2 The Architect shall designate, when necessary, a representative authorized to act in the Architect's behalf with respect to This Part of the Project. The Architect, or such authorized representative, shall examine documents submitted by the Engineer and shall render decisions pertaining thereto promptly, to avoid unreasonable delay in the progress of the Engineer's services.

2.3 If the Engineer deems it necessary for This Part of the Project, the Architect shall require the Owner to furnish, with reasonable promptness, a certified land survey and legal description of the site giving, as applicable, grades and lines of streets, alleys, pavements and adjoining property; rights-of-way, restrictions, easements, encroachments, zoning, deed restrictions, boundaries and contours of the site; locations, dimensions and complete data pertaining to existing buildings, other improvements and trees; and full information concerning available service and utility lines, both public and private, above and below grade, including inverts and depths.

2.4 If the Engineer deems it necessary for This Part of the Project, the Architect shall furnish to the Engineer, with reasonable promptness, (1) detailed layouts showing the location of connections, and (2) tabulations giving sizes and loads of equipment designed, specified or furnished by others for incorporation into other parts of the Project.

2.5 If the Engineer deems it necessary for This Part of the Project, the Architect shall require the Owner to furnish, with reasonable promptness, the services of a soil engineer or other consultant. Such services shall include test borings, test pits, soil bearing values, percolation tests, air and water pollution tests, ground corrosion and resistivity tests, including necessary operations for determining subsoil, air and water conditions, with reports and appropriate professional recommendations.

2.6 The Architect shall require the Owner to furnish structural, mechanical, chemical and other laboratory tests, inspections and reports required by law or the Contract Documents.

2.7 The services, information, surveys and reports required by Paragraphs 2.3 through 2.6, inclusive, shall be furnished at no expense to the Engineer who shall be entitled to rely upon the accuracy and completeness thereof.

2.8 If the Architect observes or otherwise becomes aware of any fault or defect with respect to This Part of the Project, or nonconformance with the Contract Documents, prompt written notice thereof shall be given by the Architect to the Engineer.

2.9 The Architect shall consult with the Engineer before issuing interpretations or clarifications of the Engineer's Drawings and Specifications and shall request the recommendation of the Engineer before acting upon Shop Drawings, Product Data, Samples or other submissions of the Contractor, or upon Change Orders affecting This Part of the Project.

2.10 The Architect shall furnish to the Engineer a copy of Statements of Probable Construction Cost or Detailed Estimates of Construction Cost as submitted to the Owner, bidding documents, bid tabulations, negotiated proposals and Contract Documents (including Change Orders as issued) to the extent that they pertain to This Part of the Project.

2.11 The Architect shall advise the Engineer of the identity of other consultants participating in the Project and the scope of their services.

2.12 The Architect shall review the Engineer's work for compliance with the Owner's program and for overall coordination with the architectural and other engineering requirements.

ARTICLE 3
CONSTRUCTION COST

3.1 The Construction Cost of the Project shall be the total cost or estimated cost to the Owner of all elements of the Project designed or specified by the Architect or the Architect's consultants. The Construction Cost of This Part of the Project shall be the total cost or estimated cost to the Owner of all elements of the Project designed or specified by the Engineer.

3.2 The Construction Cost of the Project or of This Part of the Project shall include at current market rates, including a reasonable allowance for overhead and profit, the cost of labor and materials furnished by the Owner and any equipment which has been designed, specified, selected or specially provided for by the Architect and the Architect's consultants.

3.3 Construction Cost does not include the compensation of the Architect and the Architect's consultants, the cost of the land, rights-of-way, or other costs which are the responsibility of the Architect as provided in Article 2, or which are the responsibility of the Owner as defined in the Prime Agreement.

3.4 Evaluations of the Owner's Project budget, Statements of Probable Construction Cost and Detailed Estimates of Construction Cost, if any, prepared by the Engineer, represent the Engineer's best judgment as a design professional familiar with the construction industry. It is recognized, however, that neither the Engineer nor the Architect has control over the cost of labor, materials or equipment, over the Contractor's methods of determining bid prices, or over competitive bidding, market or negotiating conditions. Accordingly, the Engineer cannot and does not warrant or represent that bids or negotiated prices for This Part of the Project will not vary from the Project budget proposed, established or approved by the Owner, if any, or from any Statement of Probable Construction Cost or other cost estimate prepared by the Engineer.

3.5 If a fixed limit of Construction Cost has been established as a condition of the Prime Agreement, the Architect and the Engineer shall establish, if practicable, a fixed limit of Construction Cost for This Part of the Project. If such a fixed limit is established, the Engineer, after consultation with the Architect, shall be permitted to include contingencies for design, bidding and price escalation, to determine what materials, equipment, component systems and types of construction are to be included in the Contract Documents with respect to This Part of the Project, and to make reasonable adjustments in the scope of This Part of the Project to bring it within the fixed limit. If required, the Engineer shall assist the Architect in including in the Contract Documents alternate bids to adjust the Construction Cost to the fixed limit. Any such fixed limit shall be increased in the amount of any increase in the Contract Sum related to This Part of the Project occurring after execution of the Contract for Construction.

3.5.1 If the Bidding or Negotiation Phase for This Part of the Project has not commenced within three months after the Architect submits the Construction Documents to the Owner, any Project budget or fixed limit of Construction Cost for This Part of the Project established as a condition of this Agreement shall be adjusted to reflect any change in the general level of prices in the construction industry between the date of submission of the Construction Documents to the Owner and the date on which bids or proposals are sought.

3.5.2 If a Project budget or fixed limit of Construction Cost for This Part of the Project (adjusted as provided in Subparagraph 3.5.1) is exceeded by the lowest bona fide bid or negotiated proposal, the Architect may require the Engineer, without additional charge, to modify the Engineer's Drawings and Specifications for This Part of the Project as necessary to bring the Construction Cost thereof within such fixed limit for This Part of the Project. If it was not practicable to establish a fixed limit of Construction Cost for This Part of the Project, and if the lowest bona fide bid or negotiated proposal, the Detailed Estimate of Construction Cost or the Statement of Probable Construction Cost established for the entire Project (including the bidding contingency) exceeds the fixed limit of Construction Cost of the entire Project, the Architect may require that the Drawings and Specifications prepared by the Engineer be modified without additional compensation as necessary to make them bear a reasonable portion of the burden of reducing the Construction Cost of This Part of the Project so that the fixed limit of Construction Cost for the entire Project is not exceeded. The providing of such service shall be the limit of the Engineer's responsibility in this regard, and having done so the Engineer shall be entitled to compensation for all services performed in accordance with this Agreement.

ARTICLE 4
DIRECT PERSONNEL EXPENSE

4.1 Direct Personnel Expense is defined as the direct salaries of all the Engineer's personnel engaged on the Project, and the portion of the cost of their mandatory and customary contributions and benefits related thereto, such as employment taxes and other statutory employee benefits, insurance, sick leave, holidays, vacations, pensions, and similar contributions and benefits.

ARTICLE 5
REIMBURSABLE EXPENSES

5.1 Reimbursable Expenses are in addition to compensation for Basic and Additional Services and include actual expenditures made by the Engineer and the Engineer's employees and consultants in the interest of the Project (to the extent they are reimbursable by the Owner under the Prime Agreement) for the expenses listed in the following Subparagraphs:

5.1.1 Expense of transportation in connection with the Project; living expenses in connection with out-of-town travel; long distance communications; and fees paid for securing approvals of authorities having jurisdiction over the Project.

5.1.2 Expense of reproductions, postage and handling of Drawings, Specifications and other documents, excluding reproductions for the office use of the Engineer, the Architect and the Architect's consultants.

5.1.3 Expense of data processing and photographic production techniques when used in connection with Additional Services.

5.1.4 If authorized in advance by the Architect, expense of overtime work requiring higher than regular rates.

5.1.5 Expense of renderings, models and mock-ups requested by the Architect.

5.1.6 Expense of any additional insurance coverage or limits, including professional liability insurance in excess of the requirements of Article 16.

ARTICLE 6
PAYMENTS TO THE ENGINEER

6.1 PAYMENTS ON ACCOUNT OF SERVICES

6.1.1 Payments for Basic Services, Additional Services and Reimbursable Expenses shall be made on the basis set forth in Article 15.

6.1.2 The Architect shall disclose to the Engineer, prior to the execution of this Agreement, any contingent or other special provisions relative to compensation that are included in the Architect's understanding with the Owner or in the Prime Agreement.

6.1.3 The Engineer shall submit, in timely fashion, invoices for Basic Services, Additional Services and Reimbursable Expenses. The Architect shall review such invoices and, if they are considered incorrect or untimely, the Architect shall review the matter with the Engineer and confirm, in writing to the Engineer within ten days from receipt of the Engineer's billing, the Architect's understanding of the disposition of the issue.

6.1.4 If and to the extent that the Contract Time initially established in the Contract for Construction is exceeded or extended through no fault of the Engineer, compensation for any Basic Services required for such extended period of Administration of the Construction Contract shall be computed as set forth in Paragraph 15.3 for Additional Services.

6.1.5 When compensation is based on a percentage of Construction Cost, and any portions of the Project are deleted or otherwise not constructed, compensation for such portions of the Project shall be payable to the extent services are performed on such portions, in accordance with the schedule set forth in Subparagraph 15.1.2, based on (1) the lowest bona fide bid or negotiated proposal, or (2) if no such bid or proposal is received, the most recent Statement of Probable Construction Cost or Detailed Estimate of Construction Cost for such portions of the Project.

6.2 PAYMENTS WITHHELD

6.2.1 No deductions shall be made from the Engineer's compensation on account of penalty, liquidated damages or other sums withheld from payments to contractors, or on account of the cost of changes in the Work other than those for which the Engineer is held legally liable.

6.3 PROJECT SUSPENSION OR ABANDONMENT

6.3.1 If the Project is suspended or abandoned in whole or in part for more than three months, for reasons other than the fault of the Engineer, the Engineer shall be compensated for all services performed prior to receipt of written notice from the Architect of such suspension or abandonment, together with Reimbursable Expenses then due and all Termination Expenses as defined in Paragraph 10.4. If the Project is resumed after being suspended for more than three months, the Engineer's compensation shall be equitably adjusted.

ARTICLE 7
ENGINEER'S ACCOUNTING RECORDS

7.1 Records of Reimbursable Expenses and expenses pertaining to Additional Services and services performed on the basis of a Multiple of Direct Personnel Expense shall be kept on the basis of generally accepted accounting principles and shall be available to the Architect or the Architect's authorized representative at mutually convenient times.

ARTICLE 8
OWNERSHIP AND USE OF DOCUMENTS

8.1 Except for reference and coordination purposes in connection with future additions or alterations to the Work, Drawings and Specifications prepared by the Engineer as instruments of service are and shall be the property of the Architect whether the Project for which they are made is executed or not. The Engineer shall be permitted to retain copies, including reproducible copies, of Drawings and Specifications for such information and reference. The Drawings and Specifications shall not be used by the Architect on other projects, or for completion of this Project by others provided the Engineer is not in default under this Agreement, except by agreement in writing and with appropriate compensation to the Engineer.

8.2 Neither the Architect nor the Engineer shall make changes in the other's Drawings and Specifications without written permission of the other party.

8.3 The Engineer shall maintain on file, and make available to the Architect, design calculations for This Part of the Project, and shall furnish copies thereof to the Architect on request.

8.4 Submission or distribution to meet official regulatory requirements, or for other purposes in connection with the Project, is not to be construed as publication in derogation of the Architect's or the Engineer's rights.

ARTICLE 9
ARBITRATION

9.1 All claims, disputes and other matters in question between the parties to this Agreement arising out of or relating to this Agreement or the breach thereof, shall be decided by arbitration in accordance with the Construction Industry Arbitration Rules of the American Arbitration Association then obtaining unless the parties mutually agree otherwise. No arbitration arising out of or relating to this Agreement shall include, by consolidation, joinder or in any other manner, any additional person not a party to this Agreement except by written consent containing a specific reference to this Agreement and signed by the Architect, the Engineer and any other person sought to be joined. Any consent to arbitration involving an additional person or persons shall not constitute consent to arbitration of any dispute not described therein or with any person not named or described therein. This agreement to arbitrate and any agreement to arbitrate with an additional person or persons duly consented to by the parties to this Agreement shall be specifically enforceable under the prevailing arbitration law.

9.2 Notice of the demand for arbitration shall be filed in writing with the other party to this Agreement and with the American Arbitration Association. The demand shall be made within a reasonable time after the claim, dispute or other matter in question has arisen. In no event shall the demand for arbitration be made after the date when institution of legal or equitable proceedings based on such claim, dispute or other matter in question would be barred by the applicable statute of limitations.

9.3 The award rendered by the arbitrators shall be final, and judgment may be entered upon it in accordance with applicable law in any court having jurisdiction thereof.

ARTICLE 10
TERMINATION OF AGREEMENT

10.1 This Agreement is terminated if and when the Prime Agreement is terminated. The Architect shall promptly notify the Engineer of such termination.

10.2 This Agreement may be terminated by either party upon seven days' written notice should the other party fail substantially to perform in accordance with its terms through no fault of the party initiating the termination.

10.3 Subject to the provisions of Article 15, and in the event of termination which is not the fault of the Engineer, the Engineer shall be compensated for all services performed to the termination date.

10.4 Termination Expenses include expenses directly attributable to termination for which the Engineer is not otherwise compensated, plus an amount computed as a percentage of the total compensation for Basic and Additional Services earned to the time of termination, as follows:

.1 20 percent if termination occurs during the Schematic Design Phase; or

.2 10 percent if termination occurs during the Design Development Phase; or

.3 5 percent if termination occurs during any subsequent phase.

ARTICLE 11
INSURANCE

11.1 The Architect and the Engineer shall each effect and maintain insurance to protect themselves from claims arising out of the performance of professional services under this Agreement and caused by any error, omission or negligent act for which they are legally liable. The Architect and the Engineer shall each maintain this insurance in force, if available, after the completion of professional services under this Agreement until the expiration of any applicable statutes of limitations. In the event there is no such statute specifically applicable to design and construction of improvements to real property, this insurance, if available, shall be maintained in force by both the Architect and the Engineer for a reasonable period after the Date of Substantial Completion of the Project as agreed to by the Architect and the Engineer.

11.2 Unless otherwise agreed, the Architect and the Engineer shall each effect and maintain insurance to protect themselves from claims under workers' or workmen's compensation acts; from claims for damages because of bodily injury, including personal injury, sickness, disease, or death of any employees or of any other person; from claims for damages because of injury to or destruction of property including loss of use resulting therefrom; and from damage to or destruction of property including valuable papers and records coverage and including loss of use resulting therefrom.

11.3 The insurance required by Paragraphs 11.1 and 11.2 shall be in the minimum limits required by law and as provided in Article 16.

11.4 The Architect and the Engineer shall, upon request, furnish to each other Certificates of Insurance evidencing the insurance carried in compliance with the requirements of Paragraphs 11.1 and 11.2, including appropriate evidence that each type of insurance has been properly amended to include coverage for this specific Project. These Certificates shall contain a provision that at least 30 days' prior written notice will be given to each other in the event of cancellation, reduction in or nonrenewal of the insurance.

ARTICLE 12
MISCELLANEOUS PROVISIONS

12.1 Unless otherwise specified, this Agreement shall be governed by the law of the principal place of business of the Architect.

12.2 Terms in this Agreement shall have the same meanings as those in AIA Document A201, General Conditions of the Contract for Construction, current as of the date identified in the Prime Agreement.

12.3 As between the parties to this Agreement: as to all acts or failures to act by either party to this Agreement, any applicable statute of limitations shall commence to run and any alleged cause of action shall be deemed to have accrued in any and all events not later than the relevant Date of Substantial Completion of the Work, and as to any acts or failures to act occurring after the relevant Date of Substantial Completion, not later than the date of issuance of the final Certificate for Payment.

12.4 The Architect and the Engineer waive all rights against each other and against the contractors, consultants, agents and employees of the other for damages covered by any property insurance during construction as set forth in the edition of AIA Document A201, General Conditions, current as of the date identified in the Prime Agreement. The Engineer shall require appropriate similar waivers from the Engineer's contractors, consultants and agents.

ARTICLE 13
SUCCESSORS AND ASSIGNS

13.1 The Architect and the Engineer, respectively, bind themselves, their partners, successors, assigns and legal representatives to the other party to this Agreement and to the partners, successors, assigns and legal representatives of such other party with respect to all covenants of this Agreement. Neither the Architect nor the Engineer shall assign, sublet or transfer any interest in this Agreement without the written consent of the other.

ARTICLE 14
EXTENT OF AGREEMENT

14.1 This Agreement represents the entire and integrated agreement for This Part of the Project between the Architect and the Engineer and supersedes all prior negotiations, representations or agreements, either written or oral. This Agreement may be amended only by written instrument signed by both the Architect and the Engineer.

ARTICLE 15
BASIS OF COMPENSATION

The Architect shall compensate the Engineer for the Scope of Services provided, in accordance with Article 6, Payments to the Engineer, and the other Terms and Conditions of this Agreement, as follows:

15.1 BASIC COMPENSATION

15.1.1 FOR BASIC SERVICES, as described in Paragraphs 1.1 and 1.2, and any other services included in Article 17 as part of Basic Services, Basic Compensation shall be computed as follows:

(Here insert basis of compensation, including fixed amounts, multiples or percentages, as applicable, and identify phases to which particular methods of compensation apply, as appropriate.)

15.1.2 Where compensation is based on a Stipulated Sum, Percentage of Construction Cost or Percentage of Construction Cost of This Part of the Project, payments for Basic Services shall be made as provided in Paragraph 6.1 so that Basic Compensation for each phase shall equal the following percentages of the total Basic Compensation payable:

(Include any additional phases as appropriate.)

Schematic Design Phase:	percent (%)
Design Development Phase:	percent (%)
Construction Documents Phase:	percent (%)
Bidding or Negotiation Phase:	percent (%)
Construction Phase:	percent (%)

15.2 FOR PROJECT REPRESENTATION BEYOND BASIC SERVICES, as described in Paragraph 1.3, compensation shall be computed separately in accordance with Subparagraph 1.3.2.

15.3 COMPENSATION FOR ADDITIONAL SERVICES

15.3.1 FOR ADDITIONAL SERVICES OF THE ENGINEER, as described in Paragraph 1.4, and any other services included in Article 17 as Additional Services, but excluding Additional Services of consultants for other than the normal engineering services for This Part of the Project, compensation shall be computed as follows:

(Here insert basis of compensation, including rates and/or multiples of Direct Personnel Expense for Principals and employees, and identify Principals and classify employees, if required. Identify specific services to which particular methods of compensation apply, if necessary.)

15.3.2 FOR ADDITIONAL SERVICES OF CONSULTANTS for other than normal engineering services for This Part of the Project, as provided under Subparagraph 15.3.1 or identified in Article 17 as Additional Services, a multiple of ( ) times the amounts billed to the Engineer for such services.

(Identify specific types of consultants in Article 17, if required.)

15.4 FOR REIMBURSABLE EXPENSES, as described in Article 5, and any other items included in Article 17 as Reimbursable Expenses, a multiple of ( ) times the amounts expended by the Engineer, the Engineer's employees and consultants in the interest of the Project.

15.5 The Architect and the Engineer agree in accordance with the Terms and Conditions of this Agreement that:

15.5.1 IF THE SCOPE of This Part of the Project or of the Engineer's Services is changed materially, the amounts of compensation shall be equitably adjusted.

15.5.2 IF THE SERVICES covered by this Agreement have not been completed within ( ) months of the date hereof, through no fault of the Engineer, the amounts of compensation, rates and multiples set forth herein shall be equitably adjusted.

15.6 PAYMENT PROVISIONS

(Insert provisions as to conditions, contingencies, times, manner and other particulars concerning payments, including any provisions for the payment of interest.)

ARTICLE 16
INSURANCE COVERAGES

(After consultation with insurance counsel for the Architect and for the Engineer, insert the minimum limits of insurance required for each type of insurance required in Paragraphs 11.1 and 11.2, and any other coverages which may be necessary to protect them.)

ARTICLE 17
OTHER CONDITIONS OR SERVICES

This Agreement entered into as of the day and year first written above.

ARCHITECT

ENGINEER

APPENDIX F

National Society of Professional Engineers Contracts

This document has important legal consequences; consultation with an attorney is encouraged with respect to its completion or modification.

STANDARD GENERAL CONDITIONS OF THE CONSTRUCTION CONTRACT

Jointly Issued by

PROFESSIONAL ENGINEERS IN PRIVATE PRACTICE

A practice division of the

NATIONAL SOCIETY OF PROFESSIONAL ENGINEERS

and by

AMERICAN CONSULTING ENGINEERS COUNCIL

and by

CONSTRUCTION SPECIFICATIONS INSTITUTE

2029 K Street, N.W., Washington, D.C. 20006

American Consulting Engineers Council
1155 15th Street, N.W., Washington, D.C. 20005

Construction Specifications Institute
1150 17th Street, N.W., Washington, D.C. 20036

These General Conditions have been prepared for use with the Owner-Contractor Agreements (NSPE-ACEC Document 1910-8-A-1 or 1910-8-A-2; CSI 56467, 56468, 1978 editions). Their provisions are interrelated and a change in one may necessitate a change in the others. Comments concerning their usage are contained in the Commentary to the Documents, NSPE-ACEC 1910-9, 1978 edition.

TABLE OF CONTENTS OF GENERAL CONDITIONS

Article Number	*Title*	*Page*
1	DEFINITIONS	8
2	PRELIMINARY MATTERS	9
3	CONTRACT DOCUMENTS: INTENT AND REUSE	10
4	AVAILABILITY OF LANDS; PHYSICAL CONDITIONS; REFERENCE POINTS	10
5	BONDS AND INSURANCE	11
6	CONTRACTOR'S RESPONSIBILITIES	12
7	WORK BY OTHERS	16
8	OWNER'S RESPONSIBILITIES	17
9	ENGINEER'S STATUS DURING CONSTRUCTION	17
10	CHANGES IN THE WORK	18
11	CHANGE OF CONTRACT PRICE	18
12	CHANGE OF THE CONTRACT TIME	21
13	WARRANTY AND GUARANTEE; TESTS AND INSPECTIONS; CORRECTION, REMOVAL OR ACCEPTANCE OF DEFECTIVE WORK	21
14	PAYMENTS TO CONTRACTOR AND COMPLETION	22
15	SUSPENSION OF WORK AND TERMINATION	25
16	ARBITRATION	26
17	MISCELLANEOUS	27

*These page numbers refer to the Contract pages.

INDEX TO GENERAL CONDITIONS

	Article or Paragraph Number
Acceptance of Insurance	5.13
Access to the Work	13.2
Addenda—definition of (see definition of Specifications)	1
Agreement—definition of	1
All Risk Insurance	5.6
Application for Payment—definition of	1
Application for Payment, Final	14.12
Application for Progress Payment	14.2
Application for Progress Payment—review of	14.4 thru 14.7
Arbitration	16
Availability of Lands	4.1
Award—Notice of—defined	1
Before Starting Construction	2.5 thru 2.7
Bid—definition of	1
Bonds and Insurance—in general	5
Bonds—definition of	1
Bonds, Delivery of	2.1, 5.1
Bonds, Performance and Other	5.1, 5.2
Cash Allowances	11.10
Change Order—Definition of	1
Changes in the Work	10
Claims, Waiver of—on Final Payment	14.16
Clarifications and Interpretations	9.3
Cleaning	6.17
Completion	14
Completion, Substantial	14.8, 14.9
Conference—Pre-Construction	2.8
Construction Machinery, Equipment, etc.	6.4
Continuing Work	6.29
Contract Documents—definition of	1
Contract Documents—intent and reuse	3
Contract Documents—reuse of	3.5
Contract Price, Change of	11
Contract Price—definition	1
Contract Time, Change of	12
Contract Time—Commencement of	2.3
Contract Time—definition of	1
Contractor—definition of	1
Contractor May Stop Work or Terminate	15.5
Contractor's Continuing Obligation	14.15
Contractor's Duty to Report Discrepancy in Documents	2.5, 3.2
Contractor's Fee—Costs Plus	11.6
Contractor's Liability Insurance	5.3
Contractor's Responsibilities—in general	6
Contractor's Warranty of Title	14.3
Contractual Liability Insurance	5.4
Copies of Documents	2.2
Correction or Removal of Defective Work	13.11
Correction Period, One Year	13.12
Correction, Removal or Acceptance of Defective Work—in general	13.11 thru 13.14
Cost of Work	11.4, 11.5
Costs, Supplemental	11.4.5

Day—definition of	1
Defective Work, Acceptance of	13.13
Defective Work, Correction or Removal of	13.11
Defective—definition of	1
Defective Work—in general	13
Defective Work, Rejecting	9.4
Definitions	1
Delivery of Bonds	2.1
Disagreements, Decisions by Engineer	9.9, 9.10
Documents, Copies of	2.2
Documents, Record	6.19
Documents, Reuse	3.5
Drawings—definition of	1
Effective date of Agreement—definition of	1
Emergencies	6.22
Engineer—defintion of	1
Engineer's—Notice Work is Acceptable	14.3
Engineer's Responsibilities, Limitations on	9.11 thru 9.14
Engineer's Status During Construction—in general	9
Engineer's—Recommendation of Payment	14.4, 14.13
Equipment, Labor, Materials and	6.3 thru 6.6
Equivalent Materials and Equipment	6.7
Fee, Contractor's—Costs Plus	11.6
Field Order—definition of	1
Field Order—issued by Engineer	10.2
Final Application for Payment	14.12
Final Inspection	14.11
Final Payment, Recommendation of	14.13, 14.14
Final Payment and Acceptance	14.13
General Requirements—definition of	1
General Provisions	17.3, 17.4
Giving Notice	17.1
Guarantee of Work—by Contractor	13.1
Indemnification	6.30 thru 6.32
Inspection, Final	14.11
Inspection, Tests and	13.3 thru 13.7
Insurance, Bonds and—in general	5
Insurance—Certificates of	2.7 and 5
Insurance, Contractor's Liability	5.3
Insurance, Contractual Liability	5.4
Insurance, Owner's Liability	5.5
Insurance, Property	5.6 thru 5.12
Intent of Contract Documents	3.1 thru 3.4, 9.12
Interpretations and Clarifications	9.3
Investigations of Physical Conditions	4.2
Labor, Materials and Equipment	6.3 thru 6.6
Laws and Regulations	6.14
Liability Insurance—Contractors	5.3
Liability Insurance—Owners	5.5
Limitations on Engineer's Responsibilities	9.11
Materials and Equipment—furnished by Contractor	6.3
Materials or Equipment—Equivalent	6.7
Miscellaneous Provisions	17
Modification—definition of	1

Notice, Giving of	17.1
Notice of Award—definition of	1
Notice of Acceptability of Project	14.13
Notice to Proceed—definition of	1
Notice to Proceed—giving of	2.3
"Or-Equal" items	6.7
Other Contractors	7
Overtime Work—prohibition of	6.3
Owner—definition of	1
Owner May Correct Defective Work	13.14
Owner May Stop Work	13.10
Owner May Suspend Work, Terminate	15.1 thru 15.4
Owner's Duty to Execute Change Orders	11.8
Owner's Liability Insurance	5.5
Owner's Representative—Engineer to serve as	9.1
Owner's Responsibilities—in general	8
Owner's Separate Representative at Site	9.8
Partial Utilization	14.10
Partial Utilization—Property Insurance	5.14
Patent Fees and Royalties	6.12
Payments to Contractor—in general	14
Payments, Recommendation of	14.4 thru 14.7.5
Pre-construction Conference	2.8
Performance, and other Bonds	5.1 thru 5.2
Permits	6.13
Physical Conditions—Investigations and Reports	4.2
Physical Conditions, Unforeseen	4.3
Preconstruction Conference	2.8
Preliminary Matters	2
Premises, Use of	6.16, 6.17, 6.18
Price—Change of Contract	11
Price-Contract—definition of	1
Progress Payment, Applications for	14.2
Progress Schedule	2.6, 14.1
Project—definition of	1
Project Representative, Resident—definition of	1
Project Representation—Provision for	9.8
Project, Starting	2.4
Property Insurance	5.6 thru 5.12
Property Insurance—Receipt and Application of Proceeds	5.11, 5.12
Property Insurance—Partial Utilization	5.14
Protection, Safety and	6.20 thru 6.21
Recommendation of Payment	14.4, 14.13
Record Documents	6.19
Reference Points	4.4
Regulations, Laws and	6.14
Rejecting Defective Work	9.4
Remedies Not Exclusive	17.5
Removal or Correction of Defective Work	13.11
Resident Project Representative—definition of	1
Resident Project Representative—provision for	9.8
Responsibilities, Contractor's	6
Responsibilities, Owner's	8
Reuse of Documents	3.5
Royalties, Patent Fees and	6.13

Safety and Protection	6.20 thru 6.21
Samples	6.23
Schedule of Shop Drawing Submissions	2.6, 14.1
Schedule of Values	2.6, 14.1
Shop Drawings and Samples	6.23 thru 6.29
Shop Drawings—definition of	1
Site, Visits to—by Engineer	9.2
Specifications—definition of	1
Starting Construction, Before	2.5 thru 2.9
Starting the Project	2.4
Stopping Work—by Contractor	15.5
Stopping Work—by Owner	13.10
Subcontractor—definition of	1
Subcontractors—in general	6.8 thru 6.11
Substantial Completion—certification of	14.8
Substantial Completion—definition of	1
Subsurface Conditions	4.2, 4.3
Supplemental Costs	11.4.5
Surety—consent to payment	14.12, 14.14
Surety—notice of changes	10.5
Surety—qualification of	5.1, 5.2
Suspending Work, by Owner	15.1
Suspension of Work and Termination—in general	15
Superintendent—Contractor's	6.2
Supervision and Superintendence	6.1, 6.2
Taxes—Payment by Contractor	6.15
Termination—by Contractor	15.5
Termination—by Owner	15.2 thru 15.4
Termination, Suspension of Work and—in general	15
Tests and Inspections	13.3 thru 13.7
Time, Change of Contract	12
Time, Computation of	17.2
Time, Contract—definition of	1
Uncovering Work	13.8, 13.9
Unit Prices	11.3.1
Unit Prices—Adjustment of	11.9
Use of Premises	6.16, 6.17, 6.18
Values, Schedule of	14.1
Visits to Site—by Engineer	9.2
Waiver of Claims—on Final Payment	14.16
Waiver of Rights by Insured Parties	5.10
Warranty and Guarantee—by Contractor	13.1
Warranty of Title, Contractor's	14.3
Work, Access to	13.2
Work by Others—in general	7
Work, Cost of	11.4, 11.5
Work Continuing During Disputes	6.29
Work—defintion of	1
Work, Neglected by Contractor	13.14
Work, Stopping by Contractor	15.5
Work, Stopping by Owner	15.1 thru 15.4

GENERAL CONDITIONS

ARTICLE 1—DEFINITIONS

Wherever used in these General Conditions or in the other Contract Documents, the following terms have the meanings indicated which are applicable to both the singular and plural thereof:

Addenda—Written or graphic instruments issued prior to the opening of Bids which clarify, correct or change the bidding documents or the Contract Documents.

Agreement—The written agreement between OWNER and CONTRACTOR covering the Work to be performed; other Contract Documents are attached to the Agreement and made a part thereof as provided therein.

Application for Payment—The form accepted by ENGINEER which is to be used by CONTRACTOR in requesting progress or final payment and which is to include such supporting documentation as is required by the Contract Documents.

Bid—The offer or proposal of the Bidder submitted on the prescribed form setting forth the prices for the Work to be performed.

Bonds—Bid, performance and payment bonds and other instruments of security.

Change Order—A written order to CONTRACTOR signed by OWNER authorizing an addition, deletion or revision in the Work, or an adjustment in the Contract Price or the Contract Time issued after the effective date of the Agreement.

Contract Documents—The Agreement, Addenda (which pertain to the Contract Documents), CONTRACTOR's Bid (including documentation accompanying the Bid and any post-Bid documentation submitted prior to the Notice of Award) when attached as an exhibit to the Agreement, the Bonds, these General Conditions, the Supplementary Conditions, the Specifications, the Drawings as the same are more specifically identified in the Agreement, together with all Modifications issued after the execution of the Agreement.

Contract Price—The moneys payable by OWNER to CONTRACTOR under the Contract Documents as stated in the Agreement.

Contract Time—The number of days (computed as provided in paragraph 17.2) or the date stated in the Agreement for the completion of the Work.

CONTRACTOR—The person, firm or corporation with whom OWNER has entered into the Agreement.

day—A calendar day of twenty-four hours measured from midnight to the next midnight.

defective—An adjective which when modifying the word Work refers to Work that is unsatisfactory, faulty or deficient, or does not conform to the Contract Documents or does not meet the requirements of any inspection, test or approval referred to in the Contract Documents, or has been damaged prior to ENGINEER's recommendation of final payment.

Drawings—The drawings which show the character and scope of the Work to be performed and which have been prepared or approved by ENGINEER and are referred to in the Contract Documents.

effective date of the Agreement—The date indicated in the Agreement on which it becomes effective, but if no such date is indicated it means the date on which the Agreement is signed and delivered by the last of the two parties to sign and deliver.

ENGINEER—The person, firm or corporation named as such in the Agreement.

Field Order—A written order issued by ENGINEER which orders minor changes in the Work in accordance with paragraph 10.2 but which does not involve a change in the Contract Price or the Contract Time.

General Requirements—Sections of Division 1 of the Specifications.

Modification—(a) A written amendment of the Contract Documents signed by both parties, (b) a Change Order, or (c) a Field Order. A modification may only be issued after the effective date of the Agreement.

Notice of Award—The written notice by OWNER to the apparent successful Bidder stating that upon compliance by the apparent successful Bidder with the conditions precedent enumerated therein, within the time specified, OWNER will sign and deliver the Agreement.

Notice to Proceed—A written notice given by OWNER to CONTRACTOR (with a copy to ENGINEER) fixing the date on which the Contract Time will commence to run and on which CONTRACTOR shall start to perform his obligation under the Contract Documents.

OWNER—The public body or authority, corporation, association, partnership, or individual with whom CONTRACTOR has entered into the Agreement and for whom the Work is to be provided.

Project—The total construction of which the Work to be provided under the Contract Documents may be the whole, or a part as indicated elsewhere in the Contract Documents.

Resident Project Representative—The authorized representative of ENGINEER who is assigned to the site or any part thereof.

Shop Drawings—All drawings, diagrams, illustrations, schedules and other data which are specifically prepared by CONTRACTOR, a Subcontractor, manufacturer, fabricator, supplier or distributor to illustrate some portion of the Work and all illustrations, brochures, standard schedules, performance charts, instructions, diagrams and other information prepared by a manufacturer, fabricator, supplier or distributor and submitted by CONTRACTOR to illustrate material or equipment for some portion of the Work.

Specifications—Those portions of the Contract Documents consisting of written technical descriptions of materials, equipment, construction systems, standards and workmanship as applied to the Work and certain administrative details applicable thereto.

Subcontractor—An individual, firm or corporation having a direct contract with CONTRACTOR or with any other Subcontractor for the performance of a part of the Work at the site.

Substantial Completion—The Work (or a specified part thereof) has progressed to the point where, in the opinion of ENGINEER as evidenced by his definitive certificate of Substantial Completion, it is sufficiently complete, in accordance with the Contract Documents, so that the Work (or specified part) can be utilized for the purposes for which it was intended; or if there be no such certificate issued, when final payment is due in accordance with paragraph 14.13. The terms "substantially complete" and "substantially completed" as applied to any Work refer to Substantial Completion thereof.

Work—The entire completed construction or the various separately identifiable parts thereof required to be furnished under the Contract Documents. Work is the result of performing services, furnishing labor and furnishing and incorporating materials and equipment into the construction, all as required by the Contract Documents.

ARTICLE 2—PRELIMINARY MATTERS

Delivery of Bonds:

2.1. When CONTRACTOR delivers the executed Agreements to OWNER, CONTRACTOR shall also deliver to OWNER such Bonds as CONTRACTOR may be required to furnish in accordance with paragraph 5.1.

Copies of Documents:

2.2. OWNER shall furnish to CONTRACTOR up to ten copies (unless otherwise specified in the General Requirements) of the Contract Documents as are reasonably necessary for the execution of the Work. Additional copies will be furnished, upon request, at the cost of reproduction.

Commencement of Contract Time; Notice to Proceed:

2.3. The Contract Time will commence to run on the thirtieth day after the effective date of the Agreement, or, if a Notice to Proceed is given, on the day indicated in the Notice to Proceed; but in no event shall the Contract Time commence to run later than the ninetieth day after the day of Bid opening or the thirtieth day after the effective date of the Agreement. A Notice to Proceed may be given at any time within thirty days after the effective date of the Agreement.

Starting the Project:

2.4. CONTRACTOR shall start to perform the Work on the date when the Contract Time commences to run, but no Work shall be done at the site prior to the date on which the Contract Time commences to run.

Before Starting Construction:

2.5. Before undertaking each part of the Work, CONTRACTOR shall carefully study and compare the Contract Documents and check and verify pertinent figures shown thereon and all applicable field measurements. CONTRACTOR shall promptly report in writing to ENGINEER any conflict, error or discrepancy which CONTRACTOR may discover; however, CONTRACTOR shall not be liable to OWNER or ENGINEER for failure to report any conflict, error or discrepancy in the Drawings or Specifications, unless CONTRACTOR had actual knowledge thereof or should reasonably have known thereof.

2.6. Within ten days after the effective date of the Agreement (unless otherwise specified in the General Requirements), CONTRACTOR shall submit to ENGINEER for review and acceptance an estimated progress schedule indicating the starting and completion dates of the various stages of the Work, a preliminary schedule of Shop Drawing submissions, and a preliminary schedule of values of the Work.

2.7. Before any Work at the site is started, CONTRACTOR shall deliver to OWNER, with a copy to ENGINEER, certificates (and other evidence of insurance requested by OWNER) which CONTRACTOR is required to purchase and maintain in accordance with paragraphs 5.3 and 5.4, and OWNER shall deliver to CONTRACTOR certificates (and other evidence of insurance requested by CONTRACTOR) which OWNER is required to purchase and maintain in accordance with paragraphs 5.6 and 5.7.

Preconstruction Conference:

2.8. Within twenty days after the effective date of the Agreement, but before CONTRACTOR starts the Work at the site, a conference will be held for review and acceptance of the schedules referred to in paragraph 2.6, to establish procedures for handling Shop Drawings and other submittals and for processing Applications for Payment, and to establish a working understanding among the parties as to the Work.

ARTICLE 3—CONTRACT DOCUMENTS: INTENT AND REUSE

Intent:

3.1. The Contract Documents comprise the entire Agreement between OWNER and CONTRACTOR concerning the Work. They may be altered only by a Modification.

3.2. The Contract Documents are complementary; what is called for by one is as binding as if called for by all. If, during the performance of the Work, CONTRACTOR finds a conflict, error or discrepancy in the Contract Documents, he shall report it to ENGINEER in writing at once and before proceeding with the Work affected thereby; however, CONTRACTOR shall not be liable to OWNER or ENGINEER for failure to report any conflict, error or discrepancy in the Specifications or Drawings unless CONTRACTOR had actual knowledge thereof or should reasonably have known thereof.

3.3. It is the intent of the Specifications and Drawings to describe a complete project (or part thereof) to be constructed in accordance with the Contract Documents. Any Work that may reasonably be inferred from the Specifications or Drawings as being required to produce the intended result shall be supplied whether or not it is specifically called for. When words which have a well-known technical or trade meaning are used to describe Work, materials or equipment such words shall be interpreted in accordance with such meaning. Reference to standard specifications, manuals or codes of any technical society, organization or association, or to the code of any governmental authority, whether such reference be specific or by implication, shall mean the latest standard specification, manual or code in effect at the time of opening of Bids (or, on the effective date of the Agreement if there were no Bids), except as may be otherwise specifically stated. However, no provision of any referenced standard specification, manual or code (whether or not specifically incorporated by reference in the Contract Documents) shall change the duties and responsibilities of OWNER, CONTRACTOR or ENGINEER, or any of their agents or employees from those set forth in the Contract Documents. Clarifications and interpretations of the Contract Documents shall be issued by ENGINEER as provided for in paragraph 9.3.

3.4. The Contract Documents will be governed by the law of the place of the Project.

Reuse of Documents:

3.5. Neither CONTRACTOR nor any Subcontractor, manufacturer, fabricator, supplier or distributor shall have or acquire any title to or ownership rights in any of the Drawings, Specifications or other documents (or copies of any thereof) prepared by or bearing the seal of ENGINEER; and they shall not reuse any of them on extensions of the Project or any other project without written consent of OWNER and ENGINEER and specific written verification or adaptation by ENGINEER.

ARTICLE 4—AVAILABILITY OF LANDS; PHYSICAL CONDITIONS; REFERENCE POINTS

Availability of Lands:

4.1. OWNER shall furnish, as indicated in the Contract Documents, the lands upon which the Work is to be performed, rights-of-way for access thereto, and such other lands which are designated for the use of CONTRACTOR. Easements for permanent structures or permanent changes in existing facilities will be obtained and paid for by OWNER, unless otherwise provided in the Contract Documents. If CONTRACTOR believes that any delay in OWNER's furnishing these lands or easements entitles him to an extension of the Contract Time, CONTRACTOR may make a claim therefor as provided in Article 12. CONTRACTOR shall provide for all additional lands and access thereto that may be required for temporary construction facilities or storage of materials and equipment.

Physical Conditions—Investigations and Reports:

4.2. Reference is made to the Supplementary Conditions for identification of those reports of investigations and tests of subsurface and latent physical conditions at the site or otherwise affecting cost, progress or performance of the Work which have been relied upon by ENGINEER in preparation of the Drawings and Specifications. Such reports are not guaranteed as to accuracy or completeness and are not part of the Contract Documents.

Unforeseen Physical Conditions:

4.3. CONTRACTOR shall promptly notify OWNER and ENGINEER in writing of any subsurface or latent physical conditions at the site or in an existing structure differing materially from those indicated or referred to in the Contract Documents. ENGINEER will promptly review those conditions and advise OWNER in writing if further investigation or tests are necessary. Promptly thereafter, OWNER shall obtain the necessary additional investigations and tests and furnish copies to ENGINEER and CONTRACTOR. If ENGINEER finds that the results of such investigations or tests indicate that there are subsurface or latent physical conditions which differ materially from those intended in the Contract Documents, and which could not reasonably have been anticipated by CONTRACTOR, a Change Order shall be issued incorporating the necessary revisions.

Reference Points:

4.4. OWNER shall provide engineering surveys for construction to establish reference points which in his judgment are necessary to enable CONTRACTOR to proceed with the Work. CONTRACTOR shall be responsible for laying out the Work (unless otherwise specified in the General Requirements), shall protect and preserve the established reference points and shall make no changes or relocations without the prior written approval of OWNER. CONTRACTOR shall report to ENGINEER whenever any reference point is lost or destroyed or requires relocation because of necessary changes in grades or locations, and shall be responsible for replacement or relocation of such reference points by professionally qualified personnel.

ARTICLE 5—BONDS AND INSURANCE

Performance and Other Bonds:

5.1. CONTRACTOR shall furnish performance and payment Bonds, each in an amount at least equal to the Contract Price as security for the faithful performance and payment of all CONTRACTOR's obligations under the Contract Documents. These Bonds shall remain in effect at least until one year after the date of final payment, except as otherwise provided by law. CONTRACTOR shall also furnish such other Bonds as are required by the Supplementary Conditions. All Bonds shall be in the forms prescribed by the bidding documents or Supplementary Conditions and be executed by such Sureties as (i) are licensed to conduct business in the state where the Project is located, and (ii) are named in the current list of "Companies Holding Certificates of Authority as Acceptable Sureties on Federal Bonds and as Acceptable Reinsuring Companies" as published in Circular 570 (amended) by the Audit Staff Bureau of Accounts, U.S. Treasury Department. All Bonds signed by an agent must be accompanied by a certified copy of the authority to act.

5.2. If the Surety on any Bond furnished by CONTRACTOR is declared a bankrupt or becomes insolvent or its right to do business is terminated in any state where any part of the Project is located or it ceases to meet the requirements of clauses (i) and (ii) of paragraph 5.1, CONTRACTOR shall within five days thereafter substitute another Bond and Surety, both of which shall be acceptable to OWNER.

Contractor's Liability Insurance:

5.3. CONTRACTOR shall purchase and maintain such comprehensive general liability and other insurance as will provide protection from claims set forth below which may arise out of or result from CONTRACTOR's performance of the Work and CONTRACTOR's other obligations under the Contract Documents, whether such performance is by CONTRACTOR, by any Subcontractor, by anyone directly or indirectly employed by any of them, or by anyone for whose acts any of them may be liable:

5.3.1. Claims under workers' or workmen's compensation, disability benefits and other similar employee benefit acts;

5.3.2. Claims for damages because of bodily injury, occupational sickness or disease, or death of CONTRACTOR's employees;

5.3.3. Claims for damages because of bodily injury, sickness or disease, or death of any person other than CONTRACTOR's employees;

5.3.4. Claims for damages insured by personal injury liability coverage which are sustained (i) by any person as a result of an offense directly or indirectly related to the employment of such person by CONTRACTOR, or (ii) by any other person for any other reason;

5.3.5. Claims for damages, other than to the Work itself, because of injury to or destruction of tangible property, including loss of use resulting therefrom; and

5.3.6. Claims for damages because of bodily injury or death of any person or property damage arising out of the ownership, maintenance or use of any motor vehicle.

The insurance required by this paragraph 5.3 shall include the specific coverages and be written for not less than the limits of liability and coverages provided in the Supplementary Conditions, or required by law, whichever is greater. The comprehensive general liability insurance shall include completed operations insurance. All such insurance shall contain a provision that the coverage afforded will not be cancelled, materially changed or renewal refused until at least thirty days' prior written notice has been given to OWNER and ENGINEER. All such insurance shall remain in effect until final payment and at all times thereafter when CONTRACTOR may be correcting, removing or replacing defective Work in accordance with paragraph 13.12. In addition, CONTRACTOR shall maintain such completed operations insurance for at least two years after final payment and furnish OWNER with evidence of continuation of such insurance at final payment and one year thereafter.

Contractual Liability Insurance:

5.4. The comprehensive general liability insurance required by paragraph 5.3 will include contractual liability insurance applicable to CONTRACTOR's obligations under paragraphs 6.30 and 6.31.

Owner's Liability Insurance:

5.5. OWNER shall be responsible for purchasing and maintaining his own liability insurance and, at his option, may purchase and maintain such insurance as will protect OWNER against claims which may arise from operations under the Contract Documents.

Property Insurance:

5.6. Unless otherwise provided in the Supplementary Conditions, OWNER shall purchase and maintain property insurance upon the Work at the site to the full insurable value thereof (subject to such deductible amounts as may be provided in the Supplementary Conditions or required by law). This insurance shall include the interests of OWNER, CONTRACTOR and Subcontractors in the Work, shall insure against the perils of fire and extended coverage and shall include "all risk" insurance for physical loss and damage including theft, vandalism and malicious mischief, collapse and water damage, and such other perils as may be provided in the Supplementary Conditions, and shall include damages, losses and expenses arising out of or resulting from any insured loss or incurred in the repair or replacement of any insured property (including fees and charges of engineers, architects, attorneys and other professionals). If not covered under the "all risk" insurance or otherwise provided in the Supplementary Conditions, CONTRACTOR shall purchase and maintain similar property insurance on portions of the Work stored on and off the site or in transit when such portions of the Work are to be included in an Application for Payment. The policies of insurance required to be purchased and maintained by OWNER in accordance with paragraphs 5.6 and 5.7 shall contain a provision that the coverage afforded will not be cancelled or materially changed until at least thirty days' prior written notice has been given to CONTRACTOR.

5.7. OWNER shall purchase and maintain such boiler and machinery insurance as may be required by the Supplementary Conditions or by law. This insurance shall include the interests of OWNER, CONTRACTOR and Subcontractors in the Work.

5.8. OWNER shall not be responsible for purchasing and maintaining any property insurance to protect the interests of CONTRACTOR or Subcontractors in the Work to the extent of any deductible amounts that are provided in the Supplementary Conditions. If CONTRACTOR wishes property insurance coverage within the limits of such amounts, CONTRACTOR may purchase and maintain it at his own expense.

5.9. If CONTRACTOR requests in writing that other special insurance be included in the property insurance policy, OWNER shall, if possible, include such insurance, and the cost thereof shall be charged to CONTRACTOR by appropriate Change Order. Prior to commencement of the Work at the site, OWNER will in writing advise CONTRACTOR whether or not such other insurance has been procured by OWNER.

Waiver of Rights:

5.10. OWNER and CONTRACTOR waive all rights against each other and the Subcontractors and their agents and employees and against ENGINEER and separate contractors (if any) and their subcontractors' agents and employees, for damages caused by fire or other perils to the extent covered by insurance provided under paragraphs 5.6 and 5.7, inclusive, or any other property insurance applicable to the Work, except such rights as they may have to the proceeds of such insurance held by OWNER as trustee. OWNER shall require similar written waivers by ENGINEER and from each separate contractor, and CONTRACTOR shall require similar written waivers from each Subcontractor (in accordance with paragraph 6.11 as applicable); each such waiver will be in favor of all other parties enumerated in this paragraph 5.10.

Receipt and Application of Proceeds:

5.11. Any insured loss under the policies of insurance required by paragraphs 5.6 and 5.7 shall be adjusted with OWNER and made payable to OWNER as trustee for the insureds, as their interests may appear, subject to the requirements of any applicable mortgage clause and of paragraph 5.12. OWNER shall deposit in a separate account any money so received, and he shall distribute it in accordance with such agreement as the parties in interest may reach. If no other special agreement is reached the damaged Work shall be repaired or replaced, the moneys so received applied on account thereof and the Work and the cost thereof covered by an appropriate Change Order.

5.12. OWNER as trustee shall have power to adjust and settle any loss with the insurers unless one of the parties in interest shall object in writing within fifteen days after the occurrence of loss to OWNER's exercise of this power. If such objection be made, OWNER as trustee shall make settlement with the insurers in accordance with such agreement as the parties in interest may reach. If required in writing by any party in interest, OWNER as trustee shall upon the occurrence of an insured loss, give bond for the proper performance of his duties.

Acceptance of Insurance:

5.13. If OWNER has any objection to the coverage afforded by or other provisions of the insurance required to be purchased and maintained by CONTRACTOR in accordance with paragraphs 5.3 and 5.4 on the basis of its not complying with the Contract Documents, OWNER will notify CONTRACTOR in writing thereof within ten days of the date of delivery of such certificates to OWNER in accordance with paragraph 2.7. If CONTRACTOR has any objection to the coverage afforded by or other provisions of the policies of insurance required to be purchased and maintained by OWNER in accordance with paragraphs 5.6 and 5.7 on the basis of their not complying with the Contract Documents, CONTRACTOR will notify OWNER in writing thereof within ten days of the date of delivery of such certificates to CONTRACTOR in accordance with paragraph 2.7. OWNER and CONTRACTOR will each provide to the other such additional information in respect of insurance provided by him as the other may reasonably request. Failure by OWNER or CONTRACTOR to give any such notice of objection within the time provided shall constitute acceptance of such insurance purchased by the other as complying with the Contract Documents.

Partial Utilization—Property Insurance:

5.14. If OWNER finds it necessary to occupy or use a portion or portions of the Work prior to Substantial Completion of all the Work, such use or occupancy may be accomplished in accordance with paragraph 14.10; provided that no such use or occupancy shall commence before the insurers providing the property insurance have acknowledged notice thereof and in writing effected the changes in coverage necessitated thereby. The insurers providing the property insurance shall consent by endorsement on the policy or policies, but the property insurance shall not be cancelled or lapse on account of any such partial use or occupancy.

ARTICLE 6—CONTRACTOR'S RESPONSIBILITIES

Supervision and Superintendence:

6.1. CONTRACTOR shall supervise and direct the Work competently and efficiently, devoting such attention thereto and applying such skills and expertise as may be necessary to perform the Work in accordance with the Contract Documents. CONTRACTOR shall be solely responsible for the means, methods, techniques, sequences and procedures of construction, but CONTRACTOR shall not be solely responsible for the negligence of others in the design or selection of a specific means, method, technique, sequence or procedure of construction which is indicated in and required by the Contract Documents. CONTRACTOR shall be responsible to see that the finished Work complies accurately with the Contract Documents.

6.2. CONTRACTOR shall keep on the Work at all times during its progress a competent resident superintendent, who shall not be replaced without written notice to OWNER and ENGINEER except under extraordinary circumstances. The superintendent will be CONTRACTOR's representative at the site and shall have authority to act on behalf of CONTRACTOR. All communications given to the superintendent shall be as binding as if given to CONTRACTOR.

Labor, Materials and Equipment:

6.3. CONTRACTOR shall provide competent, suitably qualified personnel to survey and lay out the Work and perform construction as required by the Contract Documents. CONTRACTOR shall at all times maintain good discipline and order at the site. Except in connection with the safety or protection of persons or the Work or property at the site or adjacent thereto, and except as otherwise indicated in the Supplementary Conditions, all Work at the site shall be performed during regular working hours, and CONTRACTOR will not permit overtime work or the performance of Work on Saturday, Sunday or any legal holiday without OWNER's written consent given after prior written notice to ENGINEER.

6.4. CONTRACTOR shall furnish all materials, equipment, labor, transportation, construction equipment and machinery, tools, appliances, fuel, power, light, heat, telephone, water and sanitary facilities and all other facilities and incidentals necessary for the execution, testing, initial operation and completion of the Work.

6.5. All materials and equipment shall be of good quality and new, except as otherwise provided in the Contract Documents. If required by ENGINEER, CONTRACTOR shall furnish satisfactory evidence (including reports of required tests) as to the kind and quality of materials and equipment.

6.6. All materials and equipment shall be applied, installed, connected, erected, used, cleaned and conditioned in accordance with the instructions of the applicable manufacturer, fabricator, supplier or distributor, except as otherwise provided in the Contract Documents.

Equivalent Materials and Equipment:

6.7. Whenever materials or equipment are specified or described in the Drawings or Specifications by using the name of a proprietary item or the name of a particular manufacturer, fabricator, supplier or distributor, the naming of the item is intended to establish the type, function and quality required. Unless the name is followed by words indicating that no substitution is permitted, materials or equipment of other manufacturers, fabricators, suppliers or distributors may be accepted by ENGINEER if sufficient information is submitted by CONTRACTOR to allow ENGINEER to determine that the material or equipment proposed is equivalent to that named. The procedure for review by ENGINEER will be as set forth in paragraphs 6.7.1 and 6.7.2 below as supplemented in the General Requirements.

6.7.1. Requests for review of substitute items of material and equipment will not be accepted by ENGINEER from anyone other than CONTRACTOR. If CONTRACTOR wishes to furnish or use a substitute item of material or equipment CONTRACTOR shall make written application to ENGINEER for acceptance thereof, certifying that the proposed substitute will perform adequately the functions called for by the general design, be similar and of equal substance to that specified and be suited to the same use and capable of performing the same function as that specified. The application will state whether or not acceptance of the substitute for use in the Work will require a change in the Drawings or Specifications to adapt the design to the substitute and whether or not incorporation or use of the substitute in connection with the Work is subject to payment of any license fee or royalty. All variations of the proposed substitute from that specified shall be identified in the application and available maintenance, repair and replacement service will be indicated. The application will also contain an itemized estimate of all costs that will result directly or indirectly from acceptance of such substitute, including costs of redesign and claims of other contractors affected by the resulting change, all of which shall be considered by ENGINEER in evaluating the proposed substitute. ENGINEER may require CONTRACTOR to furnish at CONTRACTOR's expense additional data about the proposed substitute. ENGINEER will be the sole judge of acceptability, and no substitute will be ordered or installed without ENGINEER's prior written acceptance. OWNER may require CONTRACTOR to furnish at CONTRACTOR's expense a special performance guarantee or other surety with respect to any substitute.

6.7.2. ENGINEER will record time required by ENGINEER and ENGINEER's consultants in evaluating substitutions proposed by CONTRACTOR and in making changes in the Drawings or Specifications occasioned thereby. Whether or not ENGINEER accepts a proposed substitute, CONTRACTOR shall reinburse OWNER for the charges of ENGINEER and ENGINEER's consultants for evaluating any proposed substitute.

Concerning Subcontractors:

6.8. CONTRACTOR shall not employ any Subcontractor or other person or organization (including those who are to furnish the principal items of materials or equipment), whether initially or as a substitute, against whom OWNER or ENGINEER may have reasonable objection. A Subcontractor or other person or organization identified in writing to OWNER and ENGINEER by CONTRACTOR prior to the Notice of Award and not objected to in writing by OWNER or ENGINEER prior to the Notice of Award will be deemed acceptable to OWNER and ENGINEER. Acceptance of any Subcontractor, other person or organization by OWNER or ENGINEER shall not constitute a waiver of any right of OWNER or ENGINEER to reject defective Work. If OWNER or ENGINEER after due investigation has reasonable objection to any Subcontractor, other person or organization proposed by CONTRACTOR after the Notice of Award, CONTRACTOR shall submit an acceptable substitute and the Contract Price shall be increased or decreased by the difference in cost occasioned by such substitution, and an appropriate Change Order shall be issued. CONTRACTOR shall not be required to employ any Subcontractor, other person or organization against whom CONTRACTOR has reasonable objection.

6.9. CONTRACTOR shall be fully responsible for all acts and omissions of his Subcontractors and of persons and organizations directly or indirectly employed by them and of persons and organizations for whose acts any of them may be liable to the same extent that CONTRACTOR is responsible for the acts and omissions of persons directly employed by CONTRACTOR. Nothing in the Contract Documents shall create any contractual relationship between OWNER or ENGINEER and any Subcontractor or other person or organization having a direct contact with CONTRACTOR, nor shall it create any obligation on the part of OWNER or ENGINEER to pay or to see to the payment of any moneys due any Subcontractor or other person or organization, except as may otherwise be required by law. OWNER or ENGINEER may furnish to any Subcontractor or other person or organization, to the extent practicable, evidence of amounts paid to CONTRACTOR on account of specific Work done.

6.10. The divisions and sections of the Specifications and the identifications of any Drawings shall not control CONTRACTOR in dividing the Work among Subcontractors or delineating the Work to be performed by any specific trade.

6.11. All Work performed for CONTRACTOR by a Subcontractor will be pursuant to an appropriate agreement between CONTRACTOR and the Subcontractor which specifically binds the Subcontractor to the applicable terms and conditions of the Contract Documents for the benefit of OWNER and ENGINEER and contains waiver provisions as required by paragraph 5.10. CONTRACTOR shall pay each Subcontractor a just share of any insurance moneys received by CONTRACTOR on account of losses under policies issued pursuant to paragraphs 5.6 through 5.8.

Patent Fees and Royalties:

6.12. CONTRACTOR shall pay all license fees and royalties and assume all costs incident to the use in the performance of the Work or the incorporation in the Work of any invention, design, process, product or device which is the subject of patent rights or copyrights held by others. If a particular invention, design, process, product or device is specified in the Contract Documents for use in the performance of the Work and if to the actual knowledge of OWNER or ENGINEER its use is subject to patent rights or copyrights calling for the payment of any license fee or royalty to others, the existence of such rights shall be disclosed by OWNER in the Contract Documents. CONTRACTOR shall indemnify and hold harmless OWNER and ENGINEER and anyone directly or indirectly employed by either of them from and against all claims, damages, losses and expenses (including attorneys' fees) arising out of any infringement of patent rights or copyrights incident to the use in the performance of the Work or resulting from the incorporation in the Work of any invention, design, process, product or device not specified in the Contract Documents, and shall defend all such claims in connection with any alleged infringement of such rights.

Permits:

6.13. Unless otherwise provided in the Supplementary Conditions, CONTRACTOR shall obtain and pay for all construction permits and licenses. OWNER shall assist CONTRACTOR, when necessary, in obtaining such permits and licenses. CONTRACTOR shall pay all governmental charges and inspection fees necessary for the prosecution of the Work, which are applicable at the time of opening of Bids. CONTRACTOR shall pay all charges of utility service companies for connections to the Work, and OWNER shall pay all charges of such companies for capital costs related thereto.

Laws and Regulations:

6.14. CONTRACTOR shall give all notices and comply with all laws, ordinances, rules and regulations applicable to the Work. If CONTRACTOR observes that the Specifications or Drawings are at variance therewith, CONTRACTOR shall give ENGINEER prompt written notice thereof, and any necessary changes shall be adjusted by an appropriate Modification. If CONTRACTOR performs any Work knowing or having reason to know that it is contrary to such laws, ordinances, rules and regulations, and without such notice to ENGINEER, CONTRACTOR shall bear all costs arising therefrom; however, it shall not be CONTRACTOR's primary responsibility to make certain that the Specifications and Drawings are in accordance with such laws, ordinances, rules and regulations.

Taxes:

6.15. CONTRACTOR shall pay all sales, consumer, use and other similar taxes required to paid by him in accordance with the law of the place of the Project.

Use of Premises:

6.16. CONTRACTOR shall confine construction equipment, the storage of materials and equipment and the operations of workmen to areas permitted by law, ordinances, permits or the requirements of the Contract Documents, and shall not unreasonably encumber the premises with construction equipment or other materials or equipment.

6.17. During the progress of the Work, CONTRACTOR shall keep the premises free from accumulations of waste materials, rubbish and other debris resulting from the Work. At the completion of the Work CONTRACTOR shall remove all waste materials, rubbish and debris from and about the premises as well as all tools, appliances, construction equipment and machinery, and surplus materials, and shall leave the site clean and ready for occupancy by OWNER. CONTRACTOR shall restore to their original condition those portions of the site not designated for alteration by the Contract Documents.

6.18. CONTRACTOR shall not load nor permit any part of any structure to be loaded in any manner that will endanger the structure, nor shall CONTRACTOR subject any part of the Work or adjacent property to stresses or pressures that will endanger it.

Record Documents:

6.19. CONTRACTOR shall keep one record copy of all Specifications, Drawings, Addenda, Modifications, Shop Drawings and samples at the site, in good order and annotated to show all changes made during the construction process. These shall be available to ENGINEER for examination and shall be delivered to ENGINEER for OWNER upon completion of the Work.

Safety and Protection:

6.20. CONTRACTOR shall be responsible for initiating, maintaining and supervising all safety precautions and programs in connection with the Work. CONTRACTOR shall take all necessary precautions for the safety of, and shall provide the necessary protection to prevent damage, injury or loss to:

6.20.1. all employees on the Work and other persons who may be affected thereby,

6.20.2. all the Work and all materials or equipment to be incorporated therein, whether in storage on or off the site, and

6.20.3. other property at the site or adjacent thereto, including trees, shrubs, lawns, walks, pavements, roadways, structures and utilities not designated for removal, relocation or replacement in the course of construction.

CONTRACTOR shall comply with all applicable laws, ordinances, rules, regulations and orders of any public body having jurisdiction for the safety of persons or property or to protect them from damage, injury or loss; and shall erect and maintain all necessary safeguards for such safety and protection. CONTRACTOR shall notify owners of adjacent property and utilities when prosecution of the Work may affect them. All damage, injury or loss to any property referred to in paragraph 6.20.2 or 6.20.3 caused, directly or indirectly, in whole or in part, by CONTRACTOR, any Subcontractor or anyone directly or indirectly employed by any of them or anyone for whose acts any of them may be liable, shall be remedied by CONTRACTOR (except damage or loss attributable to the fault of Drawings or Specifications or to the acts or omissions of OWNER or ENGINEER or anyone employed by either of them or anyone for whose acts either of them may be liable, and not attributable, directly or indirectly, in whole or in part, to the fault or negligence of CONTRACTOR). CONTRACTOR's duties and responsibilities for the safety and protection of the Work shall continue until such time as all the Work is completed and ENGINEER has issued a notice to OWNER and CONTRACTOR in accordance with paragraph 14.13 that the Work is acceptable.

6.21. CONTRACTOR shall designate a responsible member of his organization at the site whose duty shall be the prevention of accidents. This person shall be CONTRACTOR's superintendent unless otherwise designated in writing by CONTRACTOR to OWNER.

Emergencies:

6.22. In emergencies affecting the safety or protection of persons or the Work or property at the site or adjacent thereto, CONTRACTOR, without special instruction or authorization from ENGINEER or OWNER, is obligated to act to prevent threatened damage, injury or loss. CONTRACTOR shall give ENGINEER prompt written notice of any significant changes in the Work or deviations from the Contract Documents caused thereby.

Shop Drawings and Samples:

6.23. After checking and verifying all field measurements, CONTRACTOR shall submit to ENGINEER for review and approval, in accordance with the accepted schedule of Shop Drawing submissions (see paragraph 2.8), five copies (unless otherwise specified in the General Requirements) of all Shop Drawings, which shall have been checked by and stamped with the approval of CONTRACTOR and identified as ENGINEER may require. The data shown on the Shop Drawings will be complete with respect to dimensions, design criteria, materials of construction and like information to enable ENGINEER to review the information as required.

6.24. CONTRACTOR shall also submit to ENGINEER for review and approval with such promptness as to cause no delay in Work, all samples required by the Contract Documents. All samples will have been checked by and stamped with the approval of CONTRACTOR, identified clearly as to material, manufacturer, any pertinent catalog numbers and the use for which intended.

6.25. At the time of each submission, CONTRACTOR shall in writing call ENGINEER's attention to any deviations that the Shop Drawings or samples may have from the requirements of the Contract Documents.

6.26. ENGINEER will review and approve with reasonable promptness Shop Drawings and samples, but ENGINEER's review and approval shall be only for conformance with the design concept of the Project and for compliance with the information given in the Contract Documents and shall not extend to means, methods, sequences, techniques or procedures of construction or to safety precautions or programs incident thereto. The review and approval of a separate item as such will not indicate approval of the assembly in which the item functions. CONTRACTOR shall make any corrections required by ENGINEER and shall return the required number of corrected copies of Shop Drawings and resubmit new samples for review and approval. CONTRACTOR shall direct specific attention in writing to revisions other than the corrections called for by ENGINEER on previous submittals. CONTRACTOR's stamp of approval on any Shop Drawing or sample shall constitute a representation to OWNER and ENGINEER that CONTRACTOR has either determined and verified all quantities, dimensions, field construction criteria, materials, catalog numbers, and similar data or assumes full responsibility for doing so, and that CONTRACTOR has reviewed or coordinated each Shop Drawing or sample with the requirements of the Work and the Contract Documents.

6.27. Where a Shop Drawing or sample is required by the Specifications, no related Work shall be commenced until the submittal has been reviewed and approved by ENGINEER.

6.28. ENGINEER's review and approval of Shop Drawings or samples shall not relieve CONTRACTOR from responsibility for any deviations from the Contract Documents unless CONTRACTOR has in writing called ENGINEER's attention to such deviation at the time of submission and ENGINEER has given written concurrence and approval to the specific deviation, nor shall any concurrence or approval by ENGINEER relieve CONTRACTOR from responsibility for errors or omissions in the Shop Drawings.

Continuing the Work:

6.29. CONTRACTOR shall carry on the Work and maintain the progress schedule during all disputes or disagreements with OWNER. No Work shall be delayed or postponed pending resolution of any disputes or disagreements, except as CONTRACTOR and OWNER may otherwise agree in writing.

Indemnification:

6.30. To the fullest extent permitted by law, CONTRACTOR shall indemnify and hold harmless OWNER and ENGINEER and their agents and employees from and against all claims, damages, losses and expenses including but not limited to attorneys' fees arising out of or resulting from the performance of the Work, provided that any such claim, damage, loss or expense (a) is attributable to bodily injury, sickness, disease or death, or to injury to or destruction of tangible property (other than the Work itself) including the loss of use resulting therefrom and (b) is caused in whole or in part by any negligent act or omission of CONTRACTOR, any Subcontractor, anyone directly or indirectly employed by any of them or anyone for whose acts any of them may be liable, regardless of whether or not it is caused in part by a party indemnified hereunder.

6.31. In any and all claims against OWNER or ENGINEER or any of their agents or employees by any employee of CONTRACTOR, any Subcontractor, anyone directly or indirectly employed by any of them or anyone for whose acts any of them may be liable, the indemnification obligation under paragraph 6.30 shall not be limited in any way by any limitation on the amount or type of damages, compensation or benefits payable by or for CONTRACTOR or any Subcontractor under workers' or workmen's compensation acts, disability benefit acts or other employee benefit acts.

6.32. The obligations of CONTRACTOR under paragraph 6.30 shall not extend to the liability of ENGINEER, his agents or employees arising out of the preparation or approval of maps, drawings, opinions, reports, surveys, Change Orders, designs or specifications.

ARTICLE 7—WORK BY OTHERS

7.1. OWNER may perform additional work related to the Project by himself, or have additional work performed by utility service companies, or let other direct contracts therefor which shall contain General Conditions similar to these. CONTRACTOR shall afford the utility service companies and the other contractors who are parties to such direct contracts (or OWNER, if OWNER is performing the additional work with OWNER's employees) reasonable opportunity for the introduction and storage of materials and equipment and the execution of work, and shall properly connect and coordinate his Work with theirs.

7.2. If any part of CONTRACTOR's Work depends for proper execution or results upon the work of any such other contractor or utility service company (or OWNER), CONTRACTOR shall inspect and promptly report to ENGINEER in writing any patent or apparent defects or deficiencies in such work that render it unsuitable for such proper execution and results. CONTRACTOR's failure so to report shall constitute an acceptance of the other work as fit and proper for integration with CONTRACTOR's Work except for latent or non-apparent defects and deficiencies in the other work.

7.3. CONTRACTOR shall do all cutting, fitting and patching of his Work that may be required to make its several parts come together properly and integrate with such other work. CONTRACTOR shall not endanger any work of others by cutting, excavating or otherwise altering their work and will only cut or alter their work with the written consent of ENGINEER and the others whose work will be affected.

7.4. If the performance of additional work by other contractors or utility service companies or OWNER was not noted in the Contract Documents, written notice thereof shall be given to CONTRACTOR prior to starting any such additional work. If CONTRACTOR believes that the performance of such additional work by OWNER or others involves additional expense to CONTRACTOR or requires an extension of the Contract Time, CONTRACTOR may make a claim therefor as provided in Articles 11 and 12.

ARTICLE 8—OWNER'S RESPONSIBILITIES

8.1. OWNER shall issue all communications to CONTRACTOR through ENGINEER.

8.2. In case of termination of the employment of ENGINEER, OWNER shall appoint an engineer against whom CONTRACTOR makes no reasonable objection, whose status under the Contract Documents shall be that of the former ENGINEER. Any dispute in connection with such appoinment shall be subject to arbitration.

8.3. OWNER shall furnish the data required of OWNER under the Contract Documents promptly and shall make payments to CONTRACTOR promptly after they are due as provided in paragraphs 14.4 and 14.13.

8.4. OWNER's duties in respect of providing lands and easements and providing engineering surveys to establish reference points are set forth in paragraphs 4.1 and 4.4 Paragraph 4.2 refers to OWNER's identifying and making available to CONTRACTOR copies of reports of investigations and tests of subsurface and latent physical conditions at the site or otherwise affecting performance of the Work which have been relied upon by ENGINEER in preparing the Drawings and Specifications.

8.5. OWNER's responsibilities in respect of purchasing and maintaining liability and property insurance are set forth in paragraphs 5.5 through 5.7.

8.6. In connection with OWNER's rights to request changes in the Work in accordance with Article 10, OWNER (especially in certain instances as provided in paragraph 10.4) is obligated to execute Change Orders.

8.7. OWNER's responsibility in respect of certain inspections, tests and approvals is set forth in paragraph 13.4.

8.8. In connection with OWNER's right to stop Work or suspend Work, see paragraphs 13.10 and 15.1. Paragraph 15.2 deals with OWNER's right to terminate services of CONTRACTOR under certain circumstances.

ARTICLE 9—ENGINEER'S STATUS DURING CONSTRUCTION

Owner's Representative:

9.1. ENGINEER will be OWNER's respresentative during the construction period. The duties and responsibilities and the limitations of authority of ENGINEER as OWNER's representative during construction are set forth in the Contract Documents and shall not be extended without written consent of OWNER and ENGINEER.

Visits to Site:

9.2. ENGINEER will make visits to the site at intervals appropriate to the various stages of construction to observe the progress and quality of the executed Work and to determine, in general, if the Work is proceeding in accordance with the Contract Documents. ENGINEER will not be required to make exhaustive or continuous on-site inspections to check the quality or quantity of the Work. ENGINEER's efforts will be directed toward providing for OWNER a greater degree of confidence that the completed Work will conform to the Contract Documents. On the basis of such visits and on-site observations as an experienced and qualified design professional, ENGINEER will keep OWNER informed of the progress of the Work and will endeavor to guard OWNER against defects and deficiencies in the Work.

Clarifications and Interpretations:

9.3. ENGINEER will issue with reasonable promptness such written clarifications or interpretations of the Contract Documents (in the form of Drawings or otherwise) as ENGINEER may determine necessary, which shall be consistent with or reasonably inferable from the overall intent of the Contract Documents. If CONTRACTOR believes that a written clarification or interpretation justifies an increase in the Contract Price or Contract Time, CONTRACTOR may make a claim therefor as provided in Article 11 or Article 12.

Rejecting Defective Work:

9.4. ENGINEER will have authority to disapprove or reject Work which is defective, and will also have authority to require special inspection or testing of the Work as provided in paragraph 13.9, whether or not the Work is fabricated, installed or completed.

Shop Drawings, Change Orders and Payments:

9.5. In connection with ENGINEER's responsibility for Shop Drawings and samples, see paragraphs 6.23 through 6.29 inclusive.

9.6. In connection with ENGINEER's responsibilities as to Change Orders, see Articles 10, 11 and 12.

9.7. In connection with ENGINEER's responsibilities in respect of Applications for Payment, etc., see Article 14.

Project Representation:

9.8. If OWNER and ENGINEER agree, ENGINEER will furnish a Resident Project Representative to assist ENGINEER in observing the performance of the Work. The duties, responsibilities and limitations of authority of any such Resident Project Representative and assistants will be as provided in the Supplementary Conditions. If OWNER designates another agent to represent him at the site who is not ENGINEER's agent or employee, the duties, responsibilities and limitations of authority of such other person will be as provided in the Supplementary Conditions.

Decisions on Disagreements:

9.9. ENGINEER will be the initial interpreter of the requirements of the Contract Documents and judge of the acceptability of the Work thereunder. Claims, disputes and other matters relating to the acceptability of the Work or the interpretation of the requirements of the Contract Documents pertaining to the execution and progress of the Work shall be referred initially to ENGINEER in writing with a request for a formal decision in accordance with this paragraph, which ENGINEER will render in writing within a reasonable time. Written notice of each such claim, dispute and other matter shall be delivered by the claimant to ENGINEER and the other party to the Agreement within fifteen days of the occurrence of the event giving rise thereto, and written supporting data will be submitted to ENGINEER and the other party within forty-five days of such occurrence unless ENGINEER allows an additional period of time to ascertain more accurate data. In his capacity as interpreter and judge ENGINEER will not show partiality to OWNER or CONTRACTOR and will not be liable in connection with any interpretation or decision rendered in good faith in such capacity.

9.10. The rendering of a decision by ENGINEER pursuant to paragraph 9.9 with respect to any such claim, dispute or other matter (except any which have been waived by the making or acceptance of final payment as provided in paragraph 14.16) will be a condition precedent to any exercise by OWNER or CONTRACTOR of such rights or remedies as either may otherwise have under the Contract Documents or at law in respect of any such claim, dispute or other matter.

Limitations on ENGINEER's Responsibilities:

9.11. Neither ENGINEER's authority to act under this Article 9 or elsewhere in the Contract Documents nor any decision made by ENGINEER in good faith either to exercise or not exercise such authority shall give rise to any duty or responsibility of ENGINEER to CONTRACTOR, any Subcontractor, any manufacturer, fabricator, supplier or distributor, or any of their agents or employees or any other person performing any of the Work.

9.12. Whenever in the Contract Documents the terms "as ordered", "as directed", "as required", "as allowed" or terms of like effect or import are used, or the adjectives "reasonable", "suitable", "acceptable", "proper" or "satisfactory" or adjectives of like effect or import are used, to describe requirement, direction, review or judgment of ENGINEER as to the Work, it is intended that such requirement, direction, review or judgment will be solely to evaluate the Work for compliance with the Contract Documents (unless there is a specific statement indicating otherwise). The use of any such term or adjective never indicates that ENGINEER shall have authority to supervise or direct performance of the Work or authority to undertake responsibility contrary to the provisions of paragraphs 9.13 or 9.14.

9.13. ENGINEER will not be responsible for CONTRACTOR's means, methods, techniques, sequences or procedures of construction, or the safety precautions and programs incident thereto, and ENGINEER will not be responsible for CONTRACTOR's failure to perform the Work in accordance with the Contract Documents.

9.14. ENGINEER will not be responsible for the acts or omissions of CONTRACTOR or of any Subcontractors, or of the agents or employees of any CONTRACTOR or Subcontractor, or of any other persons at the site or otherwise performing any of the Work.

ARTICLE 10—CHANGES IN THE WORK

10.1. Without invalidating the Agreement, OWNER may, at any time or from time to time, order additions, deletions or revisions in the Work; these will be authorized by Change Orders. Upon receipt of a Change Order, CONTRACTOR shall proceed with the Work involved. All such Work shall be executed under the applicable conditions of the Contract Documents. If any Change Order causes an increase or decrease in the Contract Price or an extension or shortening of the Contract Time, an equitable adjustment will be made as provided in Article 11 or Article 12 on the basis of a claim made by either party.

10.2. ENGINEER may authorize minor changes in the Work not involving an adjustment in the Contract Price or the Contract Time, which are consistent with the overall intent of the Contract Documents. These may be accomplished by a Field Order and shall be binding on OWNER, and also on CONTRACTOR who shall perform the change promptly. If CONTRACTOR believes that a Field Order justifies an increase in the Contract Price or Contract Time, CONTRACTOR may make a claim therefor as provided in Article 11 or Article 12.

10.3. Additional Work performed without authorization of a Change Order will not entitle CONTRACTOR to an increase in the Contract Price or an extension of the Contract Time, except in the case of an emergency as provided in paragraph 6.22 and except as provided in paragraphs 10.2 and 13.9.

10.4. OWNER shall execute appropriate Change Orders prepared by ENGINEER covering changes in the Work which are required by OWNER, or required because of unforeseen physical conditions or emergencies, or because of uncovering Work found not to be defective, or as provided in paragraphs 11.9 or 11.10, or because of any other claim of CONTRACTOR for a change in the Contract Time or the Contract Price which is recommended by ENGINEER.

10.5. If notice of any change affecting the general scope of the Work or change in the Contract Price is required by the provisions of any Bond to be given to the Surety, it will be CONTRACTOR's responsibility to so notify the Surety, and the amount of each applicable Bond shall be adjusted accordingly. CONTRACTOR shall furnish proof of such adjustment to OWNER.

ARTICLE 11—CHANGE OF CONTRACT PRICE

11.1. The Contract Price constitutes the total compensation (subject to authorized adjustments) payable to

CONTRACTOR for performing the Work. All duties, responsibilities and obligations assigned to or undertaken by CONTRACTOR shall be at his expense without change in the Contract Price.

11.2. The Contract Price may only be changed by a Change Order. Any claim for an increase in the Contract Price shall be based on written notice delivered to OWNER and ENGINEER within fifteen days of the occurrence of the event giving rise to the claim. Notice of the amount of the claim with supporting data shall be delivered within forty-five days of such occurrence unless ENGINEER allows an additional period of time to ascertain accurate cost data. All claims for adjustment in the Contract Price shall be determined by ENGINEER if OWNER and CONTRACTOR cannot otherwise agree on the amount involved. Any change in the Contract Price resulting from any such claim shall be incorporated in a Change Order.

11.3. The value of any Work covered by a Change Order or of any claim for an increase or decrease in the Contract Price shall be determined in one of the following ways:

11.3.1. Where the Work involved is covered by unit prices contained in the Contract Documents, by application of unit prices to the quantities of the items involved (subject to the provisions of paragraph 11.9).

11.3.2. By mutual acceptance of a lump sum.

11.3.3. On the basis of the Cost of the Work (determined as provided in paragraphs 11.4 and 11.5) plus a Contractor's Fee for overhead and profit (determined as provided in paragraph 11.6).

Cost of the Work:

11.4. The term Cost of the Work means the sum of all costs necessarily incurred and paid by CONTRACTOR in the proper performance of the Work. Except as otherwise may be agreed to in writing by OWNER, such costs shall be in amounts no higher than those prevailing in the locality of the Project, shall include only the following items and shall not include any of the costs itemized in paragraph 11.5:

11.4.1. Payroll costs for employees in the direct employ of CONTRACTOR in the performance of the Work under schedules of job classifications agreed upon by OWNER and CONTRACTOR. Payroll costs for employees not employed full time on the Work shall be apportioned on the basis of their time spent on the Work. Payroll costs shall include, but not be limited to, salaries and wages plus the cost of fringe benefits which shall include social security contributions, unemployment, excise and payroll taxes, workers' or workmen's compensation, health and retirement benefits, bonuses, sick leave, vacation and holiday pay applicable thereto. Such employees shall include superintendents and foremen at the site. The expenses of performing Work after regular working hours, on Sunday or legal holidays, shall be included in the above to the extent authorized by OWNER.

11.4.2. Cost of all materials and equipment furnished and incorporated in the Work, including costs of transportation and storage thereof, and manufacturers' field services required in connection therewith. All cash discounts shall accrue to CONTRACTOR unless OWNER deposits funds with CONTRACTOR with which to make payments, in which case the cash discounts shall accrue to OWNER. All trade discounts, rebates and refunds, and all returns from sale of surplus materials and equipment shall accrue to OWNER and CONTRACTOR shall make provisions so that they may be obtained.

11.4.3. Payments made by CONTRACTOR to the Subcontractors for Work performed by Subcontractors. If required by OWNER, CONTRACTOR shall obtain competitive bids from Subcontractors acceptable to CONTRACTOR and shall deliver such bids to OWNER who will then determine, with the advice of ENGINEER, which bids will be accepted. If a subcontract provides that the Subcontractor is to be paid on the basis of Cost of the Work Plus a Fee, the Subcontractor's Cost of the Work shall be determined in the same manner as CONTACTOR's Cost of the Work. All subcontracts shall be subject to the other provisions of the Contract Documents insofar as applicable.

11.4.4. Costs of special consultants (including, but not limited to, engineers, architects, testing laboratories, surveyors, lawyers and accountants) employed for services specifically related to the Work.

11.4.5. Supplemental costs including the following:

11.4.5.1. The proportion of necessary transportation, travel and subsistence expenses of CONTRACTOR's employees incurred in discharge of duties connected with the Work.

11.4.5.2. Cost, including transportation and maintenance, of all materials, supplies, equipment, machinery, appliances, office and temporary facilities at the site and hand tools not owned by the workmen, which are consumed in the performance of the Work, and cost less market value of such items used but not consumed which remain the property of CONTRACTOR.

11.4.5.3. Rentals of all construction equipment and machinery and the parts thereof whether rented from CONTRACTOR or others in accordance with rental agreements approved by OWNER with the advice of ENGINEER, and the costs of transportation, loading, unloading, installation, dismantling and removal thereof—all in accordance with terms of said rental agreements. The rental of any such equipment, machinery or parts shall cease when the use thereof is no longer necessary for the Work.

11.4.5.4. Sales, use or similar taxes related to the Work, and for which CONTRACTOR is liable, imposed by any governmental authority.

11.4.5.5. Deposits lost for causes other than CONTRACTOR's negligence, royalty payments and fees for permits and licenses.

11.4.5.6. Losses and damages (and related expenses), not compensated by insurance or otherwise, to the Work or otherwise sustained by CONTRACTOR in connection with the execution of the Work, provided they have resulted from causes other than the negligence of CONTRACTOR, any Subcontractor, or anyone directly or indirectly employed by any of them or for whose acts any of them may be liable. Such losses shall include settlements made with the written consent and approval of OWNER. No such losses, damages and expenses shall be included in the Cost of the Work for the purpose of determining Contractor's Fee. If, however, any such loss or damage requires reconstruction and CONTRACTOR is placed in charge thereof, CONTRACTOR shall be paid for services a fee proportionate to that stated in paragraph 11.6.2.

11.4.5.7. The cost of utilities, fuel and sanitary facilities at the site.

11.4.5.8. Minor expenses such as telegrams, long distance telephone calls, telephone service at the site, expressage and similar petty cash items in connection with the Work.

11.4.5.9. Cost of premiums for additional Bonds and insurance required because of changes in the Work.

11.5. The term Cost of the Work shall not include any of the following:

11.5.1. Payroll costs and other compensation of CONTRACTOR's officers, executives, principals (of partnership and sole proprietorships), general managers, engineers, architects, estimators, lawyers, auditors, accountants, purchasing and contracting agents, expeditors, timekeepers, clerks and other personnel employed by CONTRACTOR whether at the site or in his principal or a branch office for general administration of the Work and not specifically included in the agreed upon schedule of job classifications referred to in subparagraph 11.4.1—all of which are to be considered administrative costs covered by the Contractor's Fee.

11.5.2. Expenses of CONTRACTOR's principal and branch offices other than CONTRACTOR's office at the site.

11.5.3. Any part of CONTRACTOR's capital expenses, including interest on CONTRACTOR's capital employed for the Work and charges against CONTRACTOR for delinquent payments.

11.5.4. Cost of premiums for all Bonds and for all insurance whether or not CONTRACTOR is required by the Contract Documents to purchase and maintain the same (except for additional Bonds and insurance required because of changes in the Work).

11.5.5. Costs due to the negligence of CONTRACTOR, any Subcontractor, or anyone directly or indirectly employed by any of them or for whose acts any of them may be liable, including but not limited to, the correction of defective Work, disposal of materials or equipment wrongly supplied and making good any damage to property.

11.5.6. Other overhead or general expense costs of any kind and the costs of any item not specifically and expressly included in paragraph 11.4.

Contractor's Fee:

11.6. The Contractor's Fee allowed to CONTRACTOR for overhead and profit shall be determined as follows:

11.6.1. a mutually acceptable fixed fee; or if none can be agreed upon,

11.6.2. a fee based on the following percentages of the various portions of the Cost of the Work:

11.6.2.1. for costs incurred under paragraphs 11.4.1 and 11.4.2, the Contractor's Fee shall be ten percent,

11.6.2.2. for costs incurred under paragraph 11.4.3, the Contractor's Fee shall be five percent; and if a subcontract is on the basis of Cost of the Work Plus a Fee, the maximum allowable to the Subcontractor as a fee for overhead and profit shall be ten percent, and

11.6.2.3. no fee shall be payable on the basis of costs itemized under paragraphs 11.4.4, 11.4.5 and 11.5.

11.7. The amount of credit to be allowed by CONTRACTOR to OWNER for any such change which results in a net decrease in cost, will be the amount of the actual net decrease. When both additions and credits are involved in any one change, the combined overhead and profit shall be figured on the basis of the net increase, if any.

Adjustment of Unit Prices:

11.8. Whenever the cost of any Work is to be determined pursuant to paragraphs 11.4 and 11.5, CONTRACTOR will submit in form acceptable to ENGINEER an itemized cost breakdown together with supporting data.

11.9. Where the quantity of Work with respect to any item that is covered by a unit price differs materially and significantly from the quantity of such Work indicated in the Contract Documents, an appropriate Change Order shall be issued on recommendation of ENGINEER to adjust the unit price.

Cash Allowances:

11.10. It is understood that CONTRACTOR has included in the Contract Price all allowances so named in the Contract Documents and shall cause the Work so covered to be done by such Subcontractors, manufacturers, fabricators, suppliers or distributors and for such sums within the limit of the allowances as may be acceptable to ENGINEER. Upon final payment, the Contract Price shall be adjusted as required and an appropriate Change Order issued. CONTRACTOR agrees that the original Contract Price includes such sums as CONTRACTOR deems proper for costs and profit on account of cash allowances. No demand for additional cost or profit in connection therewith will be valid.

ARTICLE 12—CHANGE OF THE CONTRACT TIME

12.1. The Contract Time may only be changed by a Change Order. Any claim for an extension in the Contract Time shall be based on written notice delivered to OWNER and ENGINEER within fifteen days of the occurrence of the event giving rise to the claim. Notice of the extent of the claim with supporting data shall be delivered within forty-five days of such occurrence unless ENGINEER allows an additional period of time to ascertain more accurate data. All claims for adjustment in the Contract Time shall be determined by ENGINEER if OWNER and CONTRACTOR cannot otherwise agree. Any change in the Contract Time resulting from any such claim shall be incorporated in a Change Order.

12.2. The Contract Time will be extended in an amount equal to time lost due to delays beyond the control of CONTRACTOR if a claim is made therefor as provided in paragraph 12.1. Such delays shall include, but not be limited to, acts or neglect by OWNER or others performing additional Work as contemplated by Article 7, or to fires, floods, labor disputes, epidemics, abnormal weather conditions, or acts of God.

12.3. All time limits stated in the Contract Documents are of the essence of the Agreement. The provisions of this Article 12 shall not exclude recovery for damages (including compensation for additional professional services) for delay by either party.

ARTICLE 13—WARRANTY AND GUARANTEE; TESTS AND INSPECTIONS; CORRECTION, REMOVAL OR ACCEPTANCE OF DEFECTIVE WORK

Warranty and Guarantee:

13.1. CONTRACTOR warrants and guarantees to OWNER and ENGINEER that all Work will be in accordance with the Contract Documents and will not be defective. Prompt notice of all defects shall be given to CONTRACTOR. All defective Work, whether or not in place, may be rejected, corrected or accepted as provided in this Article 13.

Access to Work:

13.2. ENGINEER and ENGINEER's representatives, other representatives of OWNER, testing agencies and governmental agencies with jurisdictional interests will have access to the Work at reasonable times for their observation, inspection and testing. CONTRACTOR shall provide proper and safe conditions for such access.

Tests and Inspections:

13.3. CONTRACTOR shall give ENGINEER timely notice of readiness of the Work for all required inspections, tests or approvals.

13.4. If any law, ordinance, rule, regulation, code, or order of any public body having jurisdiction requires any Work (or part thereof) to specifically be inspected, tested or approved, CONTRACTOR shall assume full responsibility therefor, pay all costs in connection therewith and furnish ENGINEER the required certificates of inspection, testing or approval. CONTRACTOR shall also be responsible for and shall pay all costs in connection with any inspection or testing required in connection with OWNER's or ENGINEER's acceptance of a manufacturer, fabricator, supplier or distributor of materials or equipment proposed to be incorporated in the Work, or of materials or equipment submitted for approval prior to CONTRACTOR's purchase thereof for incorporation in the Work. The cost of all other inspections, tests and approvals required by the Contract Documents shall be paid by OWNER (unless otherwise specified).

13.5. All inspections, tests or approvals other than those required by law, ordinance, rule, regulation, code or order of any public body having jurisdiction shall be performed by organizations acceptable to OWNER and CONTRACTOR (or by ENGINEER if so specified).

13.6. If any Work that is to be inspected, tested or approved is covered without written concurrence of ENGINEER, it must, if requested by ENGINEER, be uncovered for observation. Such uncovering shall be at CONTRACTOR's expense unless CONTRACTOR has given ENGINEER timely notice of CONTRACTOR's intention to cover such Work and ENGINEER has not acted with reasonable promptness in response to such notice.

13.7. Neither observations by ENGINEER nor inspections, tests or approvals by others shall relieve CONTRACTOR from his obligations to perform the Work in accordance with the Contract Documents.

Uncovering Work:

13.8. If any Work is covered contrary to the written request of ENGINEER, it must, if requested by ENGINEER, be uncovered for ENGINEER's observation and replaced at CONTRACTOR's expense.

13.9. If ENGINEER considers it necessary or advisable that covered Work be observed by ENGINEER or inspected or tested by others, CONTRACTOR, at ENGINEER's request, shall uncover, expose or otherwise make available for observation, inspection or testing as ENGINEER may require, that portion of the Work in question, furnishing all necessary labor, material and equipment. If it is found that such Work is defective, CONTRACTOR shall bear all the expenses of such uncovering, exposure, observation, inspection and testing and of satisfactory reconstruction, including compensation for additional professional services, and an appropriate deductive Change Order shall be issued. If, however, such Work is not found to be defective, CONTRACTOR shall be allowed an increase in the Contract Price or an extension of the Contract Time, or both, directly attributable to such uncovering, exposure, observation, inspection, testing and reconstruction if he makes a claim therefor as provided in Articles 11 and 12.

Owner May Stop the Work:

13.10. If the Work is defective, or CONTRACTOR fails to supply sufficient skilled workmen or suitable materials or equipment, OWNER may order CONTRACTOR to stop the Work, or any portion thereof, until the cause for such order has been eliminated; however, this right of OWNER to stop the Work shall not give rise to any duty on the part of OWNER to exercise this right for the benefit of CONTRACTOR or any other party.

Correction or Removal of Defective Work:

13.11. If required by ENGINEER, CONTRACTOR shall promptly, without cost to OWNER and as specified by ENGINEER, either correct any defective Work, whether or not fabricated, installed or completed, or, if the Work has been rejected by ENGINEER, remove it from the site and replace it with nondefective Work.

One Year Correction Period:

13.12. If within one year after the date of Substantial Completion or such longer period of time as may be prescribed by law or by the terms of any applicable special guarantee required by the Contract Documents or by any specific provision of the Contract Documents, any Work is found to be defective, CONTRACTOR shall promptly, without cost to OWNER and in accordance with OWNER's written instructions, either correct such defective Work, or, if it has been rejected by OWNER, remove it from the site and replace it with nondefective Work. If CONTRACTOR does not promptly comply with the terms of such instructions, or in an emergency where delay would cause serious risk of loss or damage, OWNER may have the defective Work corrected or the rejected Work removed and replaced, and all direct and indirect costs of such removal and replacement, including compensation for additional professional services, shall be paid by CONTRACTOR.

Acceptance of Defective Work:

13.13. If, instead of requiring correction or removal and replacement of defective Work, OWNER (and, prior to ENGINEER's recommendation of final payment, also ENGINEER) prefers to accept it, OWNER may do so. In such case, if acceptance occurs prior to ENGINEER's recommendation of final payment, a Change Order shall be issued incorporating the necessary revisions in the Contract Documents, including appropriate reduction in the Contract Price; or, if the acceptance occurs after such recommendation, an appropriate amount shall be paid by CONTRACTOR to OWNER.

OWNER May Correct Defective Work:

13.14. If CONTRACTOR fails within a reasonable time after written notice of ENGINEER to proceed to correct and to correct defective Work or to remove and replace rejected Work as required by ENGINEER in accordance with paragraph 13.11, or if CONTRACTOR fails to perform the Work in accordance with the Contract Documents (including any requirements of the progress schedule), OWNER may, after seven days' written notice to CONTRACTOR, correct and remedy any such deficiency. In exercising his rights under this paragraph OWNER shall proceed expeditiously. To the extent necessary to complete corrective and remedial action, OWNER may exclude CONTRACTOR from all or part of the site, take possession of all or part of the Work, and suspend CONTRACTOR's services related thereto, take possession of CONTRACTOR's tools, appliances, construction equipment and machinery at the site and incorporate in the Work all materials and equipment stored at the site or for which OWNER has paid CONTRACTOR but which are stored elsewhere. CONTRACTOR shall allow OWNER, OWNER's representatives, agents and employees such access to the site as may be necessary to enable OWNER to exercise his rights under this paragraph. All direct and indirect costs of OWNER in exercising such rights shall be charged against CONTRACTOR in an amount verified by ENGINEER, and a Change Order shall be issued incorporating the necessary revisions in the Contract Documents and a reduction in the Contract Price. Such direct and indirect costs shall include, in particular but without limitation, compensation for additional professional services required and all costs of repair and replacement of work of others destroyed or damaged by correction, removal or replacement of CONTRACTOR's defective Work. CONTRACTOR shall not be allowed an extension of the Contract Time because of any delay in performance of the Work attributable to the exercise by OWNER of OWNER's rights hereunder.

ARTICLE 14—PAYMENTS TO CONTRACTOR AND COMPLETION

Schedules:

14.1 At least ten days prior to submitting the first Application for a progress payment, CONTRACTOR shall (except as otherwise specified in the General Requirements) submit to ENGINEER a progress schedule, a final schedule of Shop Drawing submission and where applicable a schedule of values of the Work. These schedules shall be satisfactory in form and substance to ENGINEER. The schedule of values shall include quantities and unit prices aggregating the Contract Price, and shall subdivide the Work into component parts in sufficient detail to serve as the basis for progress payments during construction. Upon acceptance of the schedule of values by ENGINEER, it shall be incorporated into a form of Application for Payment acceptable to ENGINEER.

Application for Progress Payment:

14.2. At least ten days before each progress payment falls due (but not more often than once a month), CONTRACTOR shall submit to ENGINEER for review an Application for Payment filled out and signed by CONTRACTOR covering the Work completed as of the date of the Application and accompanied by such supporting documentation as is required by the Contract Documents and also as ENGINEER may reasonably require. If payment is requested on the basis of materials and equipment not incorporated in the Work but delivered and suitably stored at the site or at another location agreed to in writing, the Application for Payment shall also be accompanied by such data, satisfactory to OWNER, as will establish OWNER's title to the material and equipment and protect OWNER's interest therein, including applicable insurance. Each subsequent Application for Payment shall include an affidavit of CONTRACTOR stating that all previous progress payments received on account of the Work have been applied to discharge in full all of CONTRACTOR's obligations reflected in prior Applications for Payment. The amount of retainage with respect to progress payments will be as stipulated in the Agreement.

CONTRACTOR's Warranty of Title:

14.3. CONTRACTOR warrants and guarantees that title to all Work, materials and equipment covered by any Application for Payment, whether incorporated in the Project or not, will pass to OWNER at the time of payment free and clear of all liens, claims, security interests and encumbrances (hereafter in these General Conditions referred to as "Liens").

Review of Applications for Progress Payment:

14.4. ENGINEER will, within ten days after receipt of each Application for Payment, either indicate in writing a recommendation of payment and present the Application to OWNER, or return the Application to CONTRACTOR indicating in writing ENGINEER's reasons for refusing to recommend payment. In the latter case, CONTRACTOR may make the necessary corrections and resubmit the Application. OWNER shall, within ten days of presentation to him of the Application for Payment with ENGINEER's recommendation pay CONTRACTOR the amount recommended.

14.5. ENGINEER's recommendation of any payment requested in an Application for Payment will constitute a representation by ENGINEER to OWNER, based on ENGINEER's on-site observations of the Work in progress as an experienced and qualified design professional and on ENGINEER's review of the Application for Payment and the accompanying data and schedules that the Work has progressed to the point indicated; that, to the best of ENGINEER's knowledge, information and belief, the quality of the Work is in accordance with the Contract Documents (subject to an evaluation of the Work as a functioning Project upon Substantial Completion, to the results of any subsequent tests called for in the Contract Documents and any qualifications stated in the recommendation: and that CONTRACTOR is entitled to payment of the amount recommended. However, by recommending any such payment ENGINEER will not thereby be deemed to have represented that exhaustive or continuous on-site inspections have been made to check the quality or the quantity of the Work, or that the means, methods, techniques, sequences, and procedures of construction have been reviewed or that any examination has been made to ascertain how or for what purpose CONTRACTOR has used the moneys paid or to be paid to CONTRACTOR on account of the Contract Price, or that title to any Work, materials or equipment has passed to OWNER free and clear of any Liens.

14.6. ENGINEER's recommendation of final payment will constitute an additional representation by ENGINEER to OWNER that the conditions precedent to CONTRACTOR's being entitled to final payment as set forth in paragraph 14.13 have been fulfilled.

14.7. ENGINEER may refuse to recommend the whole or any part of any payment if, in his opinion, it would be incorrect to make such representations to OWNER. He may also refuse to recommend any such payment, or, because of subsequently discovered evidence or the results of subsequent inspections or tests, nullify and such payment previously recommended to such extent as may be necessary in ENGINEER's opinion to protect OWNER from loss because:

14.7.1. the Work is defective, or completed Work has been damaged requiring correction or replacement,

14.7.2. written claims have been made against OWNER or Liens have been filed in connection with the Work,

14.7.3. the Contract Price has been reduced because of Modifications,

14.7.4. OWNER has been required to correct defective Work or complete the Work in accordance with paragraph 13.14,

14.7.5. of CONTRACTOR's unsatisfactory prosecution of the Work in accordance with the Contract Documents, or

14.7.6. CONTRACTOR's failure to make payment to Subcontractors, or for labor, materials or equipment.

Substantial Completion:

14.8. When CONTRACTOR considers the entire Work ready for its intended use CONTRACTOR shall, in writing to OWNER and ENGINEER, certify that the entire Work is substantially complete and request that ENGINEER issue a certificate of Substantial Completion. Within a reasonable time thereafter, OWNER, CONTRACTOR and ENGINEER shall make an inspection of the Work to determine the status of completion. If ENGINEER does not consider the Work substantially complete, ENGINEER will notify CONTRACTOR in writing giving his reasons therefor. If ENGINEER considers the Work substantially complete, ENGINEER will prepare and deliver to OWNER a tentative certificate of Substantial Completion which shall fix the date of Substantial Completion. There shall be attached to the certificate a tentative list of items to be completed or corrected before final payment. OWNER shall have seven days after receipt of the tentative certificate during which he may make written objection to ENGINEER as to any provisions of the certificate or attached list. If, after considering such objections, ENGINEER concludes that the Work is not substantially complete, ENGINEER will within fourteen days after submission of the tentative certificate to OWNER notify CONTRACTOR in writing, stating his reasons therefor. If, after consideration of OWNER's objections, ENGINEER considers the Work substantially complete, ENGINEER will within said fourteen days execute and deliver to OWNER and CONTRACTOR a definitive certificate of Substantial Completion (with a revised tentative list of items to be completed or corrected) reflecting such changes from the tentative certificate as he believes justified after consideration of any objections from OWNER. At the time of delivery of the tentative certificate of Substantial Completion ENGINEER will deliver to OWNER and CONTRACTOR a written recommendation as to division of responsibilities pending final payment between OWNER and CONTRACTOR with respect to security, operation, safety, maintenance, heat, utilities and insurance. Unless OWNER and CONTRACTOR agree otherwise in writing and so inform ENGINEER prior to his issuing the definitive certificate of Substantial Completion ENGINEER's aforesaid recommendation will be binding on OWNER and CONTRACTOR until final payment.

14.9. OWNER shall have the right to exclude CONTRACTOR from the Work after the date of Substantial Completion, but OWNER shall allow CONTRACTOR reasonable access to complete or correct items on the tentative list.

Partial Utilization:

14.10. Use by OWNER of completed portions of the Work may be accomplished prior to Substantial Completion of all the Work subject to the following:

14.10.1. OWNER at any time may request CONTRACTOR in writing to permit OWNER to use any part of the Work which OWNER believes to be substantially complete and which may be so used without significant interference with construction of the other parts of the Work. If CONTRACTOR agrees, CONTRACTOR will certify to OWNER and ENGINEER that said part of the Work is substantially complete and request ENGINEER to issue a certificate of Substantial Completion for that part of the Work. Within a reasonable time thereafter OWNER, CONTRACTOR and ENGINEER shall make an inspection of that part of the Work to determine its status of completion. If ENGINEER does not consider that part of the Work to be substantially complete, ENGINEER will notify OWNER and CONTRACTOR in writing giving his reasons therefor. If ENGINEER considers that part of the Work to be substantially complete, ENGINEER will execute and deliver to OWNER and CONTRACTOR a certificate to that effect, fixing the date of Substantial Completion as to that part of the Work, attaching thereto a tentative list of items to be completed or corrected before final payment. Prior to issuing a certificate of Substantial Completion as to part of the Work ENGINEER will deliver to OWNER and CONTRACTOR a written recommendation as to the division of responsibilities pending final payment between OWNER and CONTRACTOR with respect to security, operation, safety, maintenance, utilities and insurance for that part of the Work which shall become binding upon OWNER and CONTRACTOR at the time of issuing the definitive certificate of Substantial Completion as to that part of the Work unless OWNER and CONTRACTOR shall have otherwise agreed in writing and so informed ENGINEER. OWNER shall have the right to exclude CONTRACTOR from any part of the Work which ENGINEER has so certified to be substantially complete, but OWNER shall allow CONTRACTOR reasonable access to complete or correct items on the tentative list.

14.10.2. In lieu of the issuance of a certificate of Substantial Completion as to part of the Work, OWNER may take over operation of a facility constituting part of the Work whether or not it is substantially complete if such facility is functionally and separately useable; provided that prior to any such takeover, OWNER and CONTRACTOR have agreed as to the division of responsibilities between OWNER and CONTRACTOR for security, operation, safety, maintenance, correction period, heat, utilities and insurance with respect to such facility.

14.10.3. No occupancy of part of the Work or taking over of operations of a facility will be accomplished prior to compliance with the requirements of paragraph 5.14 in respect of property insurance.

Final Inspection:

14.11. Upon written notice from CONTRACTOR that the Work is complete, ENGINEER will make a final inspection with OWNER and CONTRACTOR and will notify CONTRACTOR in writing of all particulars in which this inspection reveals that the Work is incomplete or defective. CONTRACTOR shall immediately take such measures as are necessary to remedy such deficiencies.

Final Application for Payment:

14.12. After CONTRACTOR has completed all such corrections to the satisfaction of ENGINEER and delivered all maintenance and operating instructions, schedules, guarantees, Bonds, certificates of inspection, marked-up record documents and other documents—all as required by the Contract Documents, and after ENGINEER has indicated that the Work is acceptable (subject to the provisions of paragraph 14.16), CONTRACTOR may make application for final payment following the procedure for progress payments. The final Application for Payment shall be accompanied by all documentation called for in the Contract Documents and such other data and schedules as ENGINEER may reasonably require, together with complete and legally effective releases or waivers (satisfactory to OWNER) of all Liens arising out of or filed in connection with the Work. In lieu thereof and as approved by OWNER, CONTRACTOR may furnish receipts or releases in full; an affidavit of CONTRACTOR that the releases and receipts include all labor, services, material and equipment for which a Lien could be filed, and that all payrolls, material and equipment bills, and other indebtedness connected with the Work for which OWNER or his property might in any way be responsible, have been paid or otherwise satisfied; and consent of the Surety, if any, to final payment. If any Subcontractor, manufacturer, fabricator, supplier or distributor fails to furnish a release or receipt in full, CONTRACTOR may furnish a Bond or other collateral satisfactory to OWNER to indemnify OWNER against any Lien.

Final Payment and Acceptance:

14.13. If, on the basis of ENGINEER's observation of the Work during construction and final inspection, and ENGINEER's review of the final Application for Payment and accompanying documentation—all as required by the Contract Documents, ENGINEER is satisfied that the Work has been completed and CONTRACTOR has fulfilled all of his obligations under the Contract Documents, ENGINEER will, within ten days after receipt of the final Application for Payment, indicate in writing his recommendation of payment and present the Application to OWNER for payment. Thereupon ENGINEER will give written notice to OWNER and CONTRACTOR that the Work is acceptable subject to the provisions of paragraph 14.16. Otherwise, ENGINEER will return the Application to CONTRACTOR, indicating in writing the reasons for refusing to recommend final payment, in which case CONTRACTOR shall make the necessary corrections and resubmit the Application. If the Application and accompanying documentation are appropriate as to form and substance, OWNER shall, within thirty days after receipt thereof pay CONTRACTOR the amount recommended by ENGINEER.

14.14. If, through no fault of CONTRACTOR, final completion of the Work is significantly delayed thereof and if ENGINEER so confirms, OWNER shall, upon receipt of CONTRACTOR's final Application for Payment and recommendation of ENGINEER, and without terminating the Agreement, make payment of the balance due for that portion of the Work fully completed and accepted. If the remaining balance to be held by OWNER for Work not fully completed or corrected is less than the retainage stipulated in the Agreement, and if Bonds have been furnished as required in paragraph 5.1, the written consent of the Surety to the payment of the balance due for that portion of the Work fully completed and accepted shall be submitted by CONTRACTOR to ENGINEER with the Application for such payment. Such payment shall be made under the terms and conditions governing final payment, except that it shall not constitute a waiver of claims.

Contractor's Continuing Obligation:

14.15. CONTRACTOR's obligation to perform and complete the Work in accordance with the Contract Documents shall be absolute. Neither recommendation of any progress or final payment by ENGINEER, nor the issuance of a certificate of Substantial Completion, nor any payment by OWNER to CONTRACTOR under the Contract Documents, nor any use or occupancy of the Work or any part thereof by OWNER, nor any act of acceptance by OWNER nor any failure to do so, nor the issuance of a notice of acceptability by ENGINEER pursuant to paragraph 14.13, nor any correction of defective Work by OWNER shall constitute an acceptance of Work not in accordance with the Contract Documents or a release of CONTRACTOR's obligation to perform the Work in accordance with the Contract Documents.

Waiver of Claims:

14.16. The making and acceptance of final payment shall constitute:

14.16.1. a waiver of all claims by OWNER against CONTRACTOR, except claims arising from unsettled Liens, from defective Work appearing after final inspection pursuant to paragraph 14.11 or from failure to comply with the Contract Documents or the terms of any special guarantees specified therein; however, it shall not constitute a waiver by OWNER of any rights in respect of CONTRACTOR's continuing obligations under the Contract Documents; and

14.16.2. a waiver of all claims by CONTRACTOR against OWNER other than those previously made in writing and still unsettled.

ARTICLE 15—SUSPENSION OF WORK AND TERMINATION

Owner May Suspend Work:

15.1. OWNER may, at any time and without cause, suspend the Work or any portion thereof for a period of not more than ninety days by notice in writing to CONTRACTOR and ENGINEER which shall fix the date on which Work shall be resumed. CONTRACTOR shall resume the Work on the date so fixed. CONTRACTOR will be allowed an increase in the Contract Price or an extension of the Contract Time, or both, directly attributable to any suspension if he makes a claim therefor as provided in Articles 11 and 12.

Owner May Terminate:

15.2. Upon the occurrence of any one or more of the following events:

15.2.1. if CONTRACTOR is adjudged a bankrupt or insolvent,

15.2.2. if CONTRACTOR makes a general assignment for the benefit of creditors,

15.2.3. if a trustee or receiver is appointed for CONTRACTOR or for any of CONTRACTOR's property,

15.2.4. if CONTRACTOR files a petition to take advantage of any debtor's act, or to reorganize under the bankruptcy or similar laws,

15.2.5. if CONTRACTOR repeatedly fails to supply sufficient skilled workmen or suitable materials or equipment,

15.2.6. if CONTRACTOR repeatedly fails to make prompt payments to Subcontractors or for labor, materials or equipment,

15.2.7. if CONTRACTOR disregards laws, ordinances, rules, regulations or orders of any public body having jurisdiction,

15.2.8. if CONTRACTOR disregards the authority of ENGINEER, or

15.2.9. if CONTRACTOR otherwise violates in any substantial way any provisions of the Contract Documents,

OWNER may after giving CONTRACTOR and his Surety seven days' written notice, terminate the services of CONTRACTOR, exclude CONTRACTOR from the site and take possession of the Work and of all CONTRACTOR's tools, appliances, construction equipment and machinery at the site and use the same to the full extent they could be used by CONTRACTOR (without liability to CONTRACTOR for trespass or conversion), incorporate in the Work all materials and equipment stored at the site or for which OWNER has paid CONTRACTOR but which are stored elsewhere, and finish the Work as OWNER may deem expedient. In such case CONTRACTOR shall not be entitled to receive any further payment until the Work is finished. If the unpaid balance of the Contract Price exceeds the direct and indirect costs of completing the Work, including compensation for additional professional services, such excess shall be paid to CONTRACTOR. If such costs exceed such unpaid balance, CONTRACTOR shall pay the difference to OWNER. Such costs incurred by OWNER shall be verified by ENGINEER and incorporated in a Change Order, but in finishing the Work OWNER shall not be required to obtain the lowest figure for the Work performed.

15.3. Where CONTRACTOR's services have been so terminated by OWNER, the termination shall not affect any rights of OWNER against CONTRACTOR then existing or which may thereafter accrue. Any retention or payment of moneys due CONTRACTOR by OWNER will not release CONTRACTOR from liability.

15.4. Upon seven days' written notice to CONTRACTOR and ENGINEER, OWNER may, without cause and without prejudice to any other right or remedy, elect to abandon the Work and terminate the Agreement. In such case, CONTRACTOR shall be paid for all Work executed and any expense sustained plus reasonable termination expenses.

Contractor May Stop Work or Terminate:

15.5. If, through no act or fault of CONTRACTOR, the Work is suspended for a period of more than ninety days by OWNER or under an order of court or other public authority, or ENGINEER fails to act on any Application for Payment within thiry days after it is submitted, or OWNER fails for thirty days to pay CONTRACTOR any sum finally determined to be due, then CONTRACTOR may, upon seven days' written notice to OWNER and ENGINEER, terminate the Agreement and recover from OWNER payment for all Work executed and any expense sustained plus reasonable termination expenses. In addition and in lieu of terminating the Agreement, if ENGINEER has failed to act on an Application for Payment or OWNER has failed to make any payment as aforesaid, CONTRACTOR may upon seven days' notice to OWNER and ENGINEER stop the Work until payment of all amounts then due. The provisions of this paragraph shall not relieve CONTRACTOR of his obligations under paragraph 6.29 to carry on the Work in accordance with the progress schedule and without delay during disputes and disagreements with OWNER.

ARTICLE 16—ARBITRATION

16.1. All claims, disputes and other matters in question between OWNER and CONTRACTOR arising out of, or relating to the Contract Documents or the breach thereof except for claims which have been waived by the making or acceptance of final payment as provided by paragraph 14.16, shall be decided by arbitration in accordance with the Construction Industry Arbitration Rules of the American Arbitration Association then obtaining subject to the limitations of this Article 16. This agreement so to arbitrate and any other agreement or consent to arbitrate entered into in accordance herewith as provided in this Article 16 will be specifically enforceable under the prevailing arbitration law of any court having jurisdiction.

16.2. No demand for arbitration of any claim, dispute or other matter that is required to be referred to ENGINEER initially for decision in accordance with paragraph 9.9 shall be made until the earlier of (a) the date on which ENGINEER has rendered a decision or (b) the tenth day after the parties have presented their evidence to ENGINEER if a written decision has not been rendered by ENGINEER before that date. No demand for arbitration of any such claim, dispute or other matter shall be made later than thirty days after the date on which ENGINEER has rendered a written decision in respect thereof in accordance with paragraph 9.9; and the failure to demand arbitration within said thirty days' period shall result in ENGINEER's decision being final and binding upon OWNER and CONTRACTOR. If ENGINEER renders a decision after arbitration proceedings have been initiated, such decision may be entered as evidence but shall not supersede the arbitration proceedings, except where the decision is acceptable to the parties concerned.

16.3. Notice of the demand for arbitration shall be filed in writing with the other party to the Agreement and with the

American Arbitration Association, and a copy shall be sent to ENGINEER for information. The demand for arbitration shall be made within the thirty-day period specified in paragraph 16.2 where applicable, and in all other cases within a reasonable time after the claim, dispute or other matter in question has arisen, and in no event shall any such demand be made after institution of legal or equitable proceedings based on such claim, dispute or other matter in question would be barred by the applicable statute of limitations.

16.4. No arbitration arising out of or relating to the Contract Documents shall include by consolidation, joinder or in any other manner any other person or entity (including ENGINEER, his agents, employees or consultants) who is not a party to this Agreement unless:

16.4.1. the inclusion of such other person or entity is necessary if complete relief is to be afforded among those who are already parties to the arbitration,

16.4.2. such other person or entity is substantially involved in a question of law or fact which is common to those who are already parties to the arbitration and which will arise in such proceedings, and

16.4.3. the written consent of the other person or entity sought to be included and of OWNER and CONTRACTOR has been obtained for such inclusion, which consent shall make specific reference to this paragraph; but no such consent shall constitute consent to arbitration of any dispute not specifically described in such consent or to arbitration with any party not specifically identified in such consent.

16.5. The award rendered by the arbitrators will be final, judgment may be entered upon it in any court having jurisdiction thereof, and will not be subject to modification or appeal except to the extent permitted by Sections 10 and 11 of the Federal Arbitration Act (9 U.S.C. §§10, 11).

ARTICLE 17—MISCELLANEOUS

Giving Notice:

17.1. Whenever any provision of the Contract Documents requires the giving of written notice it shall be deemed to have been validly given if delivered in person to the individual or to a member of the firm or to an officer of the corporation for whom it is intended, or if delivered at or sent by registered or certified mail, postage prepaid, to the last business address known to the giver of the notice.

Computation of Time:

17.2. When any period of time is referred to in the Contract Documents by days, it shall be computed to exclude the first and include the last day of such period. If the last day of any such period falls on a Saturday or Sunday or on a day made a legal holiday by the law of the applicable jurisdiction, such day shall be omitted from the computation.

General:

17.3. Should OWNER or CONTRACTOR suffer injury or damage to his person or property because of any error, omission or act of the other party or of any of the other party's employees or agents or others for whose acts the other party is legally liable, claim shall be made in writing to the other party within a reasonable time of the first observance of such injury or damage.

17.4. The duties and obligations imposed by these General Conditions and the rights and remedies available hereunder to the parties hereto, and, in particular but without limitation, the warranties, guarantees and obligations imposed upon CONTRACTOR by paragraphs 6.30, 13.1, 13.11, 13.14, 14.3 and 15.2 and all of the rights and remedies available to OWNER and ENGINEER thereunder, shall be in addition to, and shall not be construed in any way as a limitation of, any rights and remedies available to any or all of them which are otherwise imposed or available by law or contract, by special warranty or guarantee or by other provisions of the Contract Documents, and the provisions of this paragraph shall be as effective as if repeated specifically in the Contract Documents in connection with each particular duty, obligation, right and remedy to which they apply. All representations, warranties and guarantees made in the Contract Documents shall survive final payment and termination or completion of this Agreement.

This document has important legal consequences; consultation with an attorney is encouraged with respect to its completion or modification.

STANDARD FORM OF AGREEMENT
BETWEEN

OWNER AND ENGINEER

FOR

PROFESSIONAL SERVICES

Prepared by

ENGINEERS' JOINT CONTRACT DOCUMENTS COMMITTEE

and

Issued and Published Jointly by

PROFESSIONAL ENGINEERS IN PRIVATE PRACTICE
A practice division of the
NATIONAL SOCIETY OF PROFESSIONAL ENGINEERS

and by

AMERICAN CONSULTING ENGINEERS COUNCIL

and by

AMERICAN SOCIETY OF CIVIL ENGINEERS

1910-1 (1979 Edition)
Reprinted Jan. 1980
10M CORP.

Guide Sheet for Completing Standard Form of Agreement Between Owner and Engineer for Professional Services

1. **Page 1**—Insert proper name of partnership, corporation or governmental body which is Owner on first page and signature page. Make certain person signing for Owner has authority to do so and that there is an indication of the capacity in which he signs. (See Commentary, 1974 edit., paragraph 7.)

2. **Page 1**—Insert as complete a description of the Project as possible. Include, to the extent known, a description of the land where the Project is to be located, any special requirements as to performance, capacity or function, budgetary limitations and any special source of funds for which the Project must qualify. Identify studies, reports or analyses previously prepared which are being furnished by Owner to Engineer for his guidance, such as reports and studies referred to in paragraph 3.3. Identify other special aspects or peculiarities of the Project. (See Commentary, 1974 edit., paragraph 8.)

3. **Section 1—Basic Services**—Supplemental description in Exhibit A "Further Description of Basic Engineering Services and Related Matters." The Agreement contemplates that an exhibit will be prepared to describe in more detail the Basic Services to be rendered during Phases 1 through 5 (see language following paragraphs **1.2.6, 1.3.4, 1.4.5** and **1.5.4;** also paragraphs **1.6.1** and **2.1**). A suggested format for such an exhibit appears on the reverse of this Guide Sheet. The paragraphs of the exhibit follow the sequence of the description of Basic Services and should contain a further or expanded description of each phase of Basic Services as applicable to the particular Project undertaken (the printed language of the Agreement being generally applicable to most Projects). Exhibit A should also include an agreed time for performance of Phase 1, 2 and 3 Basic Services as indicated in paragraphs **4.2, 4.3** and **4.4**. If Basic Services are expanded to include services normally treated as Additional Services, care should be taken to strike out or modify comparable language in Section 2—Additional Services of Engineer. Post construction services should also be described in Exhibit A and paragraphs **2.1.14** and **2.1.15** modified accordingly. If Engineer rather than Owner is to furnish services of the types described in paragraphs **3.3** and **3.4,** this information should also be included in Exhibit A. Reference is also made to Exhibit A in paragraph **5.1.1** of each alternate method of payment. In those places where supplemental or amending data will not appear in Exhibit A reference to the Exhibit should be crossed out and initialed by both parties.

4. **Paragraph 2.2.2—Resident Project Representative** (see Commentary, 1974 edit., paragraph 20). If it is understood at the time of signing the Agreement that a Resident Project Representative is to be employed, this should be reflected in Exhibit A. However, paragraph **2.2.1** contemplates that a decision in this regard will not be made until later in which case the understanding should be reduced to writing and attached as an exhibit to the Agreement. Regardless of when the decision is reached, the job site responsibilities of the Resident Project Representative should be clarified in writing and this should be attached to the Agreement. This is identified as Exhibit B in paragraph **8.2.2**. The Engineers' Joint Contract Documents Committee has prepared a form entitled "Duties, Responsibilities and Limitations of Authority of Resident Project Representative" No. 1910-1-A (1979 edit.) which may be used for this purpose or as a check-off list in preparing a document specially tailored to the particular assignment.

5. **Section 4—Period of Service.** This Section has been prepared in recognition of the thought expressed in paragraph 4.1 that there will be a continuous period of service through completion of the Construction Phase with timely responses from Owner to Engineer's submission and with prompt authority to proceed with each Phase of services after the preceding Phase has been completed. The stipulated time for performing Phase 1, 2 and 3 services referred to in paragraphs **4.2, 4.3** and **4.4** should be determined in recognizing this understanding, and the agreed upon time periods should be set forth in Exhibit A. The blank space in paragraph **4.9** should be filled in with whatever is considered a reasonable time in the particular Project for the taking of bids or receiving proposals, awarding the contract and starting construction. The three-month and one-year periods provided in paragraph **4.10** may not be appropriate for all Projects and should be modified as circumstances dictate. Paragraph **4.11** recognizes the possibility of construction being performed under several prime contracts with possibly different starting times with the result that engineering services during Phases 3, 4 and 5 may have to be specially sequenced. In such cases it is expected that the special timing required for services rendered during Phases 3, 4 and 5 will be set forth in Exhibit A.

6. **Section 5—Methods of Payment.** Five different methods of payment are covered by the attached form of Agreement: these are Lump Sum, Percentage, Professional Fee plus expense, Direct Labor Costs times a factor and Payroll Costs times a factor. Each appears on a separate page so that the page for the method selected may be retained and the others destroyed without affecting the page numbers of the printed form. If some other method or combination of methods of payment is to be used, a separately-prepared page may be inserted in lieu of the printed ones. Each printed form contains blank spaces that should be filled in.

When payment is on the basis of a lump sum or a percentage fee, it is customary to provide for higher compensation for Engineer in a Project involving several prime contracts or in the case of construction contracts which contain cost-plus or incentive-savings provisions. Accordingly, blank spaces and alternate choices have been provided. The blank spaces for the number of prime contractors is intended to be filled in with the number normally anticipated for a Project of the type involved, such as a governmentally financed one where the law requires separate prime contracts. Paragraph **2.1.8** is intended to cover an additional number, if any, not anticipated at the time Engineer makes his fee commitment with Owner. In filling the blank spaces with respect to the anticipated extra services due to cost-plus or incentive-compensation arrangements with Contractors, bear in mind the provision of paragraph **3.8** which requires Owner to furnish auditing services if he wishes to ascertain how or for what purposes Contractor has used the moneys paid to him under the Construction Contract. Bear in mind also the provisions of paragraph **3.8** which require Owner to provide such inspection services as he may require to ascertain if Contractor is performing the work in accordance with legal requirements, such as safety codes, etc.

Note that the terms Construction Costs, Direct Labor Costs, Payroll Costs and Reimbursable Expenses are specifically defined and that in paragraph **5.1.1** of each alternate method of payment reference is made to Exhibit A "Further Description of Basic Engineering Services and Related Matters".

7. **Paragraph 5.1.2—Payment for Additional Services** (see Commentary, 1974 edit., paragraph 12). Space has been provided in paragraphs **5.1.2.1** through **5.1.2.4** for different rates of compensation for different types of Additional Services. Any inapplicable paragraph should be crossed out and initials inserted in the margin by both parties. If one method of compensation is to apply to all Additional Services, be certain that the cross references are correctly adjusted.

8. Note that paragraph **5.1.3** provides for reimbursement of the "actual costs" of Reimbursable Expenses without a mark up which is sometimes customary.

9. **Paragraph 5.4.1** contemplates the identification of the key personnel in Engineer's organization who are to be assigned to the Project (see Commentary, 1974 edit., paragraph 13 as to the proper meaning of the term "principals"); this applies to both the Direct Labor Cost and the Payroll Cost definitions. In developing the factor for the Payroll Cost method of compensation, bear in mind that paragraph **5.4.1** is set up so that the factor is applied to the salary, wages and fringe benefits of all personnel, and payroll costs of principals (whether they be corporate officers at fixed salaries or partners with drawing accounts who share in the profits) are to be itemized. When using Payroll Costs as a basis for compensation, instead of the detailed accounting required to substantiate the amount paid for customary and statutory benefits of personnel, a percentage of salaries and wages may be agreed to in advance and inserted in the blank space at the end of paragraph **5.4.1**; otherwise the sentence should be crossed out and initialed by both parties.

10. The blank space in paragraph **5.3.2** should be filled in after consultation with Owner.

11. **Paragraph 7.5—Arbitration.** This provides for compulsory and binding arbitration of all disputes between Owner and Engineer where the claimed amount at issue is not more than $200,000. It also restricts joinder in the arbitration proceedings of others who are not a party to the Agreement. The Engineers' Joint Contract Documents Committee believes that arbitration of such disputes under the Construction Industry Arbitration Rules of the American Arbitration Association is in the best interest of both Owner and Engineer, but also recognizes that others may differ, and the laws of all states are not similar. Accordingly, the form of Agreement has been prepared so that paragraph **7.5**, Arbitration, may be eliminated from the Agreement simply by tearing out the separate page (page 13) on which it appears AND completing the information at the bottom of the last page to indicate the page number as well as the total number of pages that make up the entire Agreement.

12. **Paragraph 8.1** contains space for additional or special provisions required by Owner or Engineer. These include any particular requirements as to insurance to be carried by Engineer or for project insurance. It is the practice of some engineers to include in their agreements language limiting their liability for negligence to a fixed dollar amount or a certain percentage of total fee for services. Any such language should be specifically negotiated with Owner after consultation with insurance counsel as to recommended verbiage and legal counsel as to enforceability. The provisions agreed to should be inserted as a subparagraph of paragraph 8.1.

13. **Paragraph 8.2** identifies two Exhibits: Exhibit A "Further Description of Basic Engineering Services and Related Matters" and Exhibit B "Duties, Responsibilities and Limitations of Authority of Resident Project Representative". The number of pages of each Exhibit should be inserted in the blank spaces. Other Exhibits should be listed in the space provided. If either Exhibit A or B is not to be attached, all reference to the document should be crossed out of the Agreement.

EXHIBIT A TO AGREEMENT BETWEEN OWNER AND ENGINEER FOR PROFESSIONAL SERVICES, dated ________________________, 19____ (for use with No. 1910-1, 1979 Edition).

Further Description of Basic Engineering Services and Related Matters

1. This is an exhibit attached to, made a part of and incorporated by reference into the Agreement made on ________________________, 19____ between ____ (Owner) and (Engineer) providing for professional engineering services. The Basic Services of Engineer as described in Section 1 of the Agreement are amended or supplemented as indicated below and the time periods for the performance of certain services as indicated in Section 4 of the Agreement are stipulated as indicated below.

2. During the Study and Report Phase ENGINEER shall:

(insert amendments or supplements to paragraph 1.2.)

The Study and Report Phase Services will be completed and the Report submitted within ______ calendar days following written authorization from OWNER to ENGINEER to proceed with that phase of services.

3. During the Preliminary Design Phase ENGINEER shall:

(insert amendments or supplements to paragraph 1.3.)

The Preliminary Design Phase Services will be completed and ENGINEER's documentation and opinion of costs submitted within ______ calendar days following written authorization from OWNER to ENGINEER to proceed with that phase of services.

4. During the Final Design Phase ENGINEER shall:

(insert amendments or supplements to paragraph 1.4.)

The Final Design Phase Services will be completed and Contract Documents and ENGINEER's opinion of costs submitted within ______ calendar days following written authorization from OWNER to ENGINEER to proceed with that phase of services.

[If the construction work is to be performed under more than one prime contract, the provisions of paragraph 4.11 will probably apply and it will be necessary to prepare a special schedule for the Final Design Phase Services.]

5. During the Bidding or Negotiating Phase ENGINEER shall:

(insert amendments or supplements to paragraph 1.5.)

[If the construction work is to be performed under more than one prime contract, the provisions of paragraph 4.11 will probably apply and it will be necessary to prepare a special schedule for the Bidding or Negotiating Phase Services.]

6. During the Construction Phase ENGINEER shall:

(insert amendments or supplements to paragraph 1.6.)

[If the construction work is to be performed under more than one prime contract, the provisions of paragraph 4.11 will probably apply and it will be necessary to prepare a special schedule for the Construction Phase Services.]

TABLE OF CONTENTS

*Page No.**

IDENTIFICATION OF THE PARTIES 1
DESCRIPTION OF THE PROJECT 1

SECTION 1—BASIC SERVICES OF ENGINEER 1
1.1. General 1
1.2. Study and Report Phase 1
1.3. Preliminary Design Phase 2
1.4. Final Design Phase 2
1.5. Bidding or Negotiating Phase 2
1.6. Construction Phase 3

SECTION 2—ADDITIONAL SERVICES OF ENGINEER 4
2.1. General 4
2.2. Resident Services During Construction 5

SECTION 3—OWNER'S RESPONSIBILITIES 6

SECTION 4—PERIOD OF SERVICE 7

SECTION 5—PAYMENTS TO ENGINEER 9
5.1. Methods of Payment for Services and Expenses of Engineer 9
5.2. Times of Payment 9
5.3. Other Provisions Concerning Payments 9
5.4. Definitions 10

SECTION 6—CONSTRUCTION COSTS AND OPINIONS OF COST 11
6.1. Construction Cost 11
6.2. Opinions of Cost 11

SECTION 7—GENERAL CONSIDERATIONS 12
7.1. Termination 12
7.2. Reuse of Documents 12
7.3. Controlling Law 12
7.4. Successors and Assigns 12
7.5. Arbitration 13

SECTION 8—SPECIAL PROVISIONS, EXHIBITS AND SCHEDULES 15

*Page numbers refer to the Form Agreement pages.

This document has important legal consequences; consultation with an attorney is encouraged with respect to its completion or modification.

STANDARD FORM OF AGREEMENT BETWEEN OWNER AND ENGINEER FOR PROFESSIONAL SERVICES

THIS IS AN AGREEMENT made as of ____________________, 19____

between ____________________

____________________ (OWNER) and

____

____________________ (ENGINEER).

OWNER intends to ____________________

____

____

____

____

____________________ (hereinafter called the Project).

OWNER and ENGINEER in consideration of their mutual covenants herein agree in respect of the performance of professional engineering services by ENGINEER and the payment for those services by OWNER as set forth below.

ENGINEER shall provide professional engineering services for OWNER in all phases of the Project to which this Agreement applies, serve as OWNER's professional engineering representative for the Project as set forth below and shall give professional engineering consultation and advice to OWNER during the performance of services hereunder.

SECTION 1—BASIC SERVICES OF ENGINEER

1.1. General.

1.1.1. ENGINEER shall perform professional services as hereinafter stated which include customary civil, structural, mechanical and electrical engineering services and customary architectural services incidental thereto.

1.2. Study and Report Phase.

After written authorization to proceed, ENGINEER shall:

1.2.1. Consult with OWNER to clarify and define OWNER's requirements for the Project and review available data.

1.2.2. Advise OWNER as to the necessity of OWNER's providing or obtaining from others data or services of the types described in paragraph 3.3, and act as OWNER's representative in connection with any such services.

1.2.3. Provide analyses of OWNER's needs, planning surveys, site evaluations and comparative studies of prospective sites and solutions.

1.2.4. Provide a general economic analysis of OWNER's requirements applicable to various alternatives.

1.2.5. Prepare a Report containing schematic layouts, sketches and conceptual design criteria with appropriate exhibits to indicate clearly the considerations involved and the alternative solutions available to OWNER and setting forth ENGINEER's findings and recommendations with opinions of probable costs for the Project, including Construction Cost, contingencies, allowances for charges of all professionals and consultants, allowances for the cost of land and rights-of-way, compensation for or damages to properties and interest and financing charges (all of which are hereinafter called "Project Costs").

1.2.6. Furnish five copies of the Report and present and review it in person with OWNER.

The duties and responsibilities of ENGINEER during the Study and Report Phase are amended and supplemented as indicated in paragraph 2 of Exhibit A "Further Description of Basic Engineering Services and Related Matters".

1.3. Preliminary Design Phase.

After written authorization to proceed with the Preliminary Design Phase, ENGINEER shall:

1.3.1. In consultation with OWNER and on the basis of the accepted Report, determine the extent of the Project.

1.3.2. Prepare preliminary design documents consisting of final design criteria, preliminary drawings and outline specifications.

1.3.3. Based on the information contained in the preliminary design documents, submit a revised opinion of probable Project Costs.

1.3.4. Furnish five copies of the above preliminary design documents and present and review them in person with OWNER.

The duties and responsibilities of ENGINEER during the Preliminary Design Phase are amended and supplemented as indicated in paragraph 3 of Exhibit A "Further Description of Basic Engineering Services and Related Matters".

1.4. Final Design Phase.

After written authorization to proceed with the Final Design Phase, ENGINEER shall:

1.4.1. On the basis of the accepted preliminary design documents and the revised opinion of probable Project Cost, prepare for incorporation in the Contract Documents final drawings to show the character and extent of the Project (hereinafter called "Drawings") and Specifications.

1.4.2. Furnish to OWNER such documents and design data as may be required for, and assist in the preparation of, the required documents so that OWNER may apply for approvals of such governmental authorities as have jurisdiction over design criteria applicable to the Project, and assist in obtaining such approvals by participating in submissions to and negotiations with appropriate authorities.

1.4.3. Advise OWNER of any adjustments to the latest opinion of probable Project Cost caused by changes in extent or design requirements of the Project or Construction Costs and furnish a revised opinion of probable Project Cost based on the Drawings and Specifications.

1.4.4. Prepare for review and approval by OWNER, his legal counsel and other advisors contract agreement forms, general conditions and supplementary conditions, and (where appropriate) bid forms, invitations to bid and instructions to bidders, and assist in the preparation of other related documents.

1.4.5. Furnish five copies of the above documents and present and review them in person with OWNER.

The duties and responsibilities of ENGINEER during the Final Design Phase are amended and supplemented as indicated in paragraph 4 of Exhibit A "Further Description of Basic Engineering Services and Related Matters".

1.5. Bidding or Negotiating Phase.

After written authorization to proceed with the Bidding or Negotiating Phase, ENGINEER shall:

1.5.1. Assist OWNER in obtaining bids or negotiating proposals for each separate prime contract for construction, materials, equipment and services.

1.5.2. Consult with and advise OWNER as to the acceptability of subcontractors and other persons and organizations proposed by the prime contractor(s) (hereinafter called "Contractor(s)") for those portions of the work as to which such acceptability is required by the bidding documents.

1.5.3. Consult with and advise OWNER as to the acceptability of substitute materials and equipment proposed by Contractor(s) when substitution prior to the award of contracts is allowed by the bidding documents.

1.5.4. Assist OWNER in evaluating bids or proposals and in assembling and awarding contracts.

The duties and responsibilities of ENGINEER during the Bidding or Negotiating Phase are amended and supplemented as indicated in paragraph 5 of Exhibit A "Further Description of Basic Engineering Services and Related Matters".

1.6. Construction Phase.

During the Construction Phase ENGINEER shall:

1.6.1. Consult with and advise OWNER and act as his representative as provided in Articles 1 through 17, inclusive, of the Standard General Conditions of the Construction Contract No. 1910-8, (1978 edition). The extent and limitations of the duties, responsibilities and authority of ENGINEER as assigned in said Standard General Conditions shall not be modified, except to the extent provided in paragraph 6 of Exhibit A "Further Description of Basic Engineering Services and Related Matters" and except as ENGINEER may otherwise agree in writing. All of OWNER's instructions to Contractor(s) will be issued through ENGINEER who will have authority to act on behalf of OWNER to the extent provided in said Standard General Conditions except as otherwise provided in writing.

1.6.2. Make visits to the site at intervals appropriate to the various stages of construction to observe as an experienced and qualified design professional the progress and quality of the executed work of Contractor(s) and to determine in general if such work is proceeding in accordance with the Contract Documents. ENGINEER shall not be required to make exhaustive or continuous on-site inspections to check the quality or quantity of such work. ENGINEER shall not be responsible for the means, methods, techniques, sequences or procedures of construction selected by Contractor(s) or the safety precautions and programs incident to the work of Contractor(s). ENGINEER's efforts will be directed toward providing a greater degree of confidence for OWNER that the completed work of Contractor(s) will conform to the Contract Documents, but ENGINEER shall not be responsible for the failure of Contractor(s) to perform the work in accordance with the Contract Documents. During such visits and on the basis of on-site observations ENGINEER shall keep OWNER informed of the progress of the work, shall endeavor to guard OWNER against defects and deficiencies in such work and may disapprove or reject work failing to conform to the Contract Documents.

1.6.3. Review and approve (or take other appropriate action in respect of) Shop Drawings (as that term is defined in the aforesaid Standard General Conditions) and samples, the results of tests and inspections and other data which each Contractor is required to submit, but only for conformance with the design concept of the Project and compliance with the information given in the Contract Documents (but such review and approval or other action shall not extend to means, methods, sequences, techniques or procedures of construction or to safety precautions and programs incident thereto); determine the acceptability of substitute materials and equipment proposed by Contractor(s); and receive and review (for general content as required by the Specifications) maintenance and operating instructions, schedules, guarantees, bonds and certificates of inspection which are to be assembled by Contractor(s) in accordance with the Contract Documents.

1.6.4. Issue all instructions of OWNER to Contractor(s); issue necessary interpretations and clarifications of the Contract Documents and in connection therewith prepare change orders as required; have authority, as OWNER's representative, to require special inspection or testing of the work; act as initial interpreter of the requirements of the Contract Documents and judge of the acceptability of the work thereunder and make decisions on all claims of OWNER and Contractor(s) relating to the acceptability of the work or the interpretation of the requirements of the Contract Documents pertaining to the execution and progress of the work; but ENGINEER shall not be liable for the results of any such interpretations or decisions rendered by him in good faith.

1.6.5. Based on ENGINEER's on-site observations as an experienced and qualified design professional and on review of applications for payment and the accompanying data and schedules, determine the amounts owing to Contractor(s) and recommend in writing payments to Contractor(s) in such amounts; such recommendations of payment will constitute a representation to OWNER, based on such observations and review, that the work has progressed to the point indicated, that, to the best of ENGINEER's knowledge, information and belief, the quality of such work is in accordance with the Contract Documents (subject to an evaluation of such work as a functioning Project upon Substantial Completion, to the results of any subsequent tests called for in the Contract Documents, and to any qualifications stated in his recommendation), and that payment of the amount recommended is due Contractor(s); but by recommending any payment ENGINEER will not thereby be deemed to have represented that continuous or exhaustive examinations have been made by ENGINEER to check the quality or quantity of the work or to review the means, methods, sequences, techniques or procedures of construction or safety precautions or programs incident thereto or that ENGINEER has made an examination to ascertain how or for what purposes any Contractor has used the moneys paid on account of the Contract Price, or that title to any of the work, materials or equipment has passed to OWNER free and clear of any lien, claims, security interests or encumbrances, or that Contractor(s) have completed their work exactly in accordance with the Contract Documents.

1.6.6. Conduct an inspection to determine if the Project is substantially complete and a final inspection to determine if the work has been completed in accordance with the Contract Documents and if each Contractor has fulfilled all of his obligations thereunder so that ENGINEER may recommend, in writing, final payment to each Contractor and may give written notice to OWNER and the Contractor(s) that the work is acceptable (subject to any conditions therein expressed), but any such recommendation and notice shall be subject to the limitations expressed in paragraph 1.6.5.

1.6.7. ENGINEER shall not be responsible for the acts or omissions of any Contractor, or subcontractor, or any of the Contractor(s)' or subcontractors' agents or employees or any other persons (except ENGINEER's own employees and agents) at the site or otherwise performing any of the Contractor(s)' work; however, nothing contained in paragraphs 1.6.1 thru 1.6.7, inclusive, shall be construed to release ENGINEER from liability for failure to properly perform duties undertaken by him in the Contract Documents.

SECTION 2—ADDITIONAL SERVICES OF ENGINEER

2.1. General

If authorized in writing by OWNER, ENGINEER shall furnish or obtain from others Additional Services of the following types which are not considered normal or customary Basic Services except to the extent provided otherwise in Exhibit A "Further Description of Basic Engineering Services and Related Matters"; these will be paid for by OWNER as indicated in Section 5.

2.1.1. Preparation of applications and supporting documents for governmental grants, loans or advances in connection with the Project; preparation or review of environmental assessments and impact statements; review and evaluation of the effect on the design requirements of the Project of any such statements and documents prepared by others; and assistance in obtaining approvals of authorities having jurisdiction over the anticipated environmental impact of the Project.

2.1.2. Services to make measured drawings of or to investigate existing conditions or facilities, or to verify the accuracy of drawings or other information furnished by OWNER.

2.1.3. Services resulting from significant changes in extent of the Project or its design including, but not limited to, changes in size, complexity, OWNER's schedule, or character of construction or method of financing; and revising previously accepted studies, reports, design documents or Contract Documents when such revisions are due to causes beyond ENGINEER's control.

2.1.4. Providing renderings or models for OWNER's use.

2.1.5. Preparing documents for alternate bids requested by OWNER for Contractor(s)' work which is not executed or documents for out-of-sequence work.

2.1.6. Investigations involving detailed consideration of operations, maintenance and overhead expenses; providing Value Engineering during the course of design; the preparation of feasibility studies, cash flow and economic evaluations, rate schedules and appraisals; assistance in obtaining financing for the Project; evaluating processes available for licensing and assisting OWNER in obtaining process licensing; detailed quantity surveys of material, equipment and labor; and audits or inventories required in connection with construction performed by OWNER.

2.1.7. Furnishing the services of special consultants for other than the normal civil, structural, mechanical and electrical engineering and normal architectural design incidental thereto, such as consultants for interior design, furniture, furnishings, communications, acoustics, kitchens and landscaping; and providing data or services of the types described in paragraph 3.3 when OWNER authorizes ENGINEER to provide such data or services in lieu of furnishing the same in accordance with paragraph 3.3.

2.1.8. Services resulting from the award of more separate prime contracts for construction, materials, equipment or services for the Project than are contemplated by paragraph 5.1.1.2, and services resulting from the arranging for performance by persons other than the principal prime contractors of services for the OWNER and administering OWNER's contracts for such services.

2.1.9. Providing any type of field surveys for design purposes and engineering surveys and staking to enable Contractor(s) to proceed with their work; and providing other special field surveys.

2.1.10. Services in connection with change orders to reflect changes requested by OWNER if the resulting change in compensation for Basic Services is not commensurate with the additional services rendered, services after the award of each contract in evaluating substitutions proposed by Contractor(s), and in making revisions to Drawings and Specifications occasioned thereby, and services resulting from significant delays, changes or price increases occurring as a direct or indirect result of material, equipment or energy shortages.

2.1.11. Services during out-of-town travel required of ENGINEER other than visits to the site as required by Section 1.

2.1.12. Preparing for OWNER, on request, a set of reproducible record prints of Drawings showing those changes made during the construction process, based on the marked-up prints, drawings and other data furnished by Contractor(s) to ENGINEER and which ENGINEER considers significant.

2.1.13. Additional or extended services during construction made necessary by (1) work damaged by fire or other cause during construction, (2) a significant amount of defective or neglected work of Contractor(s), (3) prolongation of the contract time of any prime contract by more than sixty days, (4) acceleration of the progress schedule involving services beyond normal working hours, and (5) default by Contractor(s).

2.1.14. Preparation of operating and maintenance manuals; protracted or extensive assistance in the utilization of any equipment or system (such as initial startup, testing, adjusting and balancing); and training personnel for operation and maintenance.

2.1.15. Services after completion of the Construction Phase, such as inspections during any guarantee period and reporting observed discrepancies under guarantees called for in any contract for the Project.

2.1.16. Preparing to serve or serving as a consultant or witness for OWNER in any litigation, public hearing or other legal or administrative proceeding involving the Project (except as agreed to under Basic Services).

2.1.17. Additional services in connection with the Project, including services normally furnished by OWNER and services not otherwise provided for in this Agreement.

2.2. Resident Services During Construction.

2.2.1. If requested by OWNER or recommended by ENGINEER and agreed to in writing by the other, a Resident Project Representative will be furnished and will act as directed by ENGINEER in order to assist ENGINEER in observing performance of the work of Contractor(s). Such services will be paid for by OWNER as indicated in paragraph 5.1.2.4.

2.2.2. The duties and responsibilities and the limitations on the authority of the Resident Project Representative and assistants will be set forth in Exhibit B which is to be identified, attached to and made a part of this Agreement before such services begin.

2.2.3. Through more extensive on-site observations of the work in progress and field checks of materials and equipment by the Resident Project Representative (if furnished) and assistants, ENGINEER shall endeavor to provide further protection for OWNER against defects and deficiencies in the work of Contractor(s); but the furnishing of such resident Project representation will not make ENGINEER responsible for construction means, methods, techniques, sequences or procedures or for safety precautions or programs, or for Contractor(s)' failure to perform their work in accordance with the Contract Documents.

2.2.4. If OWNER designates another person to represent OWNER at the Project site who is not ENGINEER's agent or employee, the duties, responsibilities and limitations of authority of such other person and the effect thereof on the duties and responsibilities of ENGINEER under this Agreement will be set forth in an exhibit that is to be identified, attached to and made a part of this Agreement before such services begin.

SECTION 3—OWNER'S RESPONSIBILITIES

OWNER shall:

3.1. Provide all criteria and full information as to OWNER's requirements for the Project, including design objectives and constraints, space, capacity and performance requirements, flexibility and expandability, and any budgetary limitations; and furnish copies of all design and construction standards which OWNER will require to be included in the Drawings and Specifications.

3.2. Assist ENGINEER by placing at his disposal all available information pertinent to the Project including previous reports and any other data relative to design or construction of the Project.

3.3. Furnish to ENGINEER, as required for performance of ENGINEER's Basic Services (except to the extent provided otherwise in Exhibit A "Further Description of Basic Engineering Services and Related Matters"), data prepared by or services of others, including without limitation core borings, probings and subsurface explorations, hydrographic surveys, laboratory tests and inspections of samples, materials and equipment; appropriate professional interpretations of all of the foregoing; environmental assessment and impact statements; property, boundary, easement, right-of-way, topographic and utility surveys; property descriptions; zoning, deed and other land use restriction; and other special data or consultations not covered in Section 2; all of which ENGINEER may rely upon in performing his services.

3.4. Provide field control surveys and establish reference points and base lines (except to the extent provided otherwise in Exhibit A "Further Description of Basic Engineering Services and Related Matters") to enable Contractor(s) to proceed with the layout of the work.

3.5. Arrange for access to and make all provisions for ENGINEER to enter upon public and private property as required for ENGINEER to perform his services.

3.6. Examine all studies, reports, sketches, Drawings, Specifications, proposals and other documents presented by ENGINEER, obtain advice of an attorney, insurance counselor and other consultants as OWNER deems appropriate for such examination and render in writing decisions pertaining thereto within a reasonable time so as not to delay the services of ENGINEER.

3.7. Furnish approvals and permits from all governmental authorities having jurisdiction over the Project and such approvals and consents from others as may be necessary for completion of the Project.

3.8. Provide such accounting, independent cost estimating and insurance counseling services as may be required for the Project, such legal services as OWNER may require or ENGINEER may reasonably request with regard to legal issues pertaining to the Project including any that may be raised by Contractor(s), such auditing service as OWNER may require to ascertain how or for what purpose any Contractor has used the moneys paid to him under the construction contract, and such inspection services as OWNER may require to ascertain that Contractor(s) are complying with any law, rule or regulation applicable to their performance of the work.

3.9. Designate in writing a person to act as OWNER's representative with respect to the services to be rendered under this Agreement. Such person shall have complete authority to transmit instructions, receive information, interpret and define OWNER's policies and decisions with respect to materials, equipment, elements and systems pertinent to ENGINEER's services.

3.10. Give prompt written notice to ENGINEER whenever OWNER observes or otherwise becomes aware of any development that affects the scope or timing of ENGINEER's services, or any defect in the work of Contractor(s).

3.11. Furnish, or direct ENGINEER to provide, necessary Additional Services as stipulated in Section 2 of this Agreement or other services as required.

3.12. Bear all costs incident to compliance with the requirements of this Section 3.

SECTION 4—PERIOD OF SERVICE

4.1. The provisions of this Section 4 and the various rates of compensation for ENGINEER's services provided for elsewhere in this Agreement have been agreed to in anticipation of the orderly and continuous progress of the Project through completion of the Construction Phase. ENGINEER's obligation to render services hereunder will extend for a period which may reasonably be required for the design, award of contracts and construction of the Project including extra work and required extensions thereto.

4.2. The services called for in the Study and Report Phase will be completed and the Report submitted within the stipulated period indicated in paragraph 2 of Exhibit A "Further Description of Basic Engineering Services and Related Matters" after authorization to proceed with that phase of services.

4.3. After acceptance by OWNER of the Study and Report Phase documents indicating any specific modifications or changes in the extent of the Project desired by OWNER, and upon written authorization from OWNER, ENGINEER shall proceed with the performance of the services called for in the Preliminary Design Phase, and shall submit preliminary design documents and a revised opinion of probable Project Cost within the stipulated period indicated in paragraph 3 of Exhibit A "Further Description of Basic Engineering Services and Related Matters" after authorization to proceed with that phase of services.

4.4. After acceptance by OWNER of the Preliminary Design Phase documents and revised opinion of probable Project Cost, indicating any specific modifications or changes in the extent of the Project desired by OWNER, and upon written authorization from OWNER, ENGINEER shall proceed with the performance of the services called for in the Final Design Phase; and shall deliver Contract Documents and a revised opinion of probable Project Cost for all work of Contractor(s) on the Project within the stipulated period indicated in paragraph 4 of Exhibit A "Further Description of Basic Engineering Services and Related Matters" after authorization to proceed with that phase of services.

4.5. ENGINEER's services under the Study and Report Phase, Preliminary Design Phase and Final Design Phase shall each be considered complete at the earlier of (1) the date when the submissions for that phase have been accepted by OWNER or (2) thirty days after the date when such submissions are delivered to OWNER for final acceptance, plus such additional time as may be considered reasonable for obtaining approval of governmental authorities having jurisdiction over design criteria applicable to the Project.

4.6. After acceptance by OWNER of the ENGINEER's Drawings, Specifications and other Final Design Phase documentation including the most recent opinion of probable Project Cost and upon written authorization to proceed, ENGINEER shall proceed with performance of the services called for in the Bidding or Negotiating Phase. This Phase shall terminate and the services to be rendered thereunder shall be considered complete upon commencement of the Construction Phase or upon cessation of the negotiations with prospective Contractor(s) (except as may be otherwise required to complete the services called for in paragraph 6.2.2.5).

4.7. The Construction Phase will commence with the execution of the first prime contract to be executed for the work of the Project or any part thereof, and will terminate upon written approval by ENGINEER of final payment on the last prime contract to be completed. Construction Phase services may be rendered at different times in respect of separate prime contracts if the Project involves more than one prime contract.

4.8. If OWNER has requested significant modifications or changes in the extent of the Project, the time of performance of ENGINEER's services and his various rates of compensation shall be adjusted appropriately.

4.9. If OWNER fails to give prompt written authorization to proceed with any phase of services after completion of the immediately preceding phase, or if the Construction Phase has not commenced within ______ calendar days (plus such additional time as may be required to complete the services called for under paragraph 6.2.2.5) after completion of the Final Design Phase, ENGINEER may, after giving seven days' written notice to OWNER, suspend services under this Agreement.

4.10. If ENGINEER's services for design or during construction of the Project are delayed or suspended in whole or in part by OWNER for more than three months for reasons beyond ENGINEER's control, ENGINEER shall on written demand to OWNER (but without termination of this Agreement) be paid as provided in paragraph 5.3.2. If such delay or suspension extends for more than one year for reasons beyond ENGINEER's control, or if ENGINEER for any reason is required to render services more than one year

after Substantial Completion, the various rates of compensation provided for elsewhere in this Agreement shall be subject to renegotiation.

4.11. In the event that the work designed or specified by ENGINEER is to be performed under more than one prime contract, OWNER and ENGINEER shall, prior to commencement of the Final Design Phase, develop a schedule for performance of ENGINEER's services during the Final Design, Bidding or Negotiating and Construction Phases in order to sequence and coordinate properly such services as applicable to the work under such separate contracts. This schedule is to be prepared whether or not the work under such contracts is to proceed concurrently and is to be included in Exhibit A "Further Description of Basic Engineering Services and Related Matters," and the provisions of paragraphs 4.4 through 4.10 inclusive, will be modified accordingly.

[The remainder of this page was left blank intentionally.]

LUMP SUM METHOD OF PAYMENT

SECTION 5—PAYMENTS TO ENGINEER

5.1. Methods of Payment for Services and Expenses of ENGINEER.

5.1.1. *For Basic Services.* OWNER shall pay ENGINEER for Basic Services rendered under Section 1 (as amended and supplemented by Exhibit A "Further Description of Basic Engineering Services and Related Matters") as follows:

[Delete inapplicable paragraph and initial]

5.1.1.1. *One Prime Contract.* If only one prime contract is awarded for construction, materials, equipment and services for the entire Project, a lump sum fee of $____________; but, if the prime contract contains cost-plus or incentive savings provisions for Contractor(s)' basic compensation, a lump sum fee of $____________.

5.1.1.2. *Several Prime Contracts.* If more than one but less than ______ separate prime contracts are awarded for construction, materials, equipment and services for the entire Project, a lump sum fee of $____________; but, if any prime contract contains cost-plus or incentive savings provisions for Contractor(s)' basic compensation, a lump sum fee of $____________.

5.1.2. *For Additional Services.* OWNER shall pay ENGINEER for Additional Services rendered under Section 2 as follows:

[Delete inapplicable language and initial]

5.1.2.1. *General.* For Additional Services rendered under paragraphs 2.1.1 through 2.1.17, inclusive (except services covered by paragraph 2.1.7 and services as a consultant or witness under paragraph 2.1.16), on the basis of Payroll Costs times a factor of ______ for services rendered by principals and employees assigned to the Project.

5.1.2.2. *Special Consultants.* For services and reimbursable expenses of special consultants employed by ENGINEER pursuant to paragraph 2.1.7 or 2.1.17, the amount billed to ENGINEER therefor times a factor of ______.

5.1.2.3. *Serving as a Witness.* For the services rendered by principals and employees as consultants or witnesses in any litigation, hearing or proceeding in accordance with paragraph 2.1.16, at the rate of $________ per day or any portion thereof (but compensation for time spent in preparing to appear in any such litigation, hearing or proceeding will be on the basis provided in paragraph 5.1.2.1).

5.1.2.4. *Resident Project Services.* For resident services during construction furnished under paragraph 2.2.1, on the basis of Payroll Costs times a factor of ________ for services rendered by principals and employees assigned to field offices in connection with resident Project representation.

5.1.3. *For Reimbursable Expenses.* In addition to payments provided for in paragraphs 5.1.1 and 5.1.2, OWNER shall pay ENGINEER the actual costs of all Reimbursable Expenses incurred in connection with all Basic and Additional Services.

5.1.4. The terms "Payroll Costs" and "Reimbursable Expenses" will have the meanings assigned to them in paragraph 5.4 which appears on the reverse side of this page.

5.2. Times of Payments.

5.2.1. ENGINEER shall submit monthly statements for Basic and Additional Services rendered and for Reimbursable Expenses incurred. The statements will be based upon ENGINEER's estimate of the proportion of the total services actually completed at the time of billing. OWNER shall make prompt monthly payments in response to ENGINEER's monthly statements.

5.3. Other Provisions Concerning Payments.

5.3.1. If OWNER fails to make any payment due ENGINEER for services and expenses within sixty days after receipt of ENGINEER's bill therefor, the amounts due ENGINEER shall include a charge at the rate of 1% per month from said sixtieth day, and in addition, ENGINEER may, after giving seven days' written notice to OWNER, suspend services under this Agreement until he has been paid in full all amounts due him for services and expenses.

5.3.2. In the event of termination by OWNER under paragraph 7.1 upon the completion of any phase of the Basic Services, progress payments due ENGINEER for services rendered through such phase shall constitute total payment for such services. In the event of such termination by OWNER during any phase of the Basic Services, ENGINEER will be paid for services rendered during that phase on the basis of Payroll Costs times a factor of ________ for services rendered during that phase to date of termination by principals and employees assigned to the Project. In the event of any such termination, ENGINEER will be paid for all unpaid Additional Services and unpaid Reimbursable Expenses, plus all termination expenses. Termination expenses mean Reimbursable Expenses directly attributable to termination, which shall include an amount computed as a percentage of total compensation for Basic Services earned by ENGINEER to the date of termination, as follows:

20% if termination occurs after commencement of the Preliminary Design Phase but prior to commencement of the Final Design Phase; or

10% if termination occurs after commencement of the Final Design Phase.

5.4. Definitions.

5.4.1. The Payroll Costs used as a basis for payment mean salaries and wages (basic and incentive) paid to all personnel engaged directly on the Project, including, but not limited to, engineers, architects, surveyors, designers, draftsmen, specification writers, estimators, other technical personnel, stenographers, typists and clerks; plus the cost of customary and statutory benefits including, but not limited to, social security contributions, unemployment, excise and payroll taxes, workers' compensation, health and retirement benefits, sick leave, vacation and holiday pay applicable thereto. For the purposes of this Agreement, the principals of ENGINEER and their hourly payroll costs are:

____

____

The amount of customary and statutory benefits of all other personnel will be considered equal to _______% of salaries and wages.

5.4.2. Reimbursable Expenses mean the actual expenses incurred directly or indirectly in connection with the Project for: transportation and subsistence incidental thereto; obtaining bids or proposals from Contractor(s); furnishing and maintaining field office facilities; subsistence and transportation of Resident Project Representatives and their assistants; toll telephone calls and telegrams; reproduction of reports, Drawings, Specifications, and similar Project-related items in addition to those required under Section 1; and, if authorized in advance by OWNER, overtime work requiring higher than regular rates.

[The remainder of this page was left blank intentionally.]

Page 10 of ____ pages

PERCENTAGE METHOD OF PAYMENT

SECTION 5—PAYMENTS TO ENGINEER

5.1. Methods of Payment for Services and Expenses of ENGINEER.

5.1.1. *For Basic Services.* OWNER shall pay ENGINEER for Basic Services rendered under Section 1 (as amended and supplemented by Exhibit A "Further Description of Basic Engineering Services and Related Matters") as follows:

[Delete inapplicable language and initial]

5.1.1.1. *One Prime Contract.* If only one prime contract is awarded for construction, materials, equipment and services for the entire Project, ____% of the Construction Cost; but, if the prime contract contains cost-plus or incentive savings provisions for the Contractor's basic compensation, ____% of the Construction Cost.

5.1.1.2. *Several Prime Contracts.* If more than one but less than ____ separate prime contracts are awarded for construction, materials, equipment and services for the entire Project, ____% of the Construction Cost; but, if any prime contract contains cost-plus or incentive savings provisions for Contractor(s)' basic compensation, ____% of the Construction Cost.

5.1.2. *For Additional Services.* OWNER shall pay ENGINEER for Additional Services rendered under Section 2 as follows:

[Delete inapplicable language and initial]

5.1.2.1. *General.* For Additional Services rendered under paragraphs 2.1.1 through 2.1.17, inclusive (except services covered by paragraph 2.1.7 and services as a consultant or witness under paragraph 2.1.16), on the basis of Payroll Costs times a factor of ________ for services rendered by principals and employees assigned to the Project.

5.1.2.2. *Special Consultants.* For services and reimbursable expenses of special consultants employed by ENGINEER pursuant to paragraph 2.1.7 or 2.1.17, the amount billed to ENGINEER therefor times a factor of ________.

5.1.2.3. *Serving as a Witness.* For the services rendered by principals and employees as consultants or witnesses in any litigation, hearing or proceeding in accordance with paragraph 2.1.16, at the rate of $________ per day or any portion thereof (but compensation for time spent in preparing to appear in any such litigation, hearing or proceeding will be on the basis provided in paragraph 5.1.2.1).

5.1.2.4. *Resident Project Services.* For resident services during construction furnished under paragraph 2.2.1, on the basis of Payroll Costs times a factor of ________ for services rendered by principals and employees assigned to field offices in connection with resident Project representation.

5.1.3. *For Reimbursable Expenses.* In addition to payments provided for in paragraphs 5.1.1 and 5.1.2, OWNER shall pay ENGINEER the actual costs of all Reimbursable Expenses incurred in connection with all Basic and Additional Services.

5.1.4. As used in this paragraph 5.1, the term "Construction Cost" will have the meaning assigned to it in paragraph 6.1, and terms "Payroll Costs" and "Reimbursable Expenses" will have the meanings assigned to them in paragraph 5.4 which appears on the reverse side of this page.

5.2. Times of Payments.

5.2.1. ENGINEER shall submit monthly statements for Basic and Additional Services rendered and for Reimbursable Expenses incurred. The statements will be based upon ENGINEER's estimate of the proportion of the total services actually completed at the time of billing. OWNER shall make prompt monthly payments in response to ENGINEER's monthly statements.

5.2.2. Upon conclusion of each phase of Basic Services, OWNER shall pay such additional amount, if any, as may be necessary to bring total compensation paid on account of such phase to the following percentages of total compensation payable for all phases of Basic Services:

Phase	*Insert Actual Percentage*
Study and Report	________%
Preliminary Design	________%
Final Design	________%
Bidding or Negotiating	________%
Construction	________%
TOTAL	100%

5.3. Other Provisions Concerning Payments.

5.3.1. If OWNER fails to make any payment due ENGINEER for services and expenses within sixty days after receipt of ENGINEER's bill therefor, the amounts due ENGINEER shall include a charge at the rate of 1% per month from said sixtieth day, and in addition, ENGINEER may, after giving seven days' written notice to OWNER, suspend services under this Agreement until he has been paid in full all amounts due him for services and expenses.

5.3.2. In the event of termination by OWNER under paragraph 7.1 upon the completion of any phase of the Basic Services, progress payments due ENGINEER for services rendered through such phase shall constitute total payment for such services. In the event of such termination by OWNER during any phase of the Basic Services, ENGINEER will be paid for services rendered during that phase on the basis of Payroll Costs times a factor of ________ for services rendered during that phase to date of termination by principals and employees assigned to the Project. In the event of any such termination, ENGINEER will be paid for all unpaid Additional Services and unpaid Reimbursable Expenses, plus all termination expenses. Termination expenses mean Reimbursable Expenses directly attributable to termination, which shall include an amount computed as a percentage of total compensation for Basic Services earned by ENGINEER to the date of termination, as follows:

20% if termination occurs after commencement of the Preliminary Design Phase but prior to commencement of the Final Design Phase; or

10% if termination occurs after commencement of the Final Design Phase.

5.4. Definitions.

5.4.1. The Payroll Costs used as a basis for payment mean salaries and wages (basic and incentive) paid to all personnel engaged directly on the Project, including, but not limited to, engineers, architects, surveyors, designers, draftsmen, specification writers, estimators, other technical personnel, stenographers, typists and clerks; plus the cost of customary and statutory benefits including, but not limited to, social security contributions, unemployment, excise and payroll taxes, workers' compensation, health and retirement benefits, sick leave, vacation and holiday pay applicable thereto. For the purposes of this Agreement, the principals of ENGINEER and their hourly payroll costs are:

____

____

The amount of customary and statutory benefits of all other personnel will be considered equal to ______% of salaries and wages.

5.4.2. Reimbursable Expenses mean the actual expenses incurred directly or indirectly in connection with the Project for: transportation and subsistence incidental thereto; obtaining bids or proposals from Contractor(s); furnishing and maintaining field office facilities; subsistence and transportation of Resident Project Representatives and their assistants; toll telephone calls and telegrams; reproduction of reports, Drawings, Specifications, and similar Project-related items in addition to those required under Section 1; and, if authorized in advance by OWNER, overtime work requiring higher than regular rates.

[The remainder of this page was left blank intentionally.]

PROFESSIONAL FEE PLUS EXPENSES

SECTION 5—PAYMENTS TO ENGINEER

5.1. Methods of Payment for Services and Expenses of ENGINEER.

5.1.1. *For Basic Services.* OWNER shall pay ENGINEER for Basic Services rendered under Section 1 (as amended and supplemented by Exhibit A "Further Description of Basic Engineering Services and Related Matters") a professional fee of $________ plus an amount equal to ENGINEER's Direct Labor Costs times a factor of ________ for overhead.

5.1.2. *Additional Services.* OWNER shall pay ENGINEER for Additional Services rendered under Section 2 as follows:

5.1.2.1. *General.* For Additional Services rendered under paragraphs 2.1.1 through 2.1.17, inclusive (except services covered by paragraph 2.1.7 and services as a consultant or witness under paragraph 2.1.16), on the basis of Direct Labor Costs times a factor of ________ for services rendered by principals and employees assigned to the Project.

5.1.2.2. *Special Consultants.* For services and reimbursable expenses of special consultants employed by ENGINEER pursuant to paragraph 2.1.7 or 2.1.17, the amount billed to ENGINEER therefor times a factor of ________.

5.1.2.3. *Serving as a Witness.* For the services rendered by principals and employees as consultants or witnesses in any litigation, hearing or proceeding in accordance with paragraph 2.1.16, at the rate of $________ per day or any portion thereof (but compensation for time spent in preparing to appear in any such litigation, hearing or proceeding will be on the basis provided in paragraph 5.1.2.1).

[Delete inapplicable language and initial]

5.1.2.4. *Resident Project Services.* For resident services during construction furnished under paragraph 2.2.1, on the basis of Direct Labor Costs times a factor of ________ for services rendered by principals and employees assigned to field offices in connection with resident Project representation.

5.1.3. *Reimbursable Expenses.* In addition to payments provided for in paragraphs 5.1.1 and 5.1.2, OWNER shall pay ENGINEER the actual costs of all Reimbursable Expenses incurred in connection with all Basic and Additional Services.

5.1.4. The "Direct Labor Costs" and "Reimbursable Expenses" will have the meanings assigned to them in paragraph 5.4 which appears on the reverse side of this page.

5.2. Times of Payments.

5.2.1. ENGINEER shall submit monthly statements for Basic and Additional Services rendered and for Reimbursable Expenses incurred. The amount of ENGINEER's professional fee billed will be based upon ENGINEER's estimate of the proportion of the total services actually completed at the time of billing. OWNER shall make prompt monthly payments in response to ENGINEER's monthly statements.

5.3. Other Provisions Concerning Payments.

5.3.1. If OWNER fails to make any payment due ENGINEER for services and expenses within sixty days after receipt of ENGINEER's bill therefor, the amounts due ENGINEER shall include a charge at the rate of 1% per month from said sixtieth day, and in addition, ENGINEER may, after giving seven days' written notice to OWNER, suspend services under this Agreement until he has been paid in full all amounts due him for services and expenses.

5.3.2. In the event of termination by OWNER under paragraph 7.1 upon the completion of any phase of the Basic Services, progress payments due ENGINEER for services rendered through such phase shall constitute total payment for such services. In the event of such termination by OWNER during any phase of the Basic Services, ENGINEER will be paid for services rendered during that phase on the basis of Direct Labor Costs times a factor of ________ for services rendered during that phase to date of termination by principals and employees assigned to the Project. In the event of any such termination, ENGINEER will be paid for all unpaid Additional Services and unpaid Reimbursable Expenses, plus all termination expenses. Termination expenses mean Reimbursable Expenses directly attributable to termination, which shall include an amount computed as a percentage of total compensation for Basic Services earned by ENGINEER to the date of termination, as follows:

20% if termination occurs after commencement of the Preliminary Design Phase but prior to commencement of the Final Design Phase; or

10% if termination occurs after commencement of the Final Design Phase.

5.4. Definitions.

5.4.1. Direct Labor Costs used as a basis for payment mean salaries and wages (basic and incentive) paid to all personnel engaged directly on the Project, including, but not limited to, engineers, architects, surveyors, designers, draftsmen, specification writers, estimators, other technical personnel, stenographers, typists and clerks; but does not include indirect payroll related costs or fringe benefits. For the purposes of this Agreement the principals of ENGINEER and their hourly direct labor costs are:

____

____

5.4.2. Reimbursable Expenses mean the actual expenses incurred directly or indirectly in connection with the Project for: transportation and subsistence incidental thereto; obtaining bids or proposals from Contractor(s); furnishing and maintaining field office facilities; subsistence and transportation of Resident Project Representatives and their assistants; toll telephone calls and telegrams; reproduction of reports, Drawings, Specifications, and similar Project-related items in addition to those required under Section 1; and, if authorized in advance by OWNER, overtime work requiring higher than regular rates. Reimbursable Expenses shall include the amount billed to ENGINEER by special consultants employed by ENGINEER (other than as an authorized Additional Service under Section 2) for such consultants' services and Reimbursable Expenses times a factor of ________; and shall also include expenses incurred for computer time and other highly specialized equipment, including an appropriate charge for previously established programs and expenses of photographic production techniques times a factor of ________.

[The remainder of this page was left blank intentionally.]

DIRECT LABOR COSTS METHOD OF PAYMENT

SECTION 5—PAYMENTS TO ENGINEER

5.1. Methods of Payments for Services and Expenses of ENGINEER.

5.1.1. *For Basic Services.* OWNER shall pay ENGINEER for Basic Services rendered under Section 1 (as amended and supplemented by Exhibit A "Further Description of Basic Engineering Services and Related Matters") an amount based on Direct Labor Costs times a factor of ________ for services rendered by principals and employees assigned to the Project.

5.1.2. *Additional Services.* OWNER shall pay ENGINEER for Additional Services rendered under Section 2 as follows:

5.1.2.1. *General.* For Additional Services rendered under paragraphs 2.1.1 through 2.1.17, inclusive (except services covered by paragraph 2.1.7 and services as a consultant or witness under paragraph 2.1.16), on the basis of Direct Labor Costs times a factor of ________ for services rendered by principals and employees assigned to the Project.

[Delete inapplicable language and initial]

5.1.2.2. *Special Consultants.* For services and reimbursable expenses of special consultants employed by ENGINEER pursuant to paragraph 2.1.7 or 2.1.17, the amount billed to ENGINEER therefor times a factor of ________.

5.1.2.3. *Serving as a Witness.* For the services rendered by principals and employees as consultants or witnesses in any litigation, hearing or proceeding in accordance with paragraph 2.1.16, at the rate of $ ___ per day or any portion thereof (but compensation for time spent in preparing to appear in any such litigation, hearing or proceeding will be on the basis provided in paragraph 5.1.2.1).

5.1.2.4. *Resident Project Services.* For resident services during construction furnished under paragraph 2.2.1, on the basis of Direct Labor Costs times a factor of ________________ for services rendered by principals and employees assigned to field offices in connection with resident Project representation.

5.1.3. *For Reimbursable Expenses.* In addition to payments provided for in paragraphs 5.1.1 and 5.1.2, OWNER shall pay ENGINEER the actual costs of all Reimbursable Expenses incurred in connection with all Basic and Additional Services.

5.1.4. The terms "Direct Labor Costs" and "Reimbursable Expenses" will have the meanings assigned to them in paragraph 5.4 which appears on the reverse side of this page.

5.2. Times of Payments.

5.2.1. ENGINEER shall submit monthly statements for Basic and Additional Services rendered and for Reimbursable Expenses incurred. OWNER shall make prompt monthly payments in response to ENGINEER's monthly statements.

5.3. Other Provisions Concerning Payments.

5.3.1. If OWNER fails to make any payment due ENGINEER for services and expenses within sixty days after receipt of ENGINEER's bill therefor, the amounts due ENGINEER shall include a charge at the rate of 1% per month from said sixtieth day, and in addition, ENGINEER may, after giving seven days' written notice to OWNER, suspend services under this Agreement until he has been paid in full all amounts due him for services and expenses.

5.3.2. In the event of termination by OWNER under paragraph 7.1 upon the completion of any phase of the Basic Services, progress payments due ENGINEER for services rendered through such phase shall constitute total payment for such services. In the event of such termination by OWNER during any phase of the Basic Services, ENGINEER will be paid for services rendered during that phase on the basis of Direct Labor Costs times a factor of _____ for services rendered during that phase to date of termination by principals and employees assigned to the Project. In the event of any such termination, ENGINEER will be paid for all unpaid Additional Services and unpaid Reimbursable Expenses, plus all termination expenses. Termination expenses mean Reimbursable Expenses directly attributable to termination, which shall include an amount computed as a percentage of total compensation for Basic Services earned by ENGINEER to the date of termination, as follows:

20% if termination occurs after commencement of the Preliminary Design Phase but prior to commencement of the Final Design Phase; or

10% if termination occurs after commencement of the Final Design Phase.

5.4. Definitions

5.4.1. Direct Labor Costs used as a basis for payment mean salaries and wages (basic and incentive) paid to all personnel engaged directly on the Project, including, but not limited to, engineers, architects, surveyors, designers, draftsmen, specification writers, estimators, other technical personnel, stenographers, typists and clerks; but does not include indirect payroll related costs or fringe benefits. For the purposes of this Agreement the principals of ENGINEER and their hourly direct labor costs are:

____

____

5.4.2. Reimbursable Expenses mean the actual expenses incurred directly or indirectly in connection with the Project for: transportation and subsistence incidental thereto; obtaining bids or proposals from Contractor(s); furnishing and maintaining field office facilities; subsistence and transportation of Resident Project Representatives and their assistants; toll telephone calls and telegrams; reproduction of reports, Drawings, Specifications, and similar Project-related items in addition to those required under Section 1; and, if authorized in advance by OWNER, overtime work requiring higher than regular rates. Reimbursable Expenses shall include the amount billed to ENGINEER by special consultants employed by ENGINEER (other than as an authorized Additional Service under Section 2) for such consultants' services and Reimbursable Expenses times a factor of ________; and shall also include expenses incurred for computer time and other highly specialized equipment, including an appropriate charge for previously established programs and expenses of photographic production techniques times a factor of ____________.

[The remainder of this page was left blank intentionally.]

PAYROLL COSTS METHOD OF PAYMENT

SECTION 5—PAYMENTS TO ENGINEER

5.1. Methods of Payment for Services and Expenses of ENGINEER.

5.1.1. *For Basic Services.* OWNER shall pay ENGINEER for Basic Services rendered under Section 1 (as amended and supplemented by Exhibit A "Further Description of Basic Engineering Services and Related Matters") an amount based on ENGINEER's Payroll Costs times a factor of ________ for services rendered by principals and employees assigned to the Project.

5.1.2. *For Additional Services.* OWNER shall pay ENGINEER for Additional Services rendered under Section 2 as follows:

5.1.2.1. *General.* For Additional Services rendered under paragraphs 2.1.1 through 2.1.17, inclusive (except services covered by paragraph 2.1.7 and services as a consultant or witness under paragraph 2.1.16), on the basis of Payroll Costs times a factor of ________ for services rendered by principals and employees assigned to the Project.

5.1.2.2. *Special Consultants.* For services and reimbursable expenses of special consultants employed by ENGINEER pursuant to paragraph 2.1.7 or 2.1.17, the amount billed to ENGINEER therefor times a factor of ________.

5.1.2.3. *Serving as a Witness.* For the services rendered by principals and employees as consultants or witnesses in any litigation, hearing or proceeding in accordance with paragraph 2.1.16, at the rate of $________ per day or any portion thereof (but compensation for time spent in preparing to appear in any such litigation, hearing or proceeding will be on the basis provided in paragraph 5.1.2.1).

[Delete inapplicable language and initial]

5.1.2.4. *Resident Project Services.* For resident services during construction furnished under paragraph 2.2.1, on the basis of Payroll Costs times a factor of ________ for services rendered by principals and employees assigned to field offices in connection with resident Project representation.

5.1.3. *For Reimbursable Expenses.* In addition to payments provided for in paragraphs 5.1.1 and 5.1.2, OWNER shall pay ENGINEER the actual costs of all Reimbursable Expenses incurred in connection with all Basic and Additional Services.

5.1.4. The terms "Payroll Costs" and "Reimbursable Expenses" will have the meanings assigned to them in paragraph 5.4 which appears on the reverse side of this page.

5.2. Times of Payments.

5.2.1. ENGINEER shall submit monthly statements for Basic and Additional Services rendered and for Reimbursable Expenses incurred. OWNER shall make prompt monthly payments in response to ENGINEER's monthly statements.

5.3. Other Provisions Concerning Payments.

5.3.1. If OWNER fails to make any payment due ENGINEER for services and expenses within sixty days after receipt of ENGINEER's bill therefor, the amounts due ENGINEER shall include a charge at the rate of 1% per month from said sixtieth day, and in addition, ENGINEER may, after giving seven days' written notice to OWNER, suspend services under this Agreement until he has been paid in full all amounts due him for services and expenses.

5.3.2. In the event of termination by OWNER under paragraph 7.1 upon the completion of any phase of the Basic Services, progress payments due ENGINEER for services rendered through such phase shall constitute total payment for such services. In the event of such termination by OWNER during any phase of the Basic Services, ENGINEER will be paid for services rendered during that phase on the basis of Payroll Costs times a factor of ________ for services rendered during that phase to date of termination by principals and employees assigned to the Project. In the event of any such termination, ENGINEER will be paid for all unpaid Additional Services and unpaid Reimbursable Expenses, plus all termination expenses. Termination expenses mean Reimbursable Expenses directly attributable to termination, which shall include an amount computed as a percentage of total compensation for Basic Services earned by ENGINEER to the date of termination, as follows:

20% if termination occurs after commencement of the Preliminary Design Phase but prior to commencement of the Final Design Phase; or

10% if termination occurs after commencement of the Final Design Phase.

5.4. Definitions.

5.4.1. The Payroll Costs used as a basis for payment mean salaries and wages (basic and incentive) paid to all personnel engaged directly on the Project, including, but not limited to, engineers, architects, surveyors, designers, draftsmen, specification writers, estimators, other technical personnel, stenographers, typists and clerks; plus the cost of customary and statutory benefits including, but not limited to, social security contributions, unemployment, excise and payroll taxes, workers' compensation, health and retirement benefits, sick leave, vacation and holiday pay applicable thereto. For the purposes of this Agreement, the principals of ENGINEER and their hourly payroll costs are:

____

____

The amount of customary and statutory benefits of all other personnel will be considered equal to ________% of salaries and wages.

5.4.2. Reimbursable Expenses mean the actual expenses incurred directly or indirectly in connection with the Project for: transportation and subsistence incidental thereto; obtaining bids or proposals from Contractor(s); furnishing and maintaining field office facilities; subsistence and transportation of Resident Project Representatives and their assistants; toll telephone calls and telegrams; reproduction of reports, Drawings, Specifications, and similar Project-related items in addition to those required under Section 1; and, if authorized in advance by OWNER, overtime work requiring higher than regular rates. Reimbursable Expenses shall include the amount billed to ENGINEER by special consultants employed by ENGINEER (other than as an authorized Additional Service under Section 2) for such consultants' services and Reimbursable Expenses times a factor of ________; and shall also include expenses incurred for computer time and other highly specialized equipment, including an appropriate charge for previously established programs and expenses of photographic production techniques times a factor of ________.

[The remainder of this page was left blank intentionally.]

SECTION 6—CONSTRUCTION COST AND OPINIONS OF COST

6.1. Construction Cost.

The construction cost of the entire Project (herein referred to as "Construction Cost") means the total cost of the entire Project to OWNER, but it will not include ENGINEER's compensation and expenses, the cost of land, rights-of-way, or compensation for or damages to, properties unless this Agreement so specifies, nor will it include OWNER's legal, accounting, insurance counseling or auditing services, or interest and financing charges incurred in connection with the Project. [Construction Cost is one of the items comprising Project Costs which is defined in paragraph 1.2.5.] When Construction Cost is used as a basis for payment it will be based on one of the following sources with precedence in the order listed for work designed or specified by ENGINEER:

6.1.1. For completed construction work the total costs of all work performed as designed or specified by ENGINEER.

6.1.2. For work designed or specified but not constructed, the lowest bona fide bid received from a qualified bidder for such work; or, if the work is not bid, the lowest bona fide negotiated proposal for such work.

6.1.3. For work designed or specified but not constructed upon which no such bid or proposal is received, the most recent estimate of Construction Cost, or, if none is available, ENGINEER's most recent opinion of probable Construction Cost.

Labor furnished by OWNER for the Project will be included in the Construction Cost at current market rates including a reasonable allowance for overhead and profit. Materials and equipment furnished by OWNER will be included at current market prices. No deduction is to be made from ENGINEER's compensation on account of any penalty, liquidated damages, or other amounts withheld from payments to Contractor(s).

6.2. Opinions of Cost.

6.2.1. Since ENGINEER has no control over the cost of labor, materials, equipment or services furnished by others, or over the Contractor(s)' methods of determining prices, or over competitive bidding or market conditions, his opinions of probable Project Cost and Construction Cost provided for herein are to be made on the basis of his experience and qualifications and represent his best judgment as an experienced and qualified professional engineer, familiar with the construction industry; but ENGINEER cannot and does not guarantee that proposals, bids or actual Project or Construction Cost will not vary from opinions of probable cost prepared by him. If prior to the Bidding or Negotiating Phase OWNER wishes greater assurance as to Project or Construction Cost he shall employ an independent cost estimator as provided in paragraph 3.8.

6.2.2. If a Construction Cost limit is established by written agreement between OWNER and ENGINEER, the following will apply:

6.2.2.1. The acceptance by OWNER at any time during the Basic Services of a revised opinion of probable Project or Construction Cost in excess of the then established cost limit will constitute a corresponding revision in the Construction Cost limit to the extent indicated in such revised opinion.

6.2.2.2. Any Construction Cost limit so established will include a contingency of ten percent unless another amount is agreed upon in writing.

6.2.2.3. ENGINEER will be permitted to determine what materials, equipment, component systems and types of construction are to be included in the Drawings and Specifications and to make reasonable adjustments in the extent of the Project to bring it within the cost limit.

6.2.2.4. If the Bidding or Negotiating Phase has not commenced within six months after completion of the Final Design Phase, the established Construction Cost limit will not be binding on ENGINEER, and OWNER shall consent to an adjustment in such cost limit commensurate with any applicable change in the general level of prices in the construction industry between the date of completion of the Final Design Phase and the date on which proposals or bids are sought.

6.2.2.5. If the lowest bona fide proposal or bid exceeds the established Construction Cost limit, OWNER shall (1) give written approval to increase such cost limit, (2) authorize negotiating or rebidding the Project within a reasonable time, or (3) cooperate in revising the Project's extent or quality. In the case of (3), ENGINEER shall, without additional charge, modify the Contract Documents as necessary to bring the Construction Cost within the cost limit. The providing of such service will be the limit of ENGINEER's responsibility in this regard and, having done so, ENGINEER shall be entitled to payment for his services in accordance with this Agreement.

SECTION 7—GENERAL CONSIDERATIONS

7.1. Termination.

The obligation to provide further services under this Agreement may be terminated by either party upon seven days' written notice in the event of substantial failure by the other party to perform in accordance with the terms hereof through no fault of the terminating party.

7.2. Reuse of Documents.

All documents including Drawings and Specifications prepared by ENGINEER pursuant to this Agreement are instruments of service in respect of the Project. They are not intended or represented to be suitable for reuse by OWNER or others on extensions of the Project or on any other project. Any reuse without written verification or adaptation by ENGINEER for the specific purpose intended will be at OWNER's sole risk and without liability or legal exposure to ENGINEER; and OWNER shall indemnify and hold harmless ENGINEER from all claims, damages, losses and expenses including attorneys' fees arising out of or resulting therefrom. Any such verification or adaptation will entitle ENGINEER to further compensation at rates to be agreed upon by OWNER and ENGINEER.

7.3. Controlling Law.

This Agreement is to be governed by the law of the principal place of business of ENGINEER.

7.4. Successors and Assigns.

7.4.1. OWNER and ENGINEER each binds himself and his partners, successors, executors, administrators, assigns and legal representatives to the other party to this Agreement and to the partners, successors, executors, administrators, assigns and legal representatives of such other party, in respect to all covenants, agreements and obligations of this Agreement.

7.4.2. Neither OWNER nor ENGINEER shall assign, sublet or transfer any rights under or interest in (including, but without limitation, moneys that may become due or moneys that are due) this Agreement without the written consent of the other, except as stated in paragraph 7.4.1 and except to the extent that the effect of this limitation may be restricted by law. Unless specifically stated to the contrary in any written consent to an assignment, no assignment will release or discharge the assignor from any duty or responsibility under this Agreement. Nothing contained in this paragraph shall prevent ENGINEER from employing such independent consultants, associates and subcontractors as he may deem appropriate to assist him in the performance of services hereunder.

7.4.3. Nothing herein shall be construed to give any rights or benefits hereunder to anyone other than OWNER and ENGINEER.

[The remainder of this page was left blank intentionally.]

7.5. Arbitration

7.5.1. All claims, counterclaims, disputes and other matters in question between the parties hereto arising out of or relating to this Agreement or the breach thereof will be decided by arbitration in accordance with the Construction Industry Arbitration Rules of the American Arbitration Association then obtaining, subject to the limitations and restrictions stated in paragraphs 7.5.3 and 7.5.4 below. This Agreement so to arbitrate and any other agreement or consent to arbitrate entered into in accordance herewith as provided in this paragraph 7.5 will be specifically enforceable under the prevailing arbitration law of any court having jurisdiction.

7.5.2. Notice of demand for arbitration must be filed in writing with the other parties to this Agreement and with the American Arbitration Association. The demand must be made within a reasonable time after the claim, dispute or other matter in question has arisen. In no event may the demand for arbitration be made after institution of legal or equitable proceedings based on such claim, dispute or other matter in question would be barred by the applicable statute of limitations.

7.5.3. All demands for arbitration and all answering statements thereto which include any monetary claim must contain a statement that the total sum or value in controversy as alleged by the party making such demand or answering statement is not more than $200,000 (exclusive of interest and costs). The arbitrators will not have jurisdiction, power or authority to consider, or make findings (except in denial of their own jurisdiction) concerning, any claim, counterclaim, dispute or other matter in question where the amount in controversy thereof is more than $200,000 (exclusive of interest and costs) or to render a monetary award in response thereto against any party which totals more than $200,000 (exclusive of interest and costs).

7.5.4. No arbitration arising out of, or relating to, this Agreement may include, by consolidation, joinder or in any other manner, any person or entity who is not a party to this Agreement.

7.5.5. By written consent signed by all the parties to this Agreement and containing a specific reference hereto, the limitations and restrictions contained in paragraphs 7.5.3 and 7.5.4 may be waived in whole or in part as to any claim, counterclaim, dispute or other matter specifically described in such consent. No consent to arbitration in respect of a specifically described claim, counterclaim, dispute or other matter in question will constitute consent to arbitrate any other claim, counterclaim, dispute or other matter in question which is not specifically described in such consent or in which the sum or value in controversy exceeds $200,000 (exclusive of interest and costs) or which is with any party not specifically described therein.

7.5.6. The award rendered by the arbitrators will be final, not subject to appeal and judgment may be entered upon it in any court having jurisdiction thereof.

[The remainder of this page was left blank intentionally.]

[This page was left blank intentionally.]

Revised 4/10/79

SECTION 8—SPECIAL PROVISIONS, EXHIBITS and SCHEDULES.

8.1. This Agreement is subject to the following special provisions.

8.1.1.

8.2. The following Exhibits are attached to and made a part of this Agreement:

8.2.1. Exhibit A "Further Description of Basic Engineering Services and Related Matters" consisting of ________ pages.

8.2.2. Exhibit B "Duties, Responsibilities and Limitations of Authority of Resident Project Representative" consisting of ________ pages.

8.2.3.

8.3. This Agreement (consisting of pages 1 to ________, inclusive), together with the Exhibits and schedules identified above constitute the entire agreement between OWNER and ENGINEER and supersede all prior written or oral understandings. This Agreement and said Exhibits and schedules may only be amended, supplemented, modified or cancelled by a duly executed written instrument.

IN WITNESS WHEREOF, the parties hereto have made and executed this Agreement as of the day and year first above written.

OWNER:	ENGINEER:
________________________	________________________
________________________	________________________
________________________	________________________

NAVFAC 4330/11 (Rev. 12-73)
(For use with SF 253, Aug 1970)

ADDITIONAL GENERAL PROVISIONS
(ARCHITECT-ENGINEER CONTRACT)

14. METHOD OF PAYMENT (1973 APR)

(a) Estimates shall be made monthly of the amount and value of the work and services performed by the Architect-Engineer under this contract, such estimates to be prepared by the Architect-Engineer and accompanied by such supporting data as may be required by the Contracting Officer.

(b) Upon approval of such estimate by the Contracting Officer payment upon properly certified vouchers shall be made to the Architect-Engineeer as soon as practicable of 90% of the amount as determined above, less all previous payments; *provided*, however, that if the Contracting Officer determines that the work is substantially complete and that the amount of retained percentages is in excess of the amount considered by him to be adequate for the protection of the Government, he may at his discretion release to the Architect-Engineer such excess amount.

(c) Upon satisfactory completion by the Architect-Engineer and acceptance by the Contracting Officer of the work done by the Architect-Engineer in accordance with the "Statement of Architect-Engineer Services" (Appendix A of the contract), the Architect-Engineer will be paid the unpaid balance of any money due for work under said statement, including retained percentages relating to this portion of the work. In the event that the Government exercises the option under the "Option for Supervision and Inspection Services" clause, progress payments as provided for in (a) and (b) above will be made for this portion of the contract work.

(d) Upon satisfactory completion of the construction work and its final acceptance, the Architect-Engineer shall be paid the unpaid balance of any money due hereunder. Prior to such final payment under the contract, or prior to settlement upon termination of the contract, and as a condition precedent thereto, the Architect-Engineer shall execute and deliver to the Contracting Officer a release of all claims against the Government arising under or by virtue of this contract, other than such claims, if any, as may be specifically excepted by the Architect-Engineer from the operation of the release in stated amounts to be set forth therein.

15. CONTRACTING OFFICER'S DECISIONS (1965 JAN)

The extent and character of the work to be done by the Architect-Engineer shall be subject to the general supervision, direction, control and approval of the Contracting Officer.

16. SUBCONTRACTORS AND OUTSIDE ASSOCIATES AND CONSULTANTS (1965 JAN)

Any subcontractors and outside associates or consultants required by the Architect-Engineer in connection with the services covered by the contract will be limited to such individuals or firms as were specifically identified and agreed to during negotiations. Any substitution in such subcontractors, associates, or consultants will be subject to the prior approval of the Contracting Officer.

17. RENEGOTIATION (1959 OCT)

(a) To the extent required by law, this contract is subject to the Renegotiation Act of 1951 (50 U.S.C. App. 1211, et seq.), as amended, and to any subsequent act of Congress providing for the renegotiation of Contracts. Nothing contained in this clause shall impose any renegotiation obligation with respect to this contract or any subcontract hereunder which is not imposed by an act of Congress heretofore or hereafter enacted. Subject to the foregoing this contract shall be deemed to contain all the provisions required by Section 104 of the Renegotiation Act of 1951, and by any such other act without subsequent contract amendment specifically incorporating such provisions.

(b) The Architect-Engineer agrees to insert the provisions of this clause, including this paragraph (b), in all subcontracts, as that term is defined in Section 103g of the Renegotiation Act of 1951, as amended.

18. GRATUITIES (1952 MAR)

(a) The Government may, by written notice to the Architect-Engineer, terminate the right of the Architect-Engineer to proceed under this contract if it is found, after notice and hearing, by the Secretary or his duly authorized representative, that gratuities (in the form of entertainment, gifts, or otherwise) were offered or given by the Architect-Engineer, or any agent or representative of the Architect-Engineer, to any officer or employee of the Government with a view toward securing a contract or securing favorable treatment with respect to the awarding or amending, or the making of any determinations with respect to the performing, of such contract; provided, that the existence of the facts upon which the Secretary or his duly authorized representatives makes such finding shall be in issue and may be reviewed in any competent court.

(b) In the event this contract is terminated as provided in paragraph (a) hereof, the Government shall be entitled (i) to pursue the same remedies against the Architect-Engineer as it could pursue in the event of a breach of the contract by the Architect-Engineer, and (ii) as a penalty in addition to any other damages to which it may be entitled by law, to exemplary damages in an amount (as determined by the Secretary or his duly authorized representatives) which shall not be less than three nor more than ten times the costs incurred by the Architect-Engineer in providing any such gratuities to any such officer or employee.

(c) The rights and remedies of the Government provided in this clause shall not be exclusive and are in addition to any other rights and remedies provided by law or under this contract.

19. INTEREST (1972 MAY)

Notwithstanding any other provision of this contract, unless paid within 30 days all amounts that become payable by the Contractor to the Government under this contract (net of any applicable tax credit under the Internal Revenue Code) shall bear interest from the date due until paid and shall be subject to adjustments as provided by Part 6 of Appendix E of the Armed Services Procurement Regulation, as in effect on the date of this contract. The interest rate per annum shall be the interest rate in effect which has been established by the Secretary of the Treasury pursuant to Public Law 92-41; 85 STAT 97 for the Renegotiation Board, as of the date the amount becomes due as herein provided. Amounts shall be due upon the earliest one of (i) the date fixed pursuant to this contract, (ii) the date of the first written demand for payment, consistent with this contract, including demand consequent upon default termination; (iii) the date of transmittal by the Government to the Contractor of a proposed supplemental agreement to confirm completed negotiations fixing the amount: or (iv) if this contract provides for revision of prices, the date of written notice to the Contractor stating the amount of refund payable in connection with a pricing proposal or in connection with a negotiated pricing agreement not confirmed by contract supplement.

20. COMPOSITION OF ARCHITECT-ENGINEER (1972 APR)

If the Architect-Engineer hereunder is comprised of more than one legal entity, each such entity shall be jointly and severally liable hereunder.

21. PRICING OF ADJUSTMENTS (1970 JUL)

When costs are a factor in any determination of a contract price adjustment pursuant to the "Changes" clause or any other provision of this contract, such costs shall be in accordance with Section XV of the Armed Services Procurement Regulation as in effect on the date of this contract.

22. AUDIT BY DEPARTMENT OF DEFENSE (1971 APR)

(a) General. The Contracting Officer or his representatives shall have the audit and inspection rights described in the applicable paragraphs (b), (c) and (d) below.

(b) Examination of Costs. If this is a cost reimbursement type, incentive, time and materials, labor hour, or price redeterminable contract, or any combination thereof, the Contractor shall maintain, and the Contracting Officer or his representatives shall have the right to examine books, records, documents, and other evidence and accounting procedures and practices, sufficient to reflect properly all direct and indirect costs of whatever nature claimed to have been incurred and anticipated to be incurred for the performance of this contract. Such right of examination shall include inspection at all reasonable times of the Contractor's plants, or such parts thereof, as may be engaged in the performance of this contract.

(c) Cost or Pricing Data. If the Contractor submitted cost or pricing data in connection with the pricing of this contract or any change or modification thereto, unless such pricing was based on adequate price competition, established catalog or market prices of commercial items sold in substantial quantities to the general public, or prices set by law or regulation, the Contracting Officer or his representatives who are employees of the United States Government shall have the right to examine all books, records, documents and other data of the Contractor related to the negotiation, pricing or performance of such contract, change or modification, for the purpose of evaluating the accuracy, completeness and currency of the cost or pricing data submitted. Additionally, in the case of pricing any change or modification exceeding $100,000 to formally advertised contracts, the Comptroller General of the United States or his representatives who are employees of the United States Government shall have such rights. The right of examination shall extend to all documents necessary to permit adequate evaluation of the cost or pricing data submitted, along with the computations and projections used therein.

(d) Reports. If the Contractor is required to furnish Cost Information Reports (CIR) or Contract Fund Status Reports (CFSR), the Contracting Officer or his representatives shall have the right to examine books, records, documents, and supporting materials, for the purpose of evaluating (i) the effectiveness of the Contractor's policies and procedures to produce data compatible with the objectives of these reports, and (ii) the data reported.

(e) Availability. The materials described in (b), (c) and (d) above shall be made available at the office of the Contractor, at all reasonable times, for inspection, audit, or reproduction, until the expiration of three years from the date of final payment under this contract or such lesser time specified in Appendix M of the Armed Services Procurement Regulation, and for such longer period, if any, as is required by applicable statute, or by other clauses of this contract, or by (1) and (2) below:

(1) If this contract is completely or partially terminated, the records relating to the work terminated shall be made available for a period of three years from the date of any resulting final settlement.

(2) Records which relate to appeals under the "Disputes" clause of this contract, or litigation or the settlement of claims arising out of the performance of this contract, shall be made available until such appeals, litigation, or claims have been disposed of.

(f) The Contractor shall insert a clause containing all the provisions of this clause, including this paragraph (f), in all subcontracts hereunder, except altered as necessary for proper identification of the contracting parties and the Contracting Officer under the Government prime contract.

23. DESIGN WITHIN FUNDING LIMITATIONS (1971 APR)

(The following clause applies only when the Architect-

Engineer is required to provide plans and specifications for construction at a specific site.)

(a) The Architect-Engineer shall accomplish the design services required under this contract so as to permit the award of a contract, pursuant to standard Department of Defense procedures, for the construction of the facilities designed at a price that does not exceed the estimated construction contract price set forth in this contract. When bids or proposals for the construction contract are received which exceed such estimated price, the Architect-Engineer shall perform such redesign and other services as are necessary to permit contract award within such funding limitation. These additional services shall be performed at no increase in the price of this contract. However, the Architect-Engineer shall not be required to perform such additional services at no cost to the Government if the unfavorable bids or proposals are the result of conditions beyond his reasonable control

(b) The Architect-Engineer will promptly advise the Contracting Officer if he finds that the project being designed will exceed or is likely to exceed the funding limitations and he is unable to design a usable facility within these limitations. Upon receipt of such information, the Contracting Officer will review the Architect-Engineer's revised estimate of construction cost. The Contracting Officer may, if he determines that the estimated construction contract price set forth in this contract is so low that award of a construction contract not in excess of such estimate is improbable, authorize a change in scope or materials as required to reduce the estimated construction cost to an amount within the estimated construction contract price set forth elsewhere in the contract, or he may adjust such estimated construction contract price. When bids or proposals are not solicited or where they are unreasonably delayed, the Government shall prepare an estimate of constructing the design submitted and such estimate will be used in lieu of bids or proposals to determine compliance with the funding limitation.

24. REQUIREMENTS FOR REGISTRATION OF DESIGNERS (1972 APR)

The design of architectural, structural, mechanical, electrical, civil or other engineering features of the work shall be accomplished or reviewed and approved by architects or engineers registered to practice in the particular professional field involved in a State or possession of the United States, in Puerto Rico, or in the District of Columbia.

25. PAYMENT OF INTEREST ON CONTRACTOR'S CLAIMS (1972 MAY)

(a) If an appeal is filed by the Contractor from a final decision of the Contracting Officer under the DISPUTES clause of this contract, denying a claim arising under the contract, simple interest on the amount of the claim finally determined owed by the Government shall be payable to the Contractor. Such interest shall be at the rate established by the Secretary of the Treasury pursuant to Public Law 92-41; 85 STAT 97 for the Renegotiation Board from the date the Contractor furnishes to the Contracting Officer his written appeal pursuant to the DISPUTES clause of this contract, to the date of (i) a final judgment by a court of competent jurisdiction, or (ii) mailing to the Contractor of a supplemental agreement for execution either confirming completed negotiations between the parties or carrying out a decision of a Board of Contract Appeals.

(b) Notwithstanding (a) above, (i) interest shall be applied only from the date payment was due, if such date is later than the filing of appeal; and (ii) interest shall not be paid for any period of time that the Contracting Officer determines the Contractor has unduly delayed in pursuing his remedies before a Board of Contract Appeals or a court of competent jurisdiction.

26. PRICE REDUCTION FOR DEFECTIVE COST OR PRICING DATA (1970 JAN)

If any price, including profit or fee, negotiated in connection with this contract or any cost reimbursable under this contract was increased by any significant sums because:

(i) the Contractor furnished cost or pricing data which was not complete, accurate and current as certified in the Contractor's Certificate of Current Cost or Pricing Data;

(ii) a subcontractor, pursuant to the clause of this contract entitled "Subcontractor Cost or Pricing Data" or "Subcontractor Cost or Pricing Data–Price Adjustments" or any subcontract clause therein required, furnished cost or pricing data which was not complete, accurate and current as certified in the subcontractor's Certificate of Current Cost or Pricing Data;

(iii) a subcontractor or prospective subcontractor furnished cost or pricing data which was required to be complete, accurate and current and to be submitted to support a subcontract cost estimate furnished by the Contractor but which was not complete, accurate and current as of the date certified in the Contractor's Certificate of Current Cost or Pricing Data; or

(iv) the Contractor or a subcontractor or prospective subcontractor furnished any data, not within (i), (ii) or (iii) above, which was not accurate as submitted;

the price or cost shall be reduced accordingly and the contract shall be modified in writing as may be necessary to reflect such reduction. However, any reduction in the contract price due to defective subcontract data of a prospective subcontractor when the subcontract was not subsequently awarded to such subcontractor, will be limited to the amount (plus applicable overhead and profit markup) by which the acutal subcontract, or actual cost to the Contractor if there was no subcontract, was less than the prospective subcontract cost estimate submitted by the Contractor, *provided* the actual subcontract price was not affected by defective cost or pricing data.

Note: Since the contract is subject to reduction under this clause by reason of defective cost or pricing data submitted in connection with certain subcontracts, it is expected that the contractor may wish to include a clause in each such subcontract requiring the subcontractor to appropriately indemnify the contractor. However, the inclusion of such a clause and the terms thereof are matters for negotiation and agreement between the contractor and the subcontractor, *provided* that they are consistent with ASPR 23-203 relating to Disputes provisions in subcontracts. It is also expected that any subcontractor subject to such indemnification will generally require substantially similar indemnification for defective cost or pricing data required to be submitted by his lower tier subcontractors.

27. SUBCONTRACTOR COST OR PRICING DATA (1970 JAN)

(a) The Contractor shall require subcontractors hereunder to submit, actually or by specific identification in writing, cost or pricing data under the following circumstances: (i) prior to the award of any subcontract the amount of which is expected to exceed $100,000 when entered into; (ii) prior to the pricing of any subcontract modification which involves aggregate increases and/or decreases in costs plus applicable profits expected to exceed $100,000; except where the price is based on adequate price competition, established catalog or market prices of commercial items sold in substantial quantities to the general public, or prices set by law or regulation.

(b) The Contractor shall require subcontractors to certify, in substantially the same form as that used in the certificate by the Prime Contractor to the Government, that to the best of their knowledge and belief, the cost and pricing data submitted under (a) above is accurate, complete, and current as of the date of agreement on the negotiated price of the subcontract or subcontract change or modification.

(c) The Contractor shall insert the substance of this clause including this paragraph (c) in each subcontract hereunder which exceeds $100,000 when entered into except where the price thereof is based on adequate price competition, established catalog or market prices of commercial items sold in substantial quantities to the general public, or prices set by law or regulation. In each such excepted subcontract hereunder in excess of $100,000, the Contractor shall insert the substance of the following clause:

SUBCONTRACTOR COST OR PRICING DATA–PRICE ADJUSTMENTS

(a) Paragraphs (b) and (c) of this clause shall become operative only with respect to any modification made pursuant to one or more provisions of this contract which involves aggregate increases and/or decreases in costs plus applicable profits expected to exceed $100,000. The requirements of this clause shall be limited to such contract modifications.

(b) The Contractor shall require subcontractors hereunder to submit, actually or by specific identification in writing, cost or pricing data under the following circumstances: (i) prior to award of any subcontract, the amount of which is expected to exceed $100,000 when entered into; (ii) prior to the pricing of any subcontract modification which involves aggregate increases and/or decreases in costs plus applicable profits expected to exceed $100,000; except where the price is based on adequate price competition, established catalog or market prices of commercial items sold in substantial quantities to the general public, or prices set by law or regulation.

(c) The Contractor shall require subcontractors to certify, in substantially the same form as that used in the certificate by the Prime Contractor to the Government, that to the best of their knowledge and belief the cost and pricing data submitted under (b) above is accurate, complete, and current as of the date of agreement on the negotiated price of the subcontract or subcontract change or modification.

(d) The Contractor shall insert the substance of this clause including this paragraph (d) in each subcontract hereunder which exceeds $100,000 when entered into.

28. AUTHORIZATION AND CONSENT (1964 MAR)

The Government hereby gives its authorization and consent (without prejudice to any rights of indemnification) for all use and manufacture, in the performance of this contract or any part hereof or any amendment hereto or any subcontract hereunder (including any lower-tier subcontract), of any invention described in and covered by a patent of the United States (i) embodied in the structure or composition of any article the delivery of which is accepted by the Government under this contract, or (ii) utilized in the machinery, tools, or methods the use of which necessarily results from compliance by the Contractor or the using subcontractor with (a) specifications or written provisions now or hereafter forming a part of this contract, or (b) specific written instructions given by the Contracting Officer directing the manner of performance. The entire liability to the Government for infringement of a patent of the United States shall be determined solely by the provisions of the indemnity clauses, if any, included in this contract or any subcontract hereunder (including any lower-tier subcontract), and the Government assumes liability for all other infringement to the extent of the authorization and consent hereinabove granted.

29. NOTICE AND APPROVAL OF RESTRICTED DESIGNS (1972 APR)

In the performance of this contract, the Architect-Engineer shall, to the extent practicable, make maximum use of structures, machines, products, materials, construction methods, and equipment which are readily available through Government or competitive commercial channels, or through standard or proven production techniques, methods, and processes. Unless approved by the Contracting Officer the Architect-Engineer shall not, in the performance of the work called for by this contract, produce a design or specification such as to require in this construction work the use of structures, products, materials, construction equipment, or processes which are known by the Architect-Engineer to be available only from a sole source. As to any such design or specification the Architect-Engineer shall report to the Contracting Officer giving the reason or reasons why it is considered necessary to so restrict the design or specification.

30. TECHNICAL ADEQUACY (6/72)

Approval by the Officer in Charge of drawings, designs, specifications and other incidental architectural-engineering work or materials furnished hereunder shall not in any way relieve the Architect-Engineer of responsibility for the technical adequacy of the work.

31. GOVERNMENT REPRESENTATIVES (6/72)

The work will be under the general direction of the Contracting Officer, the Commander, Naval Facilities Engineering Command, who shall designate an officer of the Civil Engineer Corps, United States Navy, or other officer or representative of the Government, as Officer in Charge, hereinafter referred to as the "OIC" who, except in connection with the Disputes clause, shall be the authorized representative of the Contracting Officer and under the direction of the Contracting Officer have complete charge of the work, and shall exercise full supervision and general direction of the work, so far as it affects the interest of the Government. For the purposes of the "Disputes" clause the Contracting Officer shall mean the Commander, Naval Facilities Engineering Command, the Acting Commander, their successors, or their representatives specially designated for this purpose.

32. GENERAL STATEMENT OF ARCHITECT-ENGINEER SERVICES, DESIGN CONTRACT (6/72)

As may be necessary to accomplish the work described in Appendix A, "Statement of Architect-Engineer Services," the Architect-Engineer shall prepare and furnish to the Government, complete and ready for use, all necessary studies, preliminary sketches, estimates, working records and other drawings (including large scale details as required), and specifications; shall check shop drawings furnished by the construction contractor; shall furnish consultation and advice as requested by the Government during the construction (but not including the supervision of the construction work); and shall furnish all other architectural and engineering services, including without limitations those specified hereinafter and required in connection with the accomplishment of Naval public works and/or utilities projects. It is agreed, without limiting the generality of the foregoing, that

(a) The Architect-Engineer shall, if necessary, visit the site and shall hold such conferences with representatives of the Government and take such other action as may be necessary to obtain the data upon which to develop the design and preliminary sketches showing the contemplated project.

(b) The preliminary sketches shall include plans, elevations and sections developed in such detail and with such descriptive specifications as will clearly indicate the scope of the work, and make possible a reasonable estimate of the cost.

(c) Preliminary sketches together with an estimate of the cost of the project shown on the sketches shall be submitted for the approval of the Officer-in-Charge.

(d) The Architect-Engineer shall change the preliminary sketches for the extent necessary to meet the requirements of the Government, and after approval by the Officer-in-Charge the Architect-Engineer shall furnish necessary prints of the approved preliminary sketches to the Officer-in-Charge.

(e) After the preliminary sketches and estimates have been approved, the Architect-Engineer shall proceed with the preparation of complete working drawings and specifications as required by the Officer-in-Charge in connection with the construction of the said project. Working drawings, specifications and estimates shall be delivered to the Officer-in-Charge in such sequence and at such times as required by the Government and as will insure that the construction work can be initiated promptly, procurement of materials made without delay, and continuous prosecution of the work promoted. Working drawings and specifications shall be revised as necessary and as required by the Officer-in-Charge. After working drawings and specifications have been approved by the Officer-in-Charge, the Architect-Engineer shall furnish such number of sets of prints of the approved working drawings and such number of sets of the approved specifications, as may be required by the Officer-in-Charge.

(f) Upon approval of final plans the Architect-Engineer shall deliver to the Government one set of tracings, in such medium and on such materials, as may be required by the Officer-in-Charge, suitable for blue-printing, showing complete approved construction requirements (not of "as-built" construction unless otherwise stipulated), provided, however, that should this contract be terminated by the Government, the Architect-Engineer shall deliver to the Government one set of such tracings as may be required by the Officer-in-Charge. Such tracings as are delivered shall be signed by the Officer-in-Charge as an indication of approval thereof, and shall become and remain the property of the Government.

(g) The Architect-Engineer shall perform all necessary architectural-engineering services of every kind stipulated, include borings, test piles and pits, or supervision of construction work executed from the drawings and specifications, provided, however, that the Architect-Engineer shall furnish upon request and without additional compensation, such amplifications and explanations and attend such conferences as may, in the opinion of the Officer-in-Charge, be necessary to clarify the intent of the drawings and specifications and shall afford the benefit of his advice on questions that may arise in connection with the construction of the project.

(h) The Architect-Engineer shall without additional fee correct or revise the drawings, specifications, or other materials furnished under this contract, if the Officer-in-Charge finds that such revision is necessary to correct errors or deficiencies for which the Architect-Engineer is responsible.

33. COST ACCOUNTING STANDARDS (1972 JUL)

(The following clause is applicable if the amount of this contract exceeds $100, 000)

(a) Unless the Cost Accounting Standards Board has prescribed rules or regulations exempting the Architect-Engineer or this contract from standards, rules, and regulations promulgated pursuant to 50 U.S.C. App. 2168 (Public Law 91-379, August 15, 1970), the Architect-Engineer, in connection with this contract shall:

(1) By submission of a Disclosure Statement, disclose in writing his cost accounting practices as required by regulations of the Cost Accounting Standards Board. The required disclosures must be made prior to contract award unless the Contracting Officer provides a written notice to the contractor authorizing post-award submission in accordance with regulations of the Cost Accounting Standards Board. The practices disclosed for this contract shall be the same as the practices currently disclosed and applied on all other contracts and subcontracts being performed by the Architect-Engineer and which contain this Cost Accounting Standards clause. If the Architect-Engineer has marked the Disclosure Statement to indicate that it contains trade secrets and commercial or financial information which is privileged and confidential, the Disclosure Statement will be protected and will not be released outside of the Government.

(2) Follow consistently the cost accounting practices disclosed pursuant to (1) above in accumulating and reporting contract performance cost data concerning this contract. If any change in disclosed practices is made for purposes of any contract or subcontract subject to Cost Accounting Standards Board requirements, the change must be applied prospectively to this contract, and the Disclosure Statement must be amended accordingly. If the contract price or cost allowance of this contract is affected by such changes, adjustment shall be made in accordance with subparagraph (a)(4) or (a)(5) below, as appropriate.

(3) Comply with all Cost Accounting Standards in effect on the date of awards of this contract or if the Architect-Engineer has submitted cost or pricing data, on the date of final agreement on price as shown on the Architiect-Engineer's signed certificate of current cost or pricing data. The Architect-Engineer shall also comply with any Cost Accounting Standard which hereafter becomes applicable to a contract or subcontract of the contractor. Such compliance shall be required prospectively from the date of applicability to such contract or subcontract.

(4) (A) Agree to an equitable adjustment as provided in the changes clause of this contract if the contract cost is affected by a Disclosure Statement change which the Architect-Engineer is required to make pursuant to (3) above. If the Architect-Engineer has not been required to file a Disclosure Statement but is required pursuant to (a)(3) above to change an established practice, then an equitable adjustment shall similarly be agreed to.

(B) Negotiate with the contracting officer to determine the terms and conditions under which any Disclosure Statement change other than changes under (4)(A) above may be made. A change to a Disclosure Statement may be proposed by either the Government or the Architect-Engineer, provided, however, that no agreement may be made under this provision, that will increase costs paid by the United States under this contract.

(5) Agree to an adjustment of the contract price or cost allowance, as appropriate, if he or a subcontractor fails to comply with an applicable Cost Accounting Standard or to follow any practice disclosed pursuant to subparagraphs (a)(1) and (a)(2) above and such failure results in any increased costs paid by the United States. Such adjustment shall provide for recovery of the increased costs to the United States together with interest thereon computed at the rate determined by the Secretary of the Treasury pursuant to Public Law 92-41, 85 Stat. 97, or 7 percent per annum, whichever is less, from the time the payment by the United States was made to the time the adjustment is effected.

(b) If the parties fail to agree whether the Architect-Engineer or a subcontractor has compiled with an applicable Cost Accounting Standard, rule, or regulation of the Cost Accounting Standards Board and as to any cost adjustment demanded by the United States, such failure to agree shall be a dispute concerning a question of fact within the meaning of the disputes clause of this contract.

(c) The Architect-Engineer shall permit any authorized representatives of the head of the agency, of the Cost Accounting Standards Board, or of the Comptroller General of the United States to examine and make copies of any documents, papers, or records relating to compliance with the requirements of this clause.

(d) The Architect-Engineer shall include in all negotiated subcontracts which he enters into the substance of this clause except paragraph (b), and shall require such inclusion in all other subcontracts of any tier, except that this requirement shall apply only to negotiated subcontracts in excess of $100, 000 where the price negotiated is not based on:

(i) Established catalog or market prices of commercial items sold in substantial quantities to the general public, or

(ii) prices set by law or regulation.

NOTE: (1) Subcontractors shall be required to submit their Disclosure Statements to the Architect-Engineer. However, if a subcontractor has previously submitted his Disclosure Statement to a Government Administrative Contracting Officer (ACO) he may satisfy that requirement by certifying to the Architect-Engineer the date of such Statement and the address of the ACO.

(2) In any case where a subcontractor determines that the Disclosure Statement information is privileged and confidential and declines

to provide it to his Architect-Engineer or higher tier subcontractor, the Architect-Engineer may authorize direct submission of that subcontractor's Disclosure Statement to the same Government offices to which the Architect-Engineer was required to make submission of his Disclosure Statement. Such authorization shall in no way relieve the Architect-Engineer of liability as provided in paragraph (a)(5) of this clause. In view of the foregoing and since the contract may be subject to adjustment under this clause by reason of any failure to comply with rules, regulations, and Standards of the Cost Accounting Standards Board in connection with covered subcontracts, it is expected that the Architect-Engineer may wish to include a clause in each such subcontract requiring the subcontractor to appropriately indemnify the Architect-Engineer. However, the inclusion of such a clause and the terms thereof are matters for negotiation and agreement between the Architect-Engineer and the subcontractor, provided that they do not conflict with the duties of the Architect-Engineer under its contract with the Government. It is also expected that any subcontractor subject to such indemnification will generally require substantially similar indemnification to be submitted by his subcontractors.

(e) The terms defined in Sec. 331.2 of Part 331 of Title 4, Code of Federal Regulations (4 CFR 331.2) shall have the same meanings herein. As there defined, "negotiated subcontract" means "any subcontract except a firm fixed-price subcontract made by an Architect-Engineer or subcontractor after receiving offers from at least two firms not associated with each other or such Architect-Engineer or subcontractor, providing (1) the solicitation to all competing firms is identical, (2) price is the only consideration in selecting the subcontractor from among the competing firms solicited, and (3) the lowest offer received in compliance with the solicitation from among those solicited is accepted." (ASPR 7-104.83 and 7-608.18)

34. LISTING OF EMPLOYMENT OPENINGS (1973 SEP)

(This clause is applicable pursuant to 41 CFR 50-250 if this contract is for $2,500 or more.)

(a) The Contractor, to provide special emphasis to the employment of qualified disabled veterans and veterans of the Vietnam era, agrees that all suitable employment openings of the Contractor which exist at the time of the execution of this contract and those which occur during the performance of this contract, including those not generated by this contract and including those occurring at an establishment of the Contractor other than the one wherein the contract is being performed but excluding those of independently operated corporate affiliates, shall be offered for listing at the appropriate office of the State employment service system wherein the opening occurs and to provide reports to such office regarding employment openings and hires as may be required.

(b) Listing of employment openings with the employment service system pursuant to this clause shall be made at least concurrently with the use of any other recruitment service or effort and shall involve the normal obligations which attach to the placing of a bona fide job order, including the acceptance of referrals of veterans and non-veterans. Listing of employment openings does not require the hiring of any particular job applicant or from any particular group of job applicants referred by the employment service system. Nothing contained herein is intended to relieve the Contractor from any requirements in any Executive Order or regulation regarding nondiscrimination in employment.

(c) (1) Reports required shall include, but not be limited to, periodic reports which shall be filed at least quarterly with the appropriate local State employment service office or, where the Contractor has more than one establishment in a State, with the central office of that State employment service. Such reports shall indicate for each establishment (i) the number of individuals who were hired during the reporting period, (ii) the number of those hired who were disabled veterans, and (iii) the number of those hired who were non-disabled veterans of the Vietnam era. The Contractor shall maintain copies of the reports submitted until the expiration of one year after final payment under the contract, during which time they shall be made available, upon request, for examination by any authorized representatives of the Contracting Officer or the Secretary of Labor.

(2) Whenever the Contractor becomes contractually bound to the listing provisions of this clause, he shall advise the employment service system in each state wherein he has establishments, of the name and location of each such establishment in the State. As long as the Contractor is contractually bound to these provisions and has so advised the State employment service system, there is no need to advise the State system of subsequent contracts. The Contractor may advise the State system when he is no longer bound by this contract clause.

(3) If the contract is for less than $10,000 or if it is with a State or local government, the procedures set forth in subparagraphs (1) and (2) of this paragraph (c) are not required.

(d) This clause does not apply to the listing of employment openings which occur and are filled outside the 50 States, the District of Columbia, Guam, Puerto Rico, and the Virgin Islands.

(e) This clause does not apply to openings which the Contractor proposes to fill from within his own organization or to fill pursuant to a customary and traditional employer-union hiring arrangement. This exclusion does not apply to a particular opening once an employer decides to consider applicants outside of his own organization or employer-union arrangements for that opening.

(f) As used in this clause:

(1) "All suitable employment openings" includes, but is not limited to, openings which occur in the following job categories: production and nonproduction; plant and office; laborers and mechanics; supervisory and nonsupervisory; technical; and executive, administrative, and professional openings which

are compensated on a basis of less than $18, 000 per year. This term includes full-time employment, temporary employment of more than three (3) days' duration, and part-time employment.

(2) "Appropriate office of the State employment service system" means the local office of the Federal-State national system of public employment offices with assigned responsibility for serving the area of the establishment where the employment opening is to be filled, including the District of Columbia, Guam, Puerto Rico, and the Virgin Islands.

(3) "Openings which the Contractor proposes to fill from within his own organization" means employment openings for which no consideration will be given to persons outside the Contractor's organization (including any affiliates, subsidiaries, and the parent companies), and includes any openings which the Contractor proposes to fill from regularly established "recall" and "rehire" lists.

(4) "Openings which the Contractor proposes to fill from within his own organization or to fill pursuant to a customary and traditional employer-union hiring arrangement" means employment openings for which no consideration will be given to persons outside of a special hiring arrangement, including openings which the Contractor proposes to fill from union hiring halls, which is part of the customary and traditional employment relationship existing between the Contractor and representatives of his employees.

(5) "Disabled veteran" means a person entitled to disability compensation under laws administered by the Veterans Administration for disability rated at thirty percent (30%) or more, or a person whose discharge or release from active duty was for a disability incurred or aggravated in line of duty.

(6) "Veteran of the Vietnam era" means a person who was discharged or released within the 48 months preceding his application for employment covered under this part and who (i) served on active duty for a period of more than 180 days, any part of which occurred after August 5, 1964, and was discharged or released therefrom with other than a dishonorable discharge, or (ii) was discharged or released from active duty for service-connected disability if any part of such duty was performed after August 5, 1964.

(g) The Contractor agrees to place this clause (excluding this paragraph (g)) in any subcontract directly under this contract provided, such subcontract is for $2, 500 or more. (Subcontracts for personal services are exempted from this requirement.)

(h) Failure of the Contractor to comply with the requirements of this clause may result in termination for default of the contract concerned.

ADDITIONAL GENERAL PROVISIONS (Architect-Engineer Contract) (Cont.):

35. ALTERATIONS

The following alterations have been made in the provisions of this contract:

(a) Delete General Provision No. 13 of SF 253 in its entirety and substitute the following provision in lieu thereof:

EQUAL OPPORTUNITY (1974 APR)

(The following clause is applicable unless this contract is exempt under the rules, regulations, and relevant orders of the Secretary of Labor (41 CFR, ch. 60).)

During the performance of this contract, the Architect-Engineer agrees as follows:

(a) The Architect-Engineer will not discriminate against any employee or applicant for employment because of race, color, religion, sex, or national origin. The Architect-Engineer will take affirmative action to ensure that applicants are employed, and that employees are treated during employment, without regard to their race, color, religion, sex, or national origin. Such action shall include, but not be limited to, the following: Employment, upgrading, demotion, or transfer; recruitment or recruitment advertising; layoff or termination; rates of pay or other forms of compensation; and selection for training, including apprenticeship. The Architect-Engineer agrees to post in conspicuous places, available to employees and applicants for employment, notices to be provided by the Contracting Officer setting forth the provisions of this Equal Opportunity clause.

(b) The Architect-Engineer will, in all solicitations or advertisements for employees placed by or on behalf of the Architect-Engineer, state that all qualified applicants will receive consideration for employment without regard to race, color, religion, sex or national origin.

(c) The Architect-Engineer will send to each labor union or representative of workers with which he has a collective bargaining agreement or other contract or understanding, a notice, to be provided by the agency Contracting Officer, advising the labor union or workers' representative of the Architect-Engineer's commitments under this Equal Opportunity clause, and shall post copies of the notice in conspicuous places available to employees and applicants for employment.

(d) The Architect-Engineer will comply with all provisions of Executive Order 11246 of September 24, 1965, as amended by Executive Order 11375 of October 13, 1967, and of the rules, regulations, and relevant orders of the Secretary of Labor.

(e) The Architect-Engineer will furnish all information and reports required by Executive Order 11264 of September 24, 1965, as amended by Executive Order 11375 of October 13, 1967, and by the rules, regulations, and orders of the Secretary of Labor, or pursuant thereto, and will permit access to his books, records, and accounts by the contracting agency and the Secretary of Labor for purposes of investigation to ascertain compliance with such rules, regulations, and orders.

(f) In the event of the Architect-Engineer's noncompliance with the Equal Opportunity clause of this contract or with any of the said rules, regulations, or orders, this contract may be canceled, terminated, or suspended, in whole or in part, and the Architect-Engineer may be declared ineligible for further Government contracts, in accordance with procedures authorized in Executive Order 11246 of September 24, 1965, as amended by Executive Order 11375 of October 13, 1967, and such other sanctions may be imposed and remedies invoked as provided in Executive Order 11246 of September 24, 1965, as amended by Executive Order 11375 of October 13, 1967, or by rule or regulation, or order of the Secretary of Labor, or as otherwise provided by law.

(g) The Architect-Engineer will include the provisions of paragraphs (a) through (g) in every subcontract or purchase order unless exempted by rules, regulations, or orders of the Secretary of Labor issued pursuant to section 204, of Executive Order 11246 of September 24, 1965, as amended by Executive Order 11375 of October 13, 1967, so that such provisions will be binding upon each subcontractor or vendor. The Architect-Engineer will take such action with respect to any subcontract or purchase order as the contracting agency may direct as a means of enforcing such provisions, including sanctions for noncompliance; *provided, however*, that in the event the Architect-Engineer becomes involved in, or is threatened with, litigation with a subcontractor or vendor as a result of such direction by the contracting agency, the Architect-Engineer may request the United States to enter into such litigation to protect the interests of the United States.

APPENDIX G

Contract for Architectural and Engineering Services

NAVFAC 4-4330/11(10-67)
Supersedes NavDocks 424 (Rev. 4—65)

SCHEDULE

It has been determined that the execution of this contract is advantageous to the national defense and that the existing facilities of the Naval Establishment are inadequate.

1. DESCRIPTION OF THE PROJECT. The Architect-Engineer shall perform all the services required under this contract for the project generally described as follows: (hereinafter refered to as the "Project")

____

____

____

having an estimated construction cost of $ ______________, and more specifically described in APPENDIX "A" which is attached hereto and made a part hereof.

2. STATEMENT OF ARCHITECT-ENGINEER SERVICES. The Architect-Engineer shall prepare and furnish to the Government, complete and ready for use, all necessary studies, preliminary sketches, estimates, working records and other drawings (including large scale details as required), and specifications; shall check shop drawings furnished by the construction contractor; shall furnish consultation and advice as requested by the Government during the construction (but not including the supervision of the construction work); and shall furnish all other architectural and engineering services, including without limitations those specified hereinafter and required in connection with the accomplishment of Naval public works and/or utilities projects.

3. PERIOD OF SERVICE. The Architect-Engineer shall complete all work and services within ____________ months after receipt of notice to proceed.

4. DESIGN WITHIN COST LIMITATION. The Architect-Engineer agrees to design a facility which can be constructed (under normal Bureau of Yards and

Docks procedures) within the estimated net construction cost, with each item of the facility to be designed so as to be constructed within applicable statutory cost limitations indicated below.

If, after receipt of competitive bids, it is found that a construction contract cannot be awarded within the aforementioned estimated construction cost and within any applicable statutory cost limitations, the Architect-Engineer will, as a part of this contract, and at no additional expense to the Government, perform such redesign, re-estimating and other services as may, in the opinion of the Officer in Charge, be required to produce a usable facility within the estimated construction cost and/or within the statutory cost limitations. In connection with the foregoing, the Architect-Engineer shall be obligated to perform such additional services at no increase in contract price only when the Navy has received competitive bids within six (6) months from the date of final approval of the plans and specifications.

STATUTORY COST LIMITS

5. ARCHITECTURAL DESIGNS AND DATA—GOVERNMENT RIGHTS (UNLIMITED) (APR.1966). The Government shall have unlimited rights, for the benefit of the Government, in all drawings, designs, specifications, architectural designs of buildings and structures, notes and other architect-engineer work produced in the performance of this contract, or in contemplation thereof, and all as-built drawings produced after completion of the work, including the right to use same on any other Government work without additional cost to the Government; and with respect thereto the Architect-Engineer agrees to and does hereby grant to the Government a royalty-free license to all such data which he may covet by copyright and to all architectural designs as to which he may assert any rights or establish any claim under the design patent or copyright laws. The Architect-Engineer for a period of three (3) years after completion of the project agrees to furnish and to provide access to the originals or copies of all such materials on the request of the Contracting Officer.

6. ADDITIONAL STATEMENT OF ARCHITECT-ENGINEER SERVICES. It is agreed without limiting the generality of the foregoing, that (a) The Architect-Engineer shall, if necessary, visit the site and shall hold such conferences with representatives of the Government and take such other action as may be necessary to obtain the data upon which to develop the design and preliminary sketches showing the contemplated project, (b) The preliminary sketches shall include plans, elevations and sections developed in such detail and with such descriptive specifications as will clearly indicate the scope of the work, and make possible a reasonable estimate of the cost, (c) Preliminary sketches together with an estimate of the cost of the project shown on the sketches shall be submitted for the approval of the Officer in Charge, (d) The Architect-Engineer shall change the preliminary sketches for the extent necessary to meet the requirements of the Government, and after approval by the Officer in Charge the Architect-Engineer shall furnish necessary prints of the approved preliminary sketches to the Officer in Charge, (e) After the preliminary sketches and estimates have been approved, the Architect-Engineer shall proceed

with the preparation of complete working drawings and specifications as required by the Officer in Charge in connection with the construction of the said project. Working drawings, specifications and estimates shall be delivered to the Officer in Charge in such sequence and at such times as required by the Government and as will insure that the construction work can be initiated promptly, procurement of materials made without delay, and continuous prosecution of the work promoted. Working drawings and specifications shall be revised as necessary and as required by the Officer in Charge. After working drawings and specifications have been approved by the Officer in Charge, the Architect-Engineer shall furnish such number of sets of prints of the approved working drawings and such number of sets of the approved specifications, as may be required by the Officer in Charge. (f) Upon approval of final plans the Architect-Engineer shall deliver to the Government one set of tracings, in such medium and on such materials, as may be required by the Officer in Charge, suitable for blue-printing, showing complete approved construction requirements (not of "as-built" construction unless otherwise stipulated), provided, however, that should this contract be terminated by the Government, the Architect-Engineer shall deliver to the Government one set of such tracings as may be required by the Officer in Charge. Such tracings as are delivered shall be signed by the Officer in Charge as an indication of approval thereof, and shall become and remain the property of the Government. (g) The Architect-Engineer shall perform all necessary architectural-engineering services of every kind required in connection with the studies, designs and the preparation of drawings and specifications, but said services shall not, unless otherwise stipulated, include borings, test piles and pits, or supervision of construction work executed from the drawings and specifications, provided, however, that the Architect-Engineer shall furnish upon request and without additional compensation, such amplifications and explanations and attend such conferences as may, in the opinion of the Officer-in-Charge, be necessary to clarify the intent of the drawings and specifications and shall afford the benefit of his advice on questions that may arise in connection with the construction of the project. (h) The Architect-Engineer shall without additional fee correct and revise the drawings, specifications, or other materials furnished under this contract, if the Officer in Charge finds that such revision is necessary to correct errors or deficiencies for which the Architect-Engineer is responsible.

7. DEFINITIONS (JUNE 1964). (a) The term "head of the agency" or "Secretary" as used herein means the Secretary, the Under Secretary, and Assistant Secretary, or any other head or assistant head of the executive or military department or other Federal agency; and the term "his duly authorized representative" means any person or persons or board (other than the Contracting Officer) authorized to act for the head of the agency or the Secretary.

(b) The term "Contracting Officer" as used herein means the person executing this contract on behalf of the Government and includes a duly appointed successor or authorized representative. (See "Government Representatives" clause).

(c) Except as otherwise provided in this contract, the term "subcontracts" includes purchase orders under this contract.

8. GOVERNMENT REPRESENTATIVES. The work will be under the general direction of the Contracting Officer. The Commander of Naval Facilities Engineering Command, who shall designate an officer of the Civil Engineer Corps, United States Navy, or other officer or representative of the Government, as Officer in Charge, hereinafter referred to as the "OIC" who, except in connection with the Disputes Clause shall be the authorized representative of the Contracting Officer and under the direction of the Contracting Officer have complete charge of the work, and shall exercise full supervision and general direction of the work, so far as it affects the interest of the Government. For the purposes of the "Disputes" clause the Contracting Officer shall mean the Commander, Naval Facilities Engineering Command, the Acting Commander, their successors, or their representatives specially designated for this purpose.

9. CONTRACTING OFFICER'S DECISIONS (JAN. 1965). The extent and character of the work to be done by the Architect-Engineer shall be subject to the general supervision, direction, control and approval of the Contracting Officer.

10. TECHNICAL ADEQUACY. Approval by the Officer in Charge of drawings, designs, specifications and other incidental architectural engineering work or materials furnished hereunder shall not in any way relieve the Architect-Engineer of responsibility of the technical adequacy of the work.

11. COMPENSATION. The Architect-Engineer shall be paid the lump sum of $ ______________ as full compensation for all services, labor and material required hereby, and for all expenditures which may be made and expenses incurred except as are otherwise expressly provided herein.

12. METHOD OF PAYMENT (JAN. 1965). (a) Estimates shall be made monthly of the amount and value of the work and services performed by the Architect-Engineer under this contract, such estimates to be prepared by the Architect-Engineer and accompanied by such supporting data as may be required by the Contracting Officer.

(b) Upon approval of such estimate by the Contracting Officer payment upon properly certified vouchers shall be made to the Architect-Engineer as soon as practicable of 90% of the amount as determined above, less all previous payments; provided, however, that if the Contracting Officer determines that the work is substantially complete and that the amount of retained percentages is in excess of the amount considered by him to be adequate for the protection of the Government, he may at his discretion release to the Architect-Engineer such excess amount.

(c) Upon satisfactory completion by the Architect-Engineer and acceptance by the Contracting Officer of the work done by the Architect-Engineer in accordance with the foregoing "Statement of Architect-Engineer Services," the Architect-Engineer will be paid the unpaid balance of any money due for work under said statement, including retained percentages relating to this portion of the work.

(d) Upon satisfactory completion of the construction work and its final acceptance, the Architect-Engineer shall be paid the unpaid balance of any money due hereunder. Prior to such final payment under this contract, or prior settlement upon termination of the contract, and as a condition precedent thereto, the Architect-Engineer shall execute and deliver to the Contracting Officer a release of all claims against the Government arising under or by virtue of this contract, other than such claims, if any, as may be specifically excepted by the Architect-Engineer from the operation of the release in stated amounts to be set forth therein.

13. CHANGES (JAN. 1965). The Contracting Officer may at any time, by written order, and without notice to the sureties, make changes within the general scope of the contract in the work and service to be performed. If any such changes cause an increase or decrease in the Architect-Engineer's cost of, or the time required for, performance of this contract, an equitable adjustment shall be made and the contract shall be modified in writing accordingly. Any claim by the Architect-Engineer for adjustment under this clause must be asserted in writing within 30 days from the date of receipt by the Architect-Engineer of the notification of change unless the Contracting Officer grants a further period of time before the date of final payment under the contract. If the parties fail to agree upon the adjustment to be made, the dispute shall be determined as provided in the "Disputes" clause of this contract; but nothing provided in this clause shall excuse the Architect-Engineer from proceeding with the prosecution of the work as changed. Except as otherwise provided in this contract, no charge for any extra work or material will be allowed.

14. FEDERAL, STATE, AND LOCAL TAXES (JUL. 1960). (a) As used throughout this clause, the term "contract date" means the date of this contract. As to additional supplies or services procured by modification of this contract, the term "contract date" means the date of such modification.

(b) Except as may be otherwise provided in this contract, the contract price includes, to the extent allocable to this contract, all Federal, State, and local taxes which, on the contract date:

(i) by Constitution, statute, or ordinance, are applicable to this contract, or to the transactions covered by this contract, or to property or interests in property; or

(ii) pursuant to written ruling or regulation, the authority charged with administering any such tax is assessing or applying to, and is not granting or honoring an exemption for, a contractor under this kind of contract, or the transactions covered by this contract, or property or interests in property.

(c) Except as may be otherwise provided in this contract, duties in effect on the contract date are included in the contract price, to the extent allocable to this contract.

(d) (1) If the Architect-Engineer is required to pay or bear the burden —

(i) of any tax or duty which either was not to be included in the contract price pursuant to the requirements of paragraphs (b) and (c), or of a tax or duty specifically excluded from the contract price by a provision of this contract; or

(ii) of an increase in rate of any tax or duty, whether or not such tax or duty was excluded from the contract price; or

(iii) of any interest or penalty on any tax or duty referred to in (i) or (ii) above; the contract price shall be increased by the amount of such tax, duty, interest, or penalty allocable to this contract; *provided,* that the Architect-Engineer warrants in writing that no amount of such tax, duty, or rate increase was included in the contract price as a contingency reserve or otherwise; and *provided further,* that liability for such tax, duty, rate increase, interest, or penalty was not incurred through the fault or negligence of the Architect-Engineer or his failure to follow instructions of the Contracting Officer.

(2) If the Architect-Engineer is not required to pay or bear the burden, or obtains a refund or drawback, in whole or in part, of any tax, duty, interest, or penalty which:

(i) was to be included in the contract price pursuant to the requirements of paragraphs (b) and (c);

(ii) was included in the contract price; or

(iii) was the basis of an increase in the contract price; the contract price shall be decreased by the amount of such relief, refund, or drawback allocable to this contract, or the allocable amount of such relief, refund, or drawback shall be paid to the Government, as directed by the Contracting Officer. The contract price also shall be similarly decreased if the Contractor, through his fault or negligence or his failure to follow instructions of the Contracting Officer, is required to pay or bear the burden, or does not obtain a refund or drawback of any such tax, duty, interest, or penalty. Interest paid or credited to the Architect-Engineer incident to a refund of taxes shall inure to the benefit of the Government to the extent that such interest was earned after the Architect-Engineer was paid or reimbursed by the Government for such taxes.

(3) Invoices or vouchers covering any adjustment of the contract price pursuant to this paragraph (d) shall set forth the amount thereof as a separate item and shall identify the particular tax or duty involved.

(4) This paragraph (d) shall not be applicable to social security taxes; income and franchise taxes, other than those levied on or measured by (i) sales or receipts from sales, or (ii) the Architect-Engineer's possession of, interest in, or use of property, title to which is in the Government; excess profits taxes; capital stock taxes; unemployment compensation taxes; or property taxes, other than such property taxes, allocable to this contract, as are assessed either on completed supplies covered by this contract, or on the Architect-Engineer's possession of, interest in, or use of property, title to which is in the Government.

(5) No adjustment of less than $100 is required to be made in the contract price pursuant to this paragraph (d).

(e) Unless there does not exist any reasonable basis to sustain an exemption, the Government upon request of the Architect-Engineer, without further liability, agrees, except as otherwise provided in this contract, to furnish evidence appropriate to establish exemption from any tax which the Architect-Engineer warrants in writing was excluded from the contract price. In addition, the Contracting Officer may furnish evidence appropriate to establish exemption from any tax that may, pursuant to this clause, give rise to either an increase or decrease in the contract price. Except as otherwise provided in this contract, evidence appropriate to establish exemption from duties will be furnished only at the discretion of the Contracting Officer.

(f) (1) The Architect-Engineer shall promptly notify the Contracting Officer on all matters pertaining to Federal, State, and local taxes, and duties, that reasonably may be expected to result in either an increase or decrease in the contract price.

(2) Whenever an increase or decrease in the contract price may be required under this clause, the Architect-Engineer shall take action as directed by the Contracting Officer, and the contract price shall be equitably adjusted to cover the costs of such action, including any interest, penalty, and reasonable attorneys' fees.

15. RESPONSIBILITY OF THE ARCHITECT-ENGINEER. (a) The Architect-Engineer shall be responsible for the professional and technical accuracy and the coordination of all designs, drawings, specifications and other work or materials furnished by the Architect-Engineer under this contract. The Architect-Engineer shall without additional cost or fee to the Government correct or revise any errors or deficiencies in his performance.

(b) Neither the Government's review, approval or acceptance of, nor payment for, any of the services required under this contract shall be construed to operate as a waiver of any rights under this contract or of any cause of action arising out of the performance of this contract, and the Architect-Engineer shall be and remain liable to the Government for all costs of any kind which were incurred by the Government as a result of the Architect-Engineer's negligent performance of the services furnished under this contract.

(c) The rights and remedies of the Government provided for under this contract are in addition to any other rights and remedies provided by law; the Government may assert its right of recovery by any appropriate means including, but not limited to, set-off, suit, withholding, recoupment, or counterclaim, either during or after performance of this contract.

16. MILITARY SECURITY REQUIREMENTS (APR. 1966). (a) The provisions of this clause shall apply to the extent that this contract involves access to information classified "Confidential" including "Confidential—Modified Handling Authorized" or higher.

(b) The Government shall notify the Architect-Engineer of the security classification of this contract and the elements thereof, and of any subsequent revisions in such security classification, by the use of a Security Requirements Check List (DD Form 254), or other written notification.

(c) To the extent the Government has indicated as of the date of this contract or thereafter indicates security classification under this contract as provided in paragraph (b) above, the Architect-Engineer shall safeguard all classified elements of this contract and shall provide and maintain a system of security controls within its own organization in accordance with the requirements of:

(i) the Security Agreement (DD Form 441), including the Department of Defense Industrial Security Manual for Safeguarding Classified Information as in effect on the date of this contract, and any modification to the Security Agreement for the purpose of adapting the Manual to the Architect-Engineer's business; and

(ii) any amendments to said Manual made after the date of this contract, notice of which has been furnished to the Architect-Engineer by the Security Office of the Military Department having security cognizance over the facility.

(d) Representatives of the Military Department having security cognizance over the facility and representatives of the contracting Military Department shall have the right to inspect at reasonable intervals the procedures, methods, and facilities utilized by the Architect-Engineer in complying with the security requirements under this contract. Should the Government, through these representatives, determine that the Architect-Engineer is not complying with the security requirements of this contract the Architect-Engineer shall be informed in writing by the Security Office of the cognizant Military Department of the proper action to be taken in order to effect compliance with such requirements.

(e) If, subsequent to the date of this contract, the security classifications or security requirements under this contract are changed by the Government as provided in this clause and the security costs or time required for delivery under this contract are thereby increased or decreased, the contract price, delivery schedule, or both and any other provision of the contract that may be affected shall be subject to an equitable adjustment by reason of such increased or decreased costs. Any equitable adjustment shall be accomplished in the same manner as if such changes were directed under the "Changes" clause in this contract.

(f) The Architect-Engineer agrees to insert, in all subcontracts hereunder which involve access to classified information, provisions which shall conform substantially to the language of this clause, including this paragraph (f) but excluding the last sentence of paragraph (e) of this clause.

(g) The Architect-Engineer also agrees that it shall determine that any subcontractor proposed by it for the furnishing of supplies and services which involve access to classified information in the Architect-Engineer's custody has been granted an appropriate facility security clearance which is still in effect, prior to being accorded access to such classified information.

17. TERMINATION FOR DEFAULT (JAN. 1965). (a) The performance of work under this contract may be terminated by the Government in accordance with this clause in whole, or from time to time in part, whenever the Architect-Ebgineer shall default in performance of this contract in accordance with its terms (including in the term "default" any such failure by the Architect-Engineer to make progress in the prosecution of the work hereunder as endangers such performance), and shall fail to cure such default within a period of ten days (or such longer periods as the Con-

tracting Officer may allow) after receipt by the Architect-Engineer of a notice specifying the default.

(b) If the contract is so terminated, the Government may take over the work and services and prosecute the same to completion by contract or otherwise, and the Architect-Engineer shall be liable to the Government for any excess cost occasioned to the Government thereby.

(c) The contract may not be so terminated if the failure to perform arises from unforeseeable causes beyond the control and without the fault or negligence of the Architect-Engineer. Such causes may include, but are not restricted to acts of God, acts of the public enemy, acts of the Government in either its sovereign or contractual capacity, fires, floods, epidemics, quarantine restrictions, strikes and unusally severe weather; but in every case the failure to perform must be beyond the control and without the fault or negligence of the Architect-Engineer; and the Architect-Engineer, within ten days from the beginning of any such delay (unless the Contracting Officer grants a further period of time before the date of final payment under the contract) notifies the Contracting Officer in writing of the causes of delay. The Contracting Office shall ascertain the facts and the extent of delay and extend the time for completing the work when, in his judgment, the findings of fact justify such an extension, and his finding of fact shall be final and conclusive on the parties, subject only to appeal as provided in the clause of this contract entitled "Disputes."

(d) If, after Notice of Termination of the contract under the provisions of this clause, it is determined for any reason that the Architect-Engineer was not in default under the provisions of this clause or that the default was excusable under the provisions of this clause, the rights and obligations of the parties shall be the same as if the Notice of Termination had been issued pursuant to the clause of this contract entitled "Termination for the Convenience of the Government."

(e) The rights and remedies of the Government provided in this clause are in addition to any other rights and remedies provided by law or under this contract.

18. DISPUTES (JUNE 1964). (a) Except as otherwise provided in this contract, any dispute concerning a question of fact arising under this contract which is not disposed of by agreement shall be decided by the Contracting Officer, who shall reduce his decision to writing and mail or otherwise furnish a copy thereof to the Architect-Engineer. The decision of the Contracting Officer shall be final and conclusive unless, within 30 days from the date of receipt of such copy, the Architect-Engineer mails or otherwise furnishes to the Contracting Officer a written appeal addressed to the head of the agency involved. The decision of the head of the agency or his duly authorized representative for the determination of such appeals shall be final and conclusive. This provision shall not be pleaded in any suit involving a question of fact arising under this contract as limiting judicial review of any such decision to cases where fraud by such official or his representative or board is alleged: Provided, however, that any such decision shall be final and conclusive unless the same is fraudulent or capricious or arbitrary or so grossly erroneous as necessarily to imply bad faith or is not supported by substantial evidence. In connection with any appeal

proceeding under this clause, the Architect-Engineer shall be afforded an opportunity to be heard and to offer evidence in support of his appeal. Pending final decision of a dispute hereunder, the Architect-Engineer shall proceed diligently with the performance of the contract and in accordance with the Contracting Officer's decision.

(b) This Disputes clause does not preclude consideration of questions of law in connection with decisions provided for in paragraph (a) above. Nothing in this contract, however, shall be construed as making final the decision of any administrative official, representative, or board on a question of law.

19. OFFICIALS NOT TO BENEFIT (JUNE 1964). No member of or delegate to Congress, or resident commissioner, shall be admitted to any share or part of this contract, or to any benefit that may arise therefrom; but this provision shall not be construed to extend to this contract if made with a corporation for its general benefit.

20. COVENANT AGAINST CONTINGENT FEES (JAN. 1958). The Architect-Engineer warrants that no person or selling agency has been employed or retained to solicit or secure this contract upon an agreement or understanding for a commission, percentage, brokerage, or contingent fee, excepting bona fide employees or bona fide established commercial or selling agencies maintained for the Architect-Engineer for the purpose of securing business. For breach or violation of this warranty the Government shall have the right to annul this contract without liability or in its discretion to deduct from the contract price or consideration the full amount of such commission, percentage, brokerage, or contingent fee.

21. ASSIGNMENT OF CLAIMS (JUNE 1964). (a) Pursuant to the provisions of the Assignment of Claims Act of 1940 as amended (31 U.S. Code 203, 41 U.S. Code 15), if this contract provides for payments aggregating $1,000 or more, claims for moneys due or to become due the Architect-Engineer from the Government under this contract may be assigned to a bank, trust company or other financing institution, including any Federal lending agency, and may thereafter be further assigned and reassigned to any such institution. Any such assignment or reassignment shall cover all amounts payable under this contract and not already paid, and shall not be made to more than one party, except that any such assignment or reassignment may be made to one party as agent or trustee for two or more parties participating in such financing. Unless otherwise provided in this contract, payments to an assignee of any moneys due or to become due under this contract shall not, to the extent provided in said Act as amended, be subject to reduction or set-off.

(b) In no event shall copies of this contract or of any plans, specifications, or other similar documents relating to work under this contract, if marked "Top Secret," "Secret," or "Confidential" be furnished to any assignee of any claim arising under this contract or to any other person not entitled to receive the same. However a copy of any part or all of this contract so marked may be furnished, or any information contained therein may be disclosed, to such assignee upon the prior written authorization of the Contracting Officer.

22. CONTRACT WORK HOURS STANDARDS ACT—OVERTIME COMPENSATION (40 U.S.C. 327-330) (APR. 1965). *(This clause is only applicable to con-*

tracts exceeding $2,500 and is required by 40 USC 327-330). (a) The Architect-Engineer shall not require or permit any laborer or mechanic in any workweek in which he is employed on any work under this contract to work in excess of eight (8) hours in any calendar day or in excess of forty (40) hours in such workweek on work subject to the provisions of the Contract Work Hours Standards Act unless such laborer or mechanic receives compensation at a rate not less than one and one-half times his basic rate of pay for all such hours worked in excess of eight (8) hours in any calendar day or in excess of forty (40) hours in such workweeks, whichever is the greater number of overtime hours. The "basic rate of pay," as used in this clause, shall be the amount paid per hour, exclusive of the Architect-Engineer contribution or cost for fringe benefits and any cash payment made in lieu of providing fringe benefits, or the basic hourly rate contained in the wage determination, whichever is greater.

(b) In the event of any violation of the provisions of paragraph (a), the Architect-Engineer shall be liable to any affected employee for any amounts due, and to the United States for liquidated damages. Such liquidated damages shall be computed with respect to each individual laborer or mechanic employed in violation of the provisions of paragraph (a) in the sum of $10 for each calendar day in which such employee was required or permitted to be employed on such work in excess of eight (8) hours or in excess of the standard workweek of forty (40) hours without payment of the overtime wages required by paragraph (a).

23. CONVICT LABOR (MAR. 1949). In connection with the performance of work under this contract, the Architect-Engineer agrees not to employ any person undergoing sentence of imprisonment at hard labor.

24. EQUAL OPPORTUNITY (APRIL 1964). (The following clause is applicable unless this contract is exempt under the rules and regulations of the President's Committee on Equal Employment Opportunity (41 C.F.R. Chapter 60). Exemptions include contracts and subcontracts (i) not exceeding $10,000, (ii) not exceeding $100,000 for standard commercial supplies or raw materials, and (iii) under which work is performed outside the United States and no recruitment of workers within the United States is involved).

During the performance of this contract, the Architect-Engineer agrees as follows:

(a) The Architect-Engineer will not discriminate against any employee or applicant for employment because of race, creed, color, or national origin. The Architect-Engineer will take affirmative action to ensure that applicants are employed, and that employees are treated during employment, without regard to their race, creed, color, or national origin. Such action shall include, but not be limited to, the following: employment, upgrading, demotion or transfer; recruitment or recruitment advertising; layoff or termination; rates of pay or other forms of compensation; and selection for training, including apprenticeship. The Architect-Engineer agrees to post in conspicuous places, available to employees and applicants for employment, notices to be provided by the Contracting Officer setting forth the provisions of this nondiscrimination clause.

(b) The Architect-Engineer will, in all solicitations or advertisements for employees placed by or on behalf of the Architect-Engineer, state that all qualified applicants will receive consideration for employment without regard to race, creed, color, or national origin.

(c) The Architect-Engineer will send to each labor union or representative of workers with which he has a collective bargaining agreement or other contract or understanding, a notice, to be provided by the Agency Contracting Officer, advising the said labor union of workers' representative of the Architect-Engineer's commitments under this non-discrimination clause, and shall post copies of the notice in conspicuous places available to employees and applicants for employment.

(d) The Architect-Engineer will comply with all provisions of Executive Order No. 10925 of March 6, 1961, as amended, and of the rules, regulations, and relevant orders of the President's Committee on Equal Employment Opportunity created thereby.

(e) The Architect-Engineer will furnish all information and reports required by Executive Order No. 10925 of March 6, 1961, as amended, and by the rules, regulations, and orders of the said Committee, or pursuant thereto, and will permit access to his books, records, and accounts by the contracting agency and the Committee for purposes of investigation to ascertain compliance with such rules, regulations, and orders.

(f) In the event of the Architect-Engineer's noncompliance with the non-discrimination clause of this contract or with any of the said rules, regulations, or orders, this contract may be cancelled, terminated, or suspended in whole or in part and the Architect-Engineer may be declared ineligible for further Government contracts in accordance with procedures authorized in Executive Order No. 10925 of March 6, 1961, as amended, and such other sanctions may be imposed and remedies invoked as provided in the said Executive Order or by rule regulation, or order of the President's Committee on Equal Employment Opportunity, or as otherwise provided by law.

(g) The Architect-Engineer will include the provisions of paragraphs (a) through (g) in every subcontract or purchase order unless exempted by rules, regulations, or orders of the President's Committee on Equal Employment Opportunity issued pursuant to Section 303 of Executive Order No. 10925 of March 6, 1961, as amended, so that such provisions will be binding upon each subcontractor or vendor.* The Architect-Engineer will take such action with respect to any subcontract or purchase order as the contracting agency may direct as a means of enforcing such provisions, including sanctions for noncompliance: *Provided, however,* that in the event

---

**Unless otherwise provided, the "Equal Opportunity" clause is not required to be inserted in subcontracts below the second tier, except for subcontracts involving the performance of "construction work" as the "site of construction" (as those terms are defined in the Committee's rules and regulations) in which case the clause must be inserted in all such contracts. Subcontracts may incorporate by reference the "Equal Opportunity" clause.*

the Architect-Engineer becomes involved in, or is threatened with, litigation with a subcontractor or vendor as a result of such direction by the contracting agency, the Architect-Engineer may request the United States to enter into such litigation to protect the interest of the United States.

25. RENEGOTIATION (OCT. 1959). (a) To the extent required by law, this contract is subject to the Renegotiation Act of 1951 (50 U.S.C. App. 1211, et seq.). as amended, and to any subsequent act of Congress providing for the renegotiation of Contracts. Nothing contained in this clause shall impose any renegotiation obligation with respect to this contract or any subcontract hereunder which is not imposed by an act of Congress heretofore or hereafter enacted. Subject to the foregoing this contract shall be deemed to contain all the provisions required by Section 104 of the Renegotiation Act of 1951, and by any such other act without subsequent contract amendment specifically incorporating such provisions.

(b) The Architect-Engineer agrees to insert the provisions of this clause, including this paragraph (b), in all subcontracts, as that term is defined in Section 103g of the Renegotiation Act of 1951, as amended.

26. EXAMINATION OF RECORDS (FEB. 1962). (a) The Architect-Engineer agrees that the Comptroller General of the United States or any of his duly authorized representatives shall, until the expiration of three years after final payment under this contract, have access to and the right to examine any directly pertinent books, documents, papers, and records of the Architect-Engineer involving transactions related to this contract.

(b) The Architect-Engineer further agrees to include in all his subcontracts hereunder a provision to the effect that the subcontractor agrees that the Comptroller General of the United States or any of his duly authorized representatives shall, until the expiration of three years after final payment under the subcontract, have access to and the right to examine any directly pertinent books, documents, papers, and records of such subcontractor involving transactions related to the subcontract. The term "subcontract" as used in this clause excludes (i) purchase orders not exceeding $2,500 and (ii) subcontracts or purchase orders for public utility services at rates established for uniform applicability to the general public.

27. GRATUITIES (MAR. 1952). (a) The Government may, by written notice to the Architect-Engineer, terminate the right of the Architect-Engineer to proceed under this contract if it is found, after notice and hearing, by the Secretary or his duly authorized representative, that gratuities (in the form of entertainment, gifts, or otherwise) were offered or given by the Architect-Engineer, or any agent or representative of the Architect-Engineer, to any officer or employee of the Government with a view toward securing a contract or securing favorable treatment with respect to the awarding or amending, or the making of any determinations with respect to the performing, of such contract; provided, that the existence of the facts upon which the Secretary or his duly authorized representatives makes such finding shall be in issue and may be reviewed in any competent court.

(b) In the event this contract is terminated as provided in paragraph (a) hereof, the Government shall be entitled (i) to pursue the same remedies against the Architect-Engineer as it could pursue in the event of a breach of the contract by the Architect-Engineer, and (ii) as a penalty in addition to any other damages to which it may be entitled by law, to exemplary damages in an amount (as determined by the Secretary or his duly authorized representatives) which shall be not less than three nor more than ten times the costs incurred by the Architect-Engineer in providing any such gratuities to any such officer or employee.

(c) The rights and remedies of the Government provided in this clause shall not be exclusive and are in addition to any other rights and remedies provided by law or under this contract.

28. INTEREST (MAY 1963). Notwithstanding any other provision of this contract, unless paid within 30 days all amounts that become payable by the Architect-Engineer to the Government under this contract (net of any applicable tax credit under the Internal Revenue Code) shall bear interest at the rate of six percent per annum from the date due until paid and shall be subject to adjustments as provided by Part 6 of Appendix E of the Armed Services Procurement Regulation, as in effect on the date of this contract. Amounts shall be due upon the earliest one of (i) the date fixed pursuant to this contract, (ii) the date of the first written demand for payment, consistent with this contract, (iii) the date of transmittal by the Government to the Architect-Engineer of a proposed supplemental agreement to confirm completed negotiations fixing the amount of (iv) if this contract provided for revision of prices, the date of written notice to the Architect-Engineer stating the amount of refund payable in connection with a pricing proposal or in connection with a negotiated pricing agreement not confirmed by contract supplement.

29. ACCIDENT PREVENTION (JAN. 1965). (a) In order to provide safety controls for protection to the life and health of employees and other persons; for prevention of damage to property, materials, supplies, and equipment; and for avoidance of work interruptions in the performance of this contract, the Architect-Engineer shall comply with all pertinent provisions of Corps of Engineers Mannual, EM385-1-1, dated 13 March 1958, entitled "General Safety Requirements," as amended, and will also take or cause to be taken such additional measures as the Contracting Officer may determine to be reasonably necessary for the purpose.

(b) The Architect-Engineer will maintain an accurate record of, and will report to the Contracting Officer in the manner and on the forms prescribed by the Contracting Officer, exposure data and all accidents resulting in death, traumatic injury, occupational disease, and damage to property, materials, supplies and equipment incident to work performed under this contract.

(c) The Contracting Officer will notify the Architect-Engineer of any noncompliance with the foregoing provisions and the action to be taken. The Architect-Engineer shall, after receipt of such notice, immediately take corrective action. Such notice, when delivered to the Architect-Engineer or his representative at the site of the work, shall be deemed sufficient for the purpose. If the Architect-Engineer fails or refuses to comply promptly the Contracting Officer may issue an order stopping all or part of the work until satisfactory corrective action has been taken. No part of the time lost due to any such stop orders shall be made the subject of claim for extension of time or for excess costs or damages by the Architect-Engineer.

(d) Compliance with the provisions of this article by subcontractors will be the responsibility of the Architect-Engineer.

30. TERMINATION FOR CONVENIENCE OF THE GOVERNMENT (JAN. 1960). *(This clause is applicable only to contracts $10,000 and under).* The Contracting Officer, by written notice, may terminate this contract, in whole or in part, when it is in the interest of the Government. If this contract is so terminated, the rights, duties and obligations of the parties hereto shall be in accordance with the applicable Sections of the Armed Services Procurement Regulation in effect on the date of this contract.

31. TERMINATION FOR CONVENIENCE OF THE GOVERNMENT (APR. 1966). *(This clause is applicable only to contracts in excess of $10,000).* (a) The performance of work under this contract may be terminated by the Government in accordance with this clause in whole, or from time to time in part, whenever the Contracting Officer shall determine that such termination is in the best interest of the Government. Any such termination shall be effected by delivery to the Architect-Engineer of a Notice of Termination specifying the extent to which performance of work under the contract is terminated, and the date upon which such termination becomes effective.

(b) After receipt of a Notice of Termination, and except as otherwise directed by the Contracting Officer, the Architect-Engineer shall:

(i) stop work under the contract on the date and to the extent specified in the Notice of Termination;

(ii) Place no further orders or subcontracts for materials, services or facilities, except as may be necessary for completion of such portion of the work under the contract as it is not terminated;

(iii) terminate all orders and subcontracts to the extent that they relate to the performance of work terminated by the Notice of Termination;

(iv) assign to the Government, in the manner, at the times and to the extent directed by the Contracting Officer, all of the right, title, and interest of the Architect-Engineer under the orders and subcontracts so terminated, in which case the Government shall have the right, in its discretion, to settle or pay any or all claims arising out of the termination of such orders and subcontracts;

(v) settle all outstanding liabilities and all claims arising out of such termination of orders and subcontracts, with the approval or ratification of the Contracting Officer, to the extent he may require which approval or ratification shall be final for all the purposes of this clause;

(vi) transfer title and deliver to the Government, in the manner, at the times, and to the extent, if any, directed by the Contracting Officer, (A) the fabricated or unfabricated parts, work in process, completed work, supplies, and other material produced as a part of, or acquired in connection with the performance of, the work terminated by the Notice of Termination, and (B) the completed or partially completed plans, drawings, information, and other property which, if the contract had been completed, would have been required to be furnished to the Government.

(vii) use his best efforts to sell, in the manner, at the times, to the extent, and at the price or prices directed or authorized by the Contracting Officer, any property of the types referred to in (vi) above; provided, however, that the Architect-Engineer (A) shall not be required to extend credit to any purchaser, and (B) may acquire any such property under the conditions prescribed by and at a price or prices approved by the Contracting Officer; and provided further that the proceeds of any such transfer or disposition shall be applied in reduction of any payments to be made by the Government to the Architect-Engineer under this contract or shall otherwise be credited to the price of cost of the work covered by this contract or paid in such other manner as the Contracting Officer may direct;

(viii) complete performance of such part of the work as shall not have been terminated by the Notice of Termination; and

(ix) take such action as may be necessary, or as the Contracting Officer may direct, for the protection and preservation of the property related to this contract which is in the possession of the Architect-Engineer and in which the Government has or may acquire an interest.

At any time after expiration of the plant clearance period, as defined in Section VIII, Armed Services Procurement Regulation, as it may be amended from time to time, the Architect-Engineer may submit to the Contracting Officer a list, certified as to quantity and quality, of any or all items of termination inventory not previously disposed of, exclusive of items the disposition of which has been directed or authorized by the Contracting Officer, and may request the Government to remove such items or enter into a storage agreement covering them. Not later than fifteen (15) days thereafter, the Government will accept title to such items and remove them or enter into a storage agreement covering the same; provided, that the list submitted shall be subject to verification by the Contracting Officer upon removal of the items, of if the items are stored, within forty-five (45) days from the date of submission of the list, and any necessary adjustment to correct the list as submitted shall be made prior to final settlement.

(c) After receipt of a Notice of Termination, the Architect-Engineer shall submit to the Contracting Officer his termination claim, in the form and with certification prescribed by the Contracting Officer. Such claim shall be submitted promptly but in no event later than one year from the effective date of termination, unless one or more extensions in writing are granted by the Contracting Officer, upon request of the Architect-Engineer made in writing within such one year period or authorized extension thereof. However, if the Contracting Officer determines that the facts justify such action, he may receive and act upon any such termination claim at any time after such one year period or any extension thereof. Upon failure of the Architect-Engineer to submit his termination claim within the time allowed, the Contracting Officer may, subject to any Settlement Review Board approvals required by Section VIII of the Armed Services Procurement Regulation in effect as of the date of execution of this contract, determine on the basis of information available to him, the amount, if any, due to the Architect-Engineer by reason of the termination and shall thereupon pay to the Architect-Engineer the amount so determined.;

(d) Subject to the provisions of paragraph (c) and subject to any Settlement Review Board approvals required by Section VIII of the Armed Services Procurement Regulation in effect as of the date of execution of this contract, the Architect-Engineer and the Contracting Officer may agree upon the whole or any part of the amount or amounts to be paid to the Architect-Engineer by reason of the total or partial termination of work pursuant to this clause, which amount or amounts may include a reasonable allowance for profit on work done; provided, that such agreed amount or amounts exclusive of settlement costs, shall not exceed the total contract price as reduced by the amount of payments otherwise made and as further reduced by the contract price of work not terminated. The contract shall be amended accordingly, and the Architect-Engineer shall be paid the agreed amount. Nothing in paragraph (e) of this clause prescribing the amount to be paid to the Architect-Engineer in the event of failure of the Architect-Engineer and the Contracting Officer to agree upon the whole amount to be paid to the Architect-Engineer by reason of the termination of work pursuant to this clause, shall be deemed to limit, restrict, or otherwise determine or affect the amount or amounts which may be agreed upon to be paid to the the Architect-Engineer pursuant to this paragraph (d).

(e) In the event of the failure of the Architect-Engineer and the Contracting Officer to agree as provided in paragraph (d), upon the whole amount to be paid to the Architect-Engineer by reason of the termination of work pursuant to this clause, the Contracting Officer shall, subject to any Settlement Review Board approvals required by Section VIII of the Armed Services Procurement Regulation in effect as of the date of execution of this contract, pay to the Architect-Engineer the amounts determined by the Contracting Officer as follows, but without duplication of any amounts agreed upon in accordance with paragraph (d):

(i) for completed work and services accepted by the Government, the price or prices specified in the contract for such work, less any payments previously made;

(ii) the total of—

(A) the costs incurred in the performance of the work and services terminated, including initial costs and preparatory expenses allocable thereto, but exclusive of any costs attributable to the work and services paid or to be paid for under paragraph (e) (i) hereof;

(B) the cost of settling and paying claims arising out of the termination of work or services under subcontracts or orders as provided in paragraph (b) (v) above, which are properly chargeable to the terminated portion of the contract exclusive of amounts paid or payable on account of work or services delivered or furnished by subcontractors prior to the effective date of termination, which amounts shall be included in the costs payable under (A) above); and

(C) a sum, as profit on (A) above, determined by the Contracting Officer pursuant to 8—303 of the Armed Services Procurement Regulation, in effect as of the date of execution of this contract, to be fair and reasonable; *provided,* that if it appears that the Architect-Engineer would have sustained a loss on the entire contract had it been completed, no profit shall be included or allowed under this subdivision (C) and an appropriate adjustment shall be made reducing the amount of settlement to reflect the indicated rate of loss; and

(iii) the reasonable cost of the preservation and protection of property incurred pursuant to paragraph (b) (ix); and any other reasonable cost incidental to the termination of work under this contract, including expense incidental to the determination of the amount due to the Architect-Engineer as a result of the termination of work under this contract.

The total sum to be paid to the Architect-Engineer under (i) and (ii) above shall not exceed the total contract price as reduced by the amount of payments otherwise made and as further reduced by the contract price of work not terminated. Except for normal spoilage, and except to the extent that the Government shall have otherwise expressly assumed the risk of loss, there shall be excluded from the amounts payable to the Architect-Engineer under (ii) above, the fair value as determined by the Contracting Officer of property which is destroyed, lost, stolen, or damaged so as to become undeliverable to the Government, or a buyer pursuant to paragraph (b) (vii).

(f) Any determination of costs under paragraph (c) or (e) hereof shall be governed by the principles for consideration of costs set forth in Section XV, Part 2, of the Armed Services Procurement Regulation, as in effect on the date of this contract.

(g) The Architect-Engineer shall have the right of appeal, under the clause of this contract entitled "Disputes," from any determination made by the Contracting Officer under paragraph (c) or (e) above, except that if the Architect-Engineer has failed to submit his claim within the time provided in paragraph (c) above and has failed to request extension of such time, he shall have no such right of appeal. In any case where the Contracting Officer has made a determination of the amount due under paragraph (c) or (e) above, the Government shall pay to the Architect-Engineer the following: (i) if there is no right of appeal hereunder or if no timely appeal has been taken, the amount so determined by the Contracting Officer, or (ii) if an appeal has been taken, the amount finally determined on such appeal.

(h) In arriving at the amount due the Architect-Engineer under this clause there shall be deducted (i) all unliquidated advance or other payments on account theretofore made to the Architect-Engineer, applicable to the terminated portion of this contract, (ii) any claim which the Government may have against the Architect-Engineer in connection with his contract, and (iii) the agreed price for, or the proceeds of sale of, any materials, supplies, or other things acquired by the Architect-Engineer or sold, pursuant to the provisions of this clause, and not otherwise recovered by or credited to the Government.

(i) If the termination hereunder be partial, prior to the settlement of the terminated portion of this contract, the Architect-Engineer may file with the Contracting Officer a request in writing for an equitable adjustment of the price or prices specified in the contract relating to the contained portion of the contract (the portion not terminated by the Notice of Termination), and such equitable adjustments as may be agreed upon shall be made in such price or prices.

(j) The Government may from time to time, under such terms and conditions as it may prescribe, make partial payments and payments on account against costs incurred by the Architect-Engineer in connection with the terminated portion of this contract whenever in the opinion of the Contracting Officer the aggregate of such payments shall be within the amount to which the Architect-Engineer will be entitled hereunder. If the total of such payments is in excess of the amount finally agreed or determined to be due under this clause, such excess shall be payable by the Architect-Engineer to the Government upon demand, together with interest computed at the rate of 6 percent per annum, for the period from the date such excess payment is received by the Architect-Engineer to the date on which such excess is repaid to the Government; provided, however, that no interest shall be charged with respect to any such excess payment attributable to a reduction in the Architect-Engineer's claim by reason of retention or other disposition of termination inventory until ten days after the date of such retention or disposition, or such later date as determined by the Contracting Officer by reason of the circumstances.

(k) Unless otherwise provided for in this contract, or by applicable statute, the Architect-Engineer from the effective date of termination and for a period of three years after final settlement under this contract, shall preserve and make available to the Government at all reasonable times at the office of the Architect-Engineer but without direct charge to the Government, all his books, records, documents, and other evidence bearing on the costs and expenses of the Architect-Engineer under this contract and relating to the work terminated hereunder, or, to the extent approved by the Contracting Officer, photographs, micro-photographs, or other authentic reproductions thereof.

32. SUBCONTRACTORS AND OUTSIDE ASSOCIATES AND CONSULTANTS (JAN. 1965). Any subcontractors and outside associates or consultants required by the Architect-Engineer in connection with the services covered by the concontract will be limited to such individuals or firms as were specifically identified and agreed to during negotiations. Any substitution in such subcontractors, associates, or consultants will be subject to the prior approval of the Contracting Officer.

33. COMPOSITION OF ARCHITECT-ENGINEER (JAN. 1965). If the Architect-Engineer hereunder is comprised of more than one legal entity, each such entity shall be jointly and severally liable hereunder.

34. AUDIT (SEP. 1964). (a) For purposes of verifying that cost or pricing data submitted, in conjunction with the negotiation of this contract or any contract change or other modification involving an amount in excess of $100,000 are accurate, complete, and current, the Contracting Officer, or his authorized representatives, shall — until the expiration of three years from the date of final payment under this contract — have the right to examine those books, records, documents and other supporting data which will permit adequate evaluation of the cost or pricing data submitted, along with the computations and projections used therein, which were available to the Architect-Engineer as of the date of execution of the Architect-Engineer's Certificate of Current Cost or Pricing Data.

(b) The Architect-Engineer agrees to insert the substance of this clause including this paragraph (b) in all subcontracts hereunder in excess of $100,000 so as to apply until three years after final payment under the subcontract, unless the price is based on adequate price competition, established catalog or market prices of commercial items sold in substantial quantities to the general public, or prices set by law or regulation. In each such excepted subcontract hereunder in excess of $100,000 the Architect-Engineer shall insert the substance of the following clause to apply until three years after final payment under the subcontract.

AUDIT-PRICE ADJUSTMENTS. (a) This clause shall become operative only with respect to any change or other modification made pursuant to one or more provisions of this contract which involves a price adjustment in excess of $100,000 unless the price adjustment is based on adequate price competition, established catalog or market prices of commercial items sold in substantial quantities to the general public, or prices set by law or regulation and further provided that such change or other modification to this contract must result from a change or other modification to the Government prime contract.

(b) For purposes of verifying that cost or pricing data submitted in conjunction with a contract change or modification involving an amount in excess of $100,000 are accurate, complete and current, the Contracting Officer or his authorized representative shall — until the expiration of three years from the data of final payment under this contract — have the right to examine those books, records, documents, and other supporting data which will permit adequate evaluation of the cost or pricing data submitted, along with the computations and projections used therein, which were available to the Contractor as of the date of execution of the Contractor's Certificate of Current Cost or Pricing Data.

(c) The Contractor agrees to insert the substance of this clause, including this paragraph (c), in all subcontracts hereunder in excess of $100,000 so as to apply until three years after final payment of the subcontract.

35. SUBCONTRACTOR COST AND PRICING DATA (SEP. 1964). (a) The Architect-Engineer shall require subcontractors hereunder to submit cost or pricing data under the following circumstances: (i) prior to award of any cost-reimbursement type, incentive, or price redeterminable subcontract; (ii) prior to the award of any subcontract the price of which is expected to exceed $100,000; (iii) prior to the pricing of any subcontract change or other modification for which the price adjustment is expected to exceed $100,000; except in the case of (ii) or (iii) where the price

is based on adequate price competition, established catalog or market prices of commercial items sold in substantial quantities to the general public, or prices set by law or regulation.

(b) The Architect-Engineer shall require subcontractors to certify, in substantially the same form as that used in the certificate by the Architect-Engineer to the Government, that to the best of their knowledge and belief, the cost and pricing data submitted under (a) above is accurate, complete, and current as of the date of execution, which date shall be as close as possible to the date of agreement on the negotiated price of the subcontract.

(c) The Architect-Engineer shall insert the substance of this clause including this paragraph (c) in each of his cost-reimbursement type, price redeterminable, or incentive subcontracts hereunder, and in any other subcontract hereunder which exceeds $100,000 except where the price thereof is based on adequate price competition, established catalog or market prices of commercial items sold in substantial quantities to the general public, or prices set by law or regulation. In each such excepted subcontract hereunder in excess of $100,000, the Architect-Engineer shall insert the substance of the following clause:

SUBCONTRACTOR COST AND PRICING DATA-PRICE ADJUSTMENTS. (a) Paragraphs (b) and (c) of this clause shall become operative only with respect to any change or other modification made pursuant to one or more provisions of this contract which involves a price adjustment in excess of $100,000. The requirements of this clause shall be limited to such price adjustments.

(b) The Contractor shall require subcontractors hereunder to submit cost or pricing data under the following circumstances: (i) prior to award of any cost-reimbursement type, incentive or price redeterminable subcontract; (ii) prior to award of any subcontract, the price of which is expected to exceed $100,000; (iii) prior to the pricing of any subcontract change or other modification for which the price adjustment is expected to exceed $100,000; except, in the case of (ii) or (iii), where the price is based on adequate price competition, established catalog or market prices of commercial items sold in substantial quantities to the general public, or prices set by law or regulation.

(c) The Contractor shall require subcontractors to certify, in substantially the same form as that used in the certificate by the Prime Contractor to the Government, that to the best of their knowledge and belief the cost and pricing data submitted under (b) above is accurate, complete, and current as of the date of execution, which date shall be as close as possible to the date of agreement of the negotiated price of the contract modification.

(d) The Contractor shall insert the substance of this clause including this paragraph (d) in each subcontract hereunder which exceeds $100,000;

36. PRICE REDUCTION FOR DEFECTIVE COST OR PRICING DATA (SEP. 1964). (a) If the Contracting Officer determines that any price, including profit or fee, negotiated in connection with this contract was increased by any significant sums because the Architect-Engineer, or any subcontractor in connection with a sub-

contract covered by (c) below, furnished incomplete or inaccurate cost or pricing data or data not current as certified in the Architect-Engineer's Certificate of Current Cost or Pricing Data, then such price shall be reduced accordingly and the contract shall be modified in writing to reflect such adjustment.

(b) Failure to agree on a reduction shall be a dispute concerning a question of fact within the meaning of the "Disputes" clause of this contract.

(c) The Architect-Engineer agrees to insert the substance of paragraphs (a) and (c) of this clause in each of his cost-reimbursement type, price redeterminable, or incentive subcontracts hereunder, and in any other subcontract hereunder, and in any other subcontract hereunder in excess of $100,000 unless the price is based on adequate price competition, established catalog or market prices of commercial items sold in substantial quantities to the general public, or prices set by law or regulation. In each such excepted subcontract hereunder which exceeds $100,000, the Architect-Engineer shall insert the substance of the following clause.

PRICE REDUCTION FOR DEFECTIVE COST OR PRICING DATA-PRICE ADJUSTMENTS. (a) This clause shall become operative only with respect to any change or other modification made pursuant to one or more provisions of this contract which involves a price adjustment in excess of $100,000, except where the price is based on adequate price competition, established catalog or market prices of commercial items sold in substantial quantities to the general public, or prices set by law or regulation. The right to price reduction under this clause shall be limited to such price adjustments.

(b) If the Contractor determines that any price, including profit or fee, negotiated in connection with any price adjustment within the purview of paragraph (a) above was increased by any significant sum because the subcontractor or any of subcontractors in connection with a subcontract covered by paragraph (c) below, furnished incomplete or inaccurate cost or pricing data, or data not current as of the date of execution of the subcontractor's certificate of current cost or pricing data, then such price shall be reduced accordingly and the subcontract shall be modified in writing to reflect such adjustment.

(c) The subcontractor agrees to insert the substance of this clause in each subcontract hereunder which exceeds $100,000.

37. EQUAL OPPORTUNITY. (a) The "Equal Opportunity" clause in paragraph 24 is amended by deleting references to the President's Committee on Equal Employment Opportunity, Executive Order 10925 of March 6, 1961, as amended, and Section 303 of Executive Order No. 10925 of March 6, 1961, as amended, and substituting therefor the Secretary of Labor, Executive Order No. 11246 of September 24, 1965, and Section 204 of Executive Order 11246 of September 24, 1965, respectively.

(b) The Equal Opportunity representation in paragraph 24 is amended to insert after the reference to "Executive Order 10925" the following: "or the clause contained in Section 201 of Executive Order No. 11114."

(c) In accordance with regulations of the Secretary of Labor, the rules, regulations, orders, instructions, designations, and other directives issued by the

the President's Committee on Equal Employment Opportunity and those issued by the heads of various departments or agencies under or pursuant to any of the Executive Orders superseded by Executive Order 11246, shall, to the extent that they are not inconsistent with Executive Order 11246, remain in full force and effect unless and until revoked or superseded by appropriate authority. References in such directives to provisions of the superseded orders shall be deemed to be references to the comparable provisions of Executive Order 11246.

APPENDIX H

Contract Disputes Act of 1978
41 U.S.C. 601 et seq.

CONTENTS

Section
601. Definition.
602. Applicability of law
  (a) Executive agency contracts.
  (b) Tennessee Valley Authority contracts.
  (c) Foreign government or international organization contracts.
603. Maritime contracts.
604. Fraudulent claims
605. Decision by contracting officer.
  (a) Contractor claims.
  (b) Review; performance of contract pending appeal.
  (c) Amount of claim; certification; notification; time of issuance; presumption.
606. Contractor's right of appeal to board of contract appeals.
607. Agency boards of contracts appeals.
  (a) Establishment; consultation; Tennessee Valley Authority.
  (b) Appointment of menbers; chairman; compensation.
  (c) Appeals; inter-agency arrangements.
  (d) Jurisdiction.
  (e) Decisions.
  (f) Accelerated appeal disposition.
  (g) Review
  (h) Procedural guidelines.
  (i) Workload studies.
608. Small claims.
  (a) Accelerated disposition of appeals.
  (b) Simplified rules of procedure.
  (c) Time of decision.
  (d) Finality of decision.
  (e) Effect of decision.
  (f) Review of requisite amount in controversy.
609. Judicial review of board decisions.
  (a) Actions in United States Court of Claims; district court actions; time for filing.
  (b) Finality of board decision.
  (c) Remand or retention of case.
  (d) Consolidation.
  (e) Judgments as to fewer than all claims.
610. Subpoena, discovery, and deposition.
611. Interest.
612. Payment of claims.
  (a) Judgments.
  (b) Monetary awards.
  (c) Reimbursement.
  (d) Tennessee Valley Authority.
613. Separability of provisions.

SECTION 601. Definitions

As used in this chapter —

(1) The term "agency head" means the head and any assistant head of an executive agency, and may "upon the designation by" the head of an executive agency include the chief official of any principal division of the agency;

(2) The term "executive agency" means an executive department as defined in section 101 of Title 5, an independent establishment as defined by section 104 of Title 5 (except that it shall not include the General Accounting Office): 1 a military department as defined by section 102 of Title 5, and a wholly owned Government corporation as defined by section 846 of Title 31, the United States Postal Service, and the Postal Rate Commission;

(3) The term "contracting officer" means any person who, by appointment in accordance with applicable regulations, has the authority to enter into and administer contracts and make determinations and findings with respect thereto. The term also includes the authorized representative of the contracting officer, acting within the limits of his authority;

(4) The term "contractor" means a party to a Government contract other than the Government;

(5) The term "Administrator" means the Administrator for Federal Procurement Policy appointed pursuant to the Office of Federal Procurement Policy Act;

(6) The term "agency board" means an agency board of contract appeals established under section 607 of this title; and

(7) The term "misrepresentation of fact" means a false statement of substantive fact, or any conduct which leads to a belief of a substantive fact material to proper understanding of the matter in hand, made with intent to deceive or mislead.

SECTION 602. Applicability of Law — Executive Agency Contracts

(a) Unless otherwise specifically provided herein, this chapter applies to any express or implied contract (including those of the nonappropriated fund activities described in sections 1346 and 1491 of Title 28) entered into by an executive agency for —

(1) the procurement of property, other than real property in being;

(2) the procurement of services;

(3) the procurement of construction, alteration, repair or maintenance of real property; or,

(4) the disposal of personal property.

TENNESSEE VALLEY AUTHORITY CONTRACTS

(b) With respect to contracts of the Tennessee Valley Authority, the provisions of this chapter shall apply only to those contracts which contain a disputes clause requiring that a contract dispute be resolved through an agency administrative process. Notwithstanding any other provision of this chapter, contracts of the Tennessee Valley Authority for the sale of fertilizer or electric power or related to the conduct or operation of the electric power system shall be excluded from the chapter.

FOREIGN GOVERNMENT OR INTERNATIONAL ORGANIZATION CONTRACTS

(c) This chapter does not apply to a contract with a foreign government, or agency thereof, or international organization, or subsidiary body thereof, if the head of the agency determines that the application of the chapter to the contract would not be in the public interest.

SECTION 603. Maritime Contracts

Appeals under paragraph (g) of section 607 of this title and suits under section 609 of this title, arising out of maritime contracts, shall be governed by section 741 to 752 or 781 to 790 of Title 46 as applicable, to the extent that those sections are not inconsistent with this chapter.

SECTION 604. Fraudulent Claims

If a contractor is unable to support any part of his claim and it is determined that such inability is attributable to misrepresentation of fact or fraud on the part of the contractor, he shall be liable to the Government for any amount equal to such unsupported part of the claim in addition to all costs to the Government attributable to the cost of reviewing said part of his claim. Liability under this section shall be determined within six years of the commission of such misrepresentation of fact or fraud.

SECTION 605. Decision by Contracting Officer — Contractor Claims

(a) All claims by a contractor against the government relating to a contract shall be in writing and shall be submitted to the contracting officer for a decision. All claims by the government against a contractor relating to a contract shall be the subject of a decision by the contracting officer. The contracting officer shall issue his decisions in writing, and shall mail or otherwise furnish a copy of the decision to the contractor. The decision shall state the reasons for the decision reached, and shall inform the contractor of his rights as provided in this chapter. Specific findings of fact are not required, but, if made, shall not be binding in any subsequent proceeding. The authority of this subsection shall not extend to a claim or dispute for penalties or forfeitures prescribed by statute or regulation which another Federal agency is specifically authorized to administer, settle or determine. This section shall not authorize any agency head to settle, compromise, pay or otherwise adjust any claim involving fraud.

REVIEW; PERFORMANCE OF CONTRACT PENDING APPEAL

(b) The contracting officer's decision on the claim shall be final and conclusive and not subject to review by any forum, tribunal, or Government agency, unless an appeal or suit is timely commenced as authorized by this chapter. Nothing in this chapter shall prohibit executive agencies from including a clause in government contracts requiring that pending final decision of an appeal, action, or final settlement, a contractor shall proceed diligently with performance of the contract in accordance with the contracting officer's decision.

AMOUNT OF CLAIM; CERTIFICATION; NOTIFICATION; TIME OF ISSUANCE; PRESUMPTION

(c) (1) A contracting officer shall issue a decision on any submitted claim of $50,000 or less within sixty days from his receipt of a written request from the contractor that a decision be rendered within that period. For claims of more than $50,000, the contractor shall certify that the claim is made in good faith, that the supporting data are accurate and complete to the best of his knowledge and belief, and that the amount requested accurately reflects the contract adjustment for which the contractor believes the government is liable.

(2) A contracting officer shall, within sixty days of receipt of a submitted certified claim over $50,000 —

(A) issue a decision; or

(B) notify the contractor of the time within which a decision will be issued.

(3) The decision of a contracting officer on submitted claims shall be issued within a reasonable time, in accordance with regulations promulgated by the agency, taking into account such factors as the size and complexity of the claim and the adequacy of the information in support of the claim provided by the contractor.

(4) A contractor may request the agency board of contract appeals to direct a contracting officer to issue a decision in a specified period of time, as determined by the board, in the event of undue delay on the part of the contracting officer.

(5) Any failure by the contracting officer to issue a decision on a contract claim within the period required will be deemed to be a decision by the contracting officer denying the claim and will authorize the commencement of the appeal or suit on the claim as otherwise provided in this chapter. However, in the event an appeal or suit is so commenced in the absence of a prior decision by the contracting officer, the tribunal concerned may, at its option, stay the proceedings to obtain a decision on the claim by the contracting officer.

SECTION 606. Contractor's Right of Appeal to Board of Contract Appeals

Within ninety days from the date of receipt of a contracting officer's decision under section 605 of this title, the contractor may appeal such decision to an agency board of contract appeals, as provided in section 607 of this title.

SECTION 607. Agency Boards of Contracts Appeals — Establishment; Consultation; Tennessee Valley Authority

(a) (1) Except as provided in paragraph (2) an agency board of contract appeals may be established within an executive agency when the agency head, after consultation with the Administrator, determines from a workload study that the volume of contract claims justifies the establishment of a full-time agency board of at least three members who shall have no other inconsistent duties. Workload studies will be updated at least once every three years and submitted to the Administrator.

(2) The Board of Directors of the Tennessee Valley Authority may establish a board of contract appeals for the Authority of an indeterminate number of members.

APPOINTMENT OF MEMBERS; CHAIRMAN; COMPENSATION

(b) (1) Except as provided in paragraph (2), the members of agency boards shall be selected and appointed to serve in the same manner as hearing examiners appointed pursuant to section 3105 of Title 5, with an additional requirement that such members shall have had not fewer than five years' experience in public contract law. Full-time members of agency boards serving as such on the effective date of this chapter shall be considered qualified. The chairman and vice chairman of each board shall be designated by the agency head from members so appointed. The chairman of each agency board shall receive compensation at a rate equal to that paid a GS-18 under the General Schedule contained in section 7332 of Title 5, the vice chairman shall receive compensation at a rate equal to that paid a GS-17 under such General Schedule, and all other members shall receive compensation at a rate equal to that paid a GS-16 under such General Schedule. Such positions shall be in addition to the number of positions which may be placed in GS-16, GS-17, and GS-18 of such General Schedule under existing law.

(2) The Board of Directors of the Tennessee Valley Authority shall establish criteria for the appointment of members to its agency board of contract appeals established in subsection (a) (2) of this section, and shall designate a chairman of such board. The chairman of such board shall receive compensation at a rate equal to the daily

rate paid a GS-18 under the General Schedule contained in section 5332 of Title 5 for each day he is engaged in the actual performance of his duties as a member of such board. All other members of such board shall receive compensation at a rate equal to the daily rate paid a GS-16 under such General Schedule for each day they are engaged in the actual performance of their duties as members of such board.

APPEALS; INTER-AGENCY ARRANGEMENTS

(c) If the volume of contract claims is not sufficient to justify an agency board under subsection (a) of this section or if he otherwise considers it appropriate, any agency head shall arrange for appeals from decisions by contracting officers of his agency to be decided by a board of contract appeals of another executive agency. In the event an agency head is unable to make such an arrangement with another agency, he shall submit the case to the Administrator for placement with an agency board. The provisions of this subsection shall not apply to the Tennessee Valley Authority.

JURISDICTION

(d) Each agency board shall have jurisdiction to decide any appeal from a decision of a contracting officer (1) relative to a contract made by its agency, and (2) relative to a contract made by any other agency when such agency or the Administrator has designated the agency board to decide the appeal. In exercising this jurisdiction, the agency board is authorized to grant any relief that would be available to a litigant asserting a contract claim in the Court of Claims.

DECISIONS

(e) An agency board shall provide to the fullest extent practicable, informal, expeditious, and inexpensive resolution of disputes, and shall issue a decision in writing or take other appropriate action on each appeal submitted, and shall mail or otherwise furnish a copy of the decision to the contractor and the contracting officer.

ACCELERATED APPEAL DISPOSITION

(f) The rules of each agency board shall include a procedure for the accelerated disposition of any appeal from a decision of a contracting officer where the amount in dispute is $50,000 or less. The accelerated procedure shall be applicable at the sole election of only the contractor. Appeals under the accelerated procedure shall be resolved, whenever possible, within one hundred and eighty days from the date the contractor elects to utilize such procedure.

REVIEW

(g) (1) The decision of an agency board of contract appeals shall be final, except that —

(A) a contractor may appeal such a decision to the Court of Claims within one hundred twenty days after the date of receipt of a copy of such decision, or

(B) the agency head, if he determines that an appeal should be taken, and with the prior approval of the Attorney General, transmits the decision of the board of contract appeals to the United States

Court of Claims for judicial review, under section 2510 of Title 28, within one hundred and twenty days from the date of the agency's receipt of a copy of the board's decision.

(2) Notwithstanding the provisions of paragraph (1), the decision of the board of contract appeals of the Tennessee Valley Authority shall be final, except that —

(A) a contractor may appeal such a decision to a United States district court pursuant to the provisions of section 1337 of Title 28 within one hundred twenty days after the date of receipt of a copy of such decision, or

(B) The Tennessee Valley Authority may appeal the decision to a United States district court pursuant to the provisions of section 1337 of Title 28 within one hundred twenty days after the date of the decision in any case.

PROCEDURAL GUIDELINES

(h) Pursuant to the authority conferred under the Office of Federal Procurement Policy Act, the Administrator is authorized and directed, as may be necessary or desirable to carry out the provisions of this chapter, to issue guidelines with respect to criteria for the establishment, functions, and procedures of the agency boards (except for a board established by the Tennessee Valley Authority).

WORKLOAD STUDIES

(i) Within one hundred and twenty days from November 1, 1978, all agency boards, except that of the Tennessee Valley Authority, of three or more full time members shall develop workload studies for approval by the agency head as specified in subsection (a) (1) of this section.

SECTION 608. Small claims — Accelerated Disposition of Appeals

(a) The rules of each agency board shall include a procedure for the expedited disposition of any appeal from a decision of a contracting officer where the amount in dispute is $10,000 or less. The small claims procedure shall be applicable at the sole election of the contractor.

SIMPLIFIED RULES OF PROCEDURE

(b) The small claims procedure shall provide for simplified rules of procedure to facilitate the decision of any appeal thereunder. Such appeals may be decided by a single member of the agency board with such concurrences as may be provided by rule or regulation.

TIME OF DECISION

(c) Appeals under the small claims procedure shall be resolved, whenever possible, within one hundred twenty days from the date on which the contractor elects to utilize such procedure.

FINALITY OF DECISION

(d) A decision against the Government or the contractor reached under the small claims procedure shall be final and conclusive and shall not be set aside except in cases of fraud.

EFFECT OF DECISION

(e) Administrative determinations and final decision under this section shall have no value as precedent for future cases under this chapter.

REVIEW OF REQUISITE AMOUNT IN CONTROVERSY

(f) The Administrator is authorized to review at least every three years, beginning with the third year after November 1, 1978, the dollar amount defined in subsection (a) of this section as a small claim, and based upon economic indexes selected by the Administrator adjust that level accordingly.

SECTION 609. Judicial Review of Board Decision — Actions in United States Court or Claims; District Court Actions; Time for Filing

(a) (1) Except as provided in paragraph (2), and in lieu of appealing the decision of the contracting officer under section 605 of this title to an agency board, a contractor may bring an action directly on the claim in the United States Court of Claims, notwithstanding any contract provision, regulation, or rule of law to the contrary.

(2) In the case of an action against the Tennessee Valley Authority, the contractor may only bring an action directly on the claim in a United States district court pursuant to section 1337 of Title 28, not withstanding any contract provision, regulation, or rule of law to the contrary.

(3) Any action under paragraph (1) or (2) shall be filed within twelve months from the date of the receipt by the contractor of the decision of the contracting officer concerning the claim, and shall proceed de novo in accordance with the rules of the appropriate court.

FINALITY OF BOARD DECISION

(b) In the event of an appeal by a contractor or the Government from a decision of any agency board pursuant to section 607 of this title, notwithstanding any contract provision, regulation, or rules of law to the contrary, the decision of the agency board on any question of law shall not be final or conclusive, but the decision on any question of fact shall be final and conclusive and shall not be set aside unless the decision is fraudulent, or arbitrary, or capricious, or so grossly erroneous as to necessarily imply bad faith, or if such decision is not supported by substantial evidence.

REMAND OR RETENTION OF CASE

(c) In any appeal by a contractor or the Government from a decision of any agency board pursuant to section 607 of this title, the court may render an opinion and judgment and remand the case for further action by the agency board or by the executive agency as appropriate, with such direction as the court considers just and proper, or, in its discretion and in lieu of remand it may retain the case and take such additional evidence or action as may be necessary for final disposition of the case.

CONSOLIDATION

(d) If two or more suits arising from one contract are filed in the Court of Claims and one or more agency boards, for the convenience of parties or witnesses or in the interest of justice, the Court of Claims may order the consolidation of such suits in that court or transfer any suits to or among the agency boards involved.

JUDGMENTS AS TO FEWER THAN ALL CLAIMS

(e) In any suit filed pursuant to this chapter involving two or more claims, counterclaims, cross-claims, or third-party claims, and where a portion of one such claim can be divided for purposes of decision or judgment, and in such suit where multiple parties are involved, the court, whenever such action is appropriate, may enter a judgment as to one or more but fewer than all of the claims, portion thereof, or parties.

SECTION 610. Subpoena, Discovery, and Deposition

A member of an agency board of contract appeals may administer oaths to witnesses, authorize depositions and discovery proceedings, and require by subpoena the attendance of witnesses, and production of books and papers, for the taking of testimony or evidence by deposition or in the hearing of an appeal by the agency board. In case of contumacy or refusal to obey a subpoena by a person who resides, is found, or transacts business within the jurisdiction of a United States district court, the court, upon application of the agency board through the Attorney General; or upon application by the board of contract appeals of the Tennessee Valley Authority, shall have jurisdiction to issue the person an order requiring him to appear before the agency board or a member thereof, to produce evidence or to give testimony, or both. Any failure of any such person to obey the order of the court may be punished by the court as a contempt thereof.

SECTION 611. Interest

Interest on amounts found due contractors on claims shall be paid to the contractor from the date the contracting officer receives the claim pursuant to section 605 (a) of this title from the contractor until payment thereof. The interest provided for in this section shall be paid at the rate established by the Secretary of the Treasury pursuant to Public Law 92-41 (85 Stat. 97) for the Renegotiation Board.

SECTION 612. Payment of Claims – Judgments

(a) Any judgment against the United States on a claim under this chapter shall be paid promptly in accordance with the procedures provided by section 724a of Title 31.

MONETARY AWARDS

(b) Any monetary award to a contractor by an agency board of contract appeals shall be paid promptly in accordance with the procedures contained in subsection (a) of this section.

REIMBURSEMENT

(c) Payments made pursuant to subsections (a) and (b) of the section shall be reimbursed to the fund provided by section 724a of Title 31, by the agency whose appropriations were used for the contract out of available funds or by obtaining additional appropriations for such purposes.

TENNESSEE VALLEY AUTHORITY

(d) (1) Notwithstanding the provisions of subsection (a) through (c) of this section, any judgment against the Tennessee Valley Authority on a claim under this chapter shall be paid promptly in accordance with the provisions of section 831h(b) of Title 16.

(2) Notwithstanding the provisions of subsection (a) through (c) of this section, any monetary award to a contractor by the board of contract appeals for the Tennessee Valley Authority shall be paid in accordance with the provisions of section 831h(b) of Title 16.

SECTION 613. Separability of Provisions

If any provision of this chapter, or the application of such provision to any persons or circumstances, is held invalid, the remainder of this chapter, or the application of such provision to persons or circumstances other than those to which it is held invalid, shall not be affected thereby.

STANDARD FORM 26, JULY 1966
GENERAL SERVICES ADMINISTRATION
FED. PROC. REG. (41CFR) 1-16.101

AWARD/CONTRACT

PAGE 1 OF

1. CONTRACT (*Proc. Inst. Ident.*) NO.	2. EFFECTIVE DATE	3. REQUISITION/PURCHASE REQUEST/PROJECT NO.	4. CERTIFIED FOR NATIONAL DEFENSE UNDER BDSA REG. 2 AND/OR DMS REG. 1. RATING:

5. ISSUED BY CODE

6. ADMINISTERED BY (*If other than block 5*) CODE

7. DELIVERY
☐ FOB DESTINATION
☐ OTHER (*See below*)

8. CONTRACTOR NAME AND ADDRESS CODE FACILITY CODE
(*Street, city, county, State, and ZIP code*)

9. DISCOUNT FOR PROMPT PAYMENT

10. SUBMIT INVOICES (*4 copies unless otherwise specified*) TO ADDRESS SHOWN IN BLOCK ______

11. SHIP TO/MARK FOR CODE

12. PAYMENT WILL BE MADE BY CODE

13. THIS PROCUREMENT WAS ☐ ADVERTISED, ☐ NEGOTIATED, PURSUANT TO: ☐ 10 U.S.C. 2304 (a)( ) ☐ 41 U.S.C. 252 (c)( )

14. ACCOUNTING AND APPROPRIATION DATA

15. ITEM NO.	16. SUPPLIES/SERVICES	17. QUANTITY	18. UNIT	19. UNIT PRICE	20. AMOUNT

21. TOTAL AMOUNT OF CONTRACT $

CONTRACTING OFFICER WILL COMPLETE BLOCK 22 OR 26 AS APPLICABLE

22. ☐ CONTRACTOR'S NEGOTIATED AGREEMENT (*Contractor is required to sign this document and return ______ copies to issuing office.*) Contractor agrees to furnish and deliver all items or perform all the services set forth or otherwise identified above and on any continuation sheets for the consideration stated herein. The rights and obligations of the parties to this contract shall be subject to and governed by the following documents: (a) this award/contract, (b) the solicitation, if any, and (c) such provisions, representations, certifications, and specifications, as are attached or incorporated by reference herein. (*Attachments are listed herein.*)

26. ☐ AWARD (*Contractor is not required to sign this document.*) Your offer on Solicitation Number ______, including the additions or changes made by you which additions or changes are set forth in full above, is hereby accepted as to the items listed above and on any continuation sheets. This award consummates the contract which consists of the following documents: (a) the Government's solicitation and your offer, and (b) this award/contract. No further contractual document is necessary.

23. NAME OF CONTRACTOR

BY ______
(Signature of person authorized to sign)

27. UNITED STATES OF AMERICA

BY ______
(Signature of Contracting Officer)

24. NAME AND TITLE OF SIGNER (*Type or print*)	25. DATE SIGNED	28. NAME OF CONTRACTING OFFICER (*Type or print*)	29. DATE SIGNED

26-105

Table of Cases

Case *Text Page*

Acoustics, Inc. v. *Trepte Construc. Co.,* 14 Cal. App. 3d 887 (1971) 80
Aetna Insurance Co. v. *Hellmuth, Obata & Kassabaum, Inc.,* 392 F.2d 472 (8th Cir. 1968) . . 76
Air-A-Plane Corp. v. *United States,* 187 Ct. Cl. 269, 408 F.2d 1030 (1969) 43
American Motorcycle Assoc. v. *Superior Court* 20 Cal. 3d 578 (1978) 108
Anno. 152 A.L.R. 1349, 1359 (1941) 41
ASPR ¶1-324-7 and ¶1-324, 1-3 46
Avent v. *Proffitt,* 95 S.E. 134 (S. Car. 1918) 45
Axel Electronics Inc., ASBCA No. 18990, 74-1 BCA ¶10,471 (1974) 43
Ball, Gordon H. Inc., ASBCA 5905, 61-2 BCA ¶3177 (1961) 40
Bay Ray Prods. Inc. v. *United States,* 167 Ct. Cl. 839 340 F.2d 343 (1964) 47
Bayshore Dev. Co. v. *Bondfoey,* 78 So. 507 (Fla. 1918) 45
Benbow Plumbing & Heating Co., ASBCA No. 13233, 69-1 BCA ¶7565 (1969) 41
Bennett, Joseph E. Co., GSBCA 2362, 72-1 BCA ¶9364 (1972) 69
Benson Tree Co., IBCA No. 812-11-69, 70-2 BCA ¶8549 (1970) 42
Blackhawk Heating and Plumbing Co., Inc., GSBCA No. 2432 76-1 BCA ¶11,649 (1976) 71, 74
Blount Brothers Construction Co. v. *United States,* 171 Ct. Cl. 478, 346 F.2d 962 (1965) 78
Brezinz Const. Co. v. *United States,* 196 Ct. Cl. 29, 33 (1971) 35
Brick v. *Okaloosa Title & Abstract Co.,* 404 F.2d 889 (5th Cir. 1968) 45
Buffalo Merchandise Warehouse Inc. v. *United States,* 115 Ct. Cl. 568, 88 F. Supp. 276 (1950) 36
Catalytic Engineering & Mfg. Corp., ASBCA No. 15257, 72-1 BCA ¶9342 (1972) 48
Chaney & James Construction Co., Inc., FAACAP No. 67-18, 66-2 BCA ¶6066 (1967) . . . 63, 68
Chapel v. *Clark,* 76 N.W. 67 (Mich. 1898) 45
Chicago & E.I. Ry. Co., 94 F.2d 296 (7th Cir. 1938) 37
City of Eveleth v. *Ruble,* 225 N.W. 2d 521 (Minn. 1974) 76, 78
City of Mounds View v. *Walijarvi,* 263 N.W. 2d 420 (Minn. 1978) 76
Coleman Engineering Co. v. *North American Aviation, Inc.,* 65 Cal. 2d 396 (1966) 78, 80
Comden v. *Superior Court,* 20 Cal. 3d 906 (1978) 105
Continental Consolidated Corporation, ENGBCA 2743, et. al., 67-2 BCA ¶6624 (1967) 58, 68, 74
Crane Const. Co. v. *Commonwealth,* 195 N.E. 110 (Mass. 1935) 84

Case *Text Page*
Craviolini v. *Scholer & Fuller Assoc. Architects,* 357 P.2d 611 (Ariz. 1961) 45
Curtis, Helene Industries, Inc., v. *United States,* 160 Ct. Cl. 437, 312 F.2d 744 (1963) 43, 89
Dale Const. Co. v. *United States,* 168 Ct. Cl. 692 (1964) 39
Dawson Construction Co., Inc., GSBCA No. 3998, 75-2 BCA ¶11,563 (1975) 60
Dobson v. *Rutgers,* 157 N.J. Super. 357, 384 A.2d 1121 (1978) 59, 72-73
Drew Brown Ltd. v. *Joseph Rugo, Inc.,* 436 F.2d 632 (1st Cir. 1971) 61, 67
Duluth Superior Dredging Company v. *United States,* 81 Ct. Cl. 912 (1935) 33
Eggers & Higgins v. *United States,* 185 Ct. Cl. 765, 403 F.2d 225 (1968) 45
Exber, Inc. v. *Sletten Constr. Co.,* 558 P.2d 517 (Nev. 1976) 108
Foster, L.W. Sportswear Co. v. *United States,* 186 Ct. Cl. 499 (1969) 38
Fuller, George A. Co. v. *United States,* 108 Ct. Cl. 70, 69 F. Supp. 409 (1949) 44
G.M. Co. Mfg. Inc., ASBCA No. 5345, 60-1 BCA ¶2576 (1960) 47
Geer v. *Bennett,* 237 So. 2d 311 (Fla. App. 1970) 45
Goslin-Birmingham Inc. ENGBCA 2800, 67-2 BCA ¶6402 (1967) 39
Great Lakes Dredge & Dock Co., ENGBCA 2654, 66-1 BCA ¶5442 (1966) 33
Green, William Const. Co. v. *United States,* 201 Ct. Cl. 616, 477 F.2d 930 (1973) 43
Gresham & Co., Inc. v. *United States,* 200 Ct. Cl. 97 (1972) 38
Guest t/a C.M. Guest & Sons, WDBCA No. 593, 2 CCF 1015 (1944) 39
Gurney Industries, Inc. v. *St. Paul Fire & Marine Ins. Co.,* 467 F.2d 588 (4th Cir. 1972) . . . 79
Heat Exchangers, Inc., ASBCA No. 8705, 1963 BCA ¶3881 (1963) 68
Hedin, J.D. Constr. Co. v. *United States,* 347 F.2d 235, 254-55 (Ct. Cl. 1965) 41
Hedla v. *McCool,* 476 F.2d 1223 (9th Cir. 1973) 77, 79
Hercules Engineering & Mfg. Co., ASBCA No. 4979, 59-2 BCA ¶2426 (1959) 47
Homestead Savings & Loan Assoc. v. *Superior Court,* 195 Cal. App. 2d 697 (1961) 107
Hosmer Inc. v. *Commonwealth of Massachusetts,* 302 Mass. 495 19 N.E. 2d 800 (1939) 41
Huber, Hunt & Nichols, Inc. v. *Moore,* 67 Cal. App. 3d 278 (1977) 76, 83
International Erectors v. *Wilhoit Steel Erectors & R. Serv.,* 400 F.2d 465 (5th Cir. 1968) 86
Johnson v. *Fenestra Inc.,* 305 F.2d 179 (3d Cir. 1962) 41
Kaplan Contractors, Inc., GSBCA No. 2747, 70-2 BCA ¶8511 (1970) 39
Keco Indus. Inc., ASBCA No. 13271, 71-1 BCA ¶8727 (1971) 46, 47
Kiewit, Peter Sons Co. v. *Iowa Southern Utilities Co.,* 355 F. Supp. 376 (S.D. Iowa 1973) 59, 65
Kiewit, Peter Sons Co. v. *United States,* 109 Ct. Cl. 390 (1974) 35
Lane-Verdugo, ASBCA No. 16327 73-2 BCA ¶10,271 (1973) 63
Leavell, C.H. & Co., ASBCA No. 16099, 72-2 BCA ¶9694 (1972) 44
Leavell, C.H. & Co., GSBCA 2901, 70-2 BCA ¶8437 (1970) 72
Lewis-Nicholson, Inc. v. *United States,* 550 F.2d 26 (Ct. Cl. 1977) 39
Ling-Temco-Vought Inc. v. *United States,* 201 Ct. Cl. 135, 475 F.2d 630 (1973) 45
Longenfelder, C.J. & Sons, WDBCA 706 (1944) 39
M.S.I. Corporation, GSBCA No. 2429, 68-2 BCA ¶7377 (1968) 41
Marden, Edward R. Corp. v. *United States,* 194 Ct. Cl. 799, 422 F.2d 364 (1971) 43
Marhoefer, E.H. Jr., Co., DOTCAB Nos. 70-17, 70-1 BCA ¶8177 (1970)
Maryland Cas. Co. v. *Tacoma,* 90 P.2d 226, 94 P.2d 217 (Wash. 1939) 42
Maryland Painting Co., ENGBCA No. 3337, 73-2 BCA ¶10,223 (1973) 43
Minmar Builders, Inc., GSBCA No. 3430 72-2 BCA ¶9599 (1972) 60

Case *Text Page*
Mix Inc. v. *Columbus,* 171 N.E. 2d 1967 (Ohio, App. 1959) 42
Montgomery-Macri Co. and Western Line Construction Co., 1963 BCA ¶3819 (1963) 68
Morrison-Knudsen Co. v. *United States,* 192 Ct. Cl. 410, 418, 427, F.2d 1181 (1970) 35
Moyer, A.R. Inc. v. *Graham,* 285 So. 2d 397 (Fla. 1973) 45
McMullan, Robert & Sons, Inc., ASBCA No. 19129, 76-2 BCA ¶12,072 (1976) 41
McShain, John, Inc. v. *United States,* 199 Ct. Cl. 364 (1972) 35
Natkin & Co. v. *George A. Fuller Co.,* 347 F. Supp. 17, 35 (W.D. Mo. 1972) 35
Natus Corp. v. *United States,* 178 Ct. Cl. 1,371 F.2d 450 (1967) 44
Ozark Dam Constructors v. *United States,* 130 Ct. Cl. 354, 127 F. Supp. 187 (1955) 88
P & B Company of Washington, Inc., ASBCA No. 9429, 1963 BCA ¶3997 (1963) 40
Pacific Car & Foundry Co., ASBCA No. 4849, 60-1 BCA ¶2505 (1960) 40
Pavement Specialists Inc., ASBCA No. 17410, 73-2 BCA ¶ 10,082 (1973) 42
Passalacqua, William Builders, Inc., GSBCA No. 4205, 77-1 BCA ¶12,406 (1977) 69
Player v. *George M. Brewster & Son, Inc.,* 18 Cal. App. 3d 999 (1975) 106
Railroad Waterproofing Corp., ASBCA No. 12384, 69-1 BCA ¶7456 (1969) 40
Randolph, Otto, Inc., ASBCA No. 11539, 66-2 BCA ¶5928 (1966) 39
Raymond International of Delaware, Inc., ASBCA No. 13121, 70-1 BCA ¶8341 (1970) . . . 40-41
Roscoe Engineering Corp. & Assocs., ASBCA 4820, 61-1 BCA ¶2919 (1961) 36-37
S & E Contr., Inc., AECBCA No. 97-12-72, 74-2 BCA ¶10,676 (1974) 41, 97
Schweigert, Inc. v. *United States,* 181 Ct. Cl. 1184 (1967) 34
Shadrick Contracting Co., Inc., ASBCA No. 14613, 71-1 BCA ¶8647 (1971) 37
Smith, William A. Contr. Co. of Mo. v. *United States,* 155 Ct. Cl. 1292 F. 2d 847 (1961). . . .43
Southern Fireproofing Co. v. *R.F. Ball Construction,* 334 F.2d 122 (8th Cir. 1964) 66-67
Southwest Mechanical Contractors, GSBCA Nos. 1310 and 1320, 1962 BCA ¶4482 (1964). . 36
Spencer Explosives, Inc., ASBCA No. 4800, 60-2 BCA ¶2795 (1960) 39
Sullivan County v. *Nezelek, Inc.,* 366 N.E. 2d 72 (N.Y. 1977) 108
Sun Shipbuilding and Dry Dock Co., ASBCA No. 11300, 68-1 BCA ¶7054 (1968) 41
Tecon Corp. v. *United States,* 188 Ct. Cl. 436, 411 F.d 1271 (1969). 44
Thompson Ramo Wooldridge, Inc. v. *United States,* 175 Ct. Cl. 527 (1966) 42
Union Paving Co. v. *United States,* 126 Ct. Cl. 478, 489 115 F. Supp. 179, 185 (1953) 36
United States ex. rel Los Angeles Testing Laboratory v. *Rogers & Rogers,*
161 F. Supp. 132 (D. Cal. 1958) 45
United States v. *Spearin,* 248 U.S. 132 (1918) 78
Varco-Pruden, Inc. v. *Hampshire Constr. Co.,* 50 Cal. App. 3d 654 (1975) 85
Walsh Bros. v. *United States,* 107 Ct. Cl. 627, 69 F. Supp. 125 (1947) 42
Watkins-Johnson Company, ASBCA No. 15457, 74-1 BCA ¶10,395 (1973) 37
W.P.C. Enterprises, Inc. v. *United States,* 163 Ct. Cl. 1, 6, 323 F.2d 874, 876-77 (1963) 35

Index

A

Acceptance, 46-48
Activity:
  and CPM schedules, 50-51
AIA 201, 21-22, 26, 45, 82, 94
  dispute, 45
Ambiguity rule, 35
  companion rule to, 35
American Society of Civil Engineers, 19
Arbitrator, 106-111
Arbitration:
  Boards of Arbitration and Review, 21
  difference from litigation, 107
  discovery as element of, 107
  use of, 22
Arbitration award, 109
Architect:
  and AIA-type dispute, 45
  liability of, 45
  obligations of, 45
Architectural concrete, 29
Article 5 of federal contracts, 45
Associated General Contractors, 22
Attorney *(see also* lawyer), 103-105
Audit, 90, 95

B

Bar charts, 49, 56-57, 60
  and CPM schedules, 60-61
Boards of Arbitration and Review, 21
Board of Contract Appeals, 90, 91, 92, 95
Budget, 76, 77, 79

C

Change order mechanism, 82
Change Study Analysis:
  and *William Passalacqua Builders, Inc.,* 69-71
Changed Claims, 42
Changed Conditions (Differing Site Conditions), clause, 38, 43, 44-45, 88
Changes, 84
  cardinal, 43
  constructive, 38-40
  express, 38
Claims:
  and federal contracts, 89-91
    denial of, 90
Claims Act of 1979, 18
Clauses:
  Changed Conditions (Differing Site Conditions), 38, 43, 44-45, 88
  exculpatory, 41-42
    disputes, 90
  inspection, 46
  no-delay-damage type, 41, 46
  Site Inspection, 42-43
  standard contract default, 34
  suspension of work, 88
  warranties, 46-48
Cold weather construction, 28
Concrete:
  architectural, 29
  handling of, 28-29
  high-strength, 29
  ready-mixed, 28-29
Consideration:
  for validity of contract, 18
Constructive acceleration rule, 40-41
Construction:
  social revolution as influence on, 29-30
Construction changes:
  categories of, 38-40
Construction contracts (*see* contracts):
  concepts basic to, 17
Construction dispute:
  and CPM and MIS, 30
  cold weather construction as cause of, 28
  rule of mobility in, 31
Construction industry:
  changes in, 25

Construction industry: *(Cont.)*
effect of travel on, 30-31
Construction litigation:
and labor problems, 30
Construction management firm, 27
Construction manager, 28
Construction problems:
minimizing, 81-86
contractor's role, 83-85
designer's role, 82-83
owner's role, 81-82
supplier's role, 86
Construction schedule, 81, 83-84
acceptance of, 73
accuracy of, 74
and the law, 49-74
approval of, 62
completeness of, 61-62
dates in, 64-67
float time, 67-69
law's view of, 58-60
mistakes in, 69-72
mutual responsibilities under, 73-74
pre-bid, 72-73
substantiation of, 62-63
types of, 49-57
updating, 57-58, 74, 84
Contract (*see also* federal contracts and federal contract disputes):
basis for claim, 33
consideration as needed for validity, 18
contractor's role in, 79-81
design phase, 75-78
designer's role, 77-78
elimination of discrepancies, 76-77
owner's role, 75-77
enforcement of; 18, 19, 20
by judges, 20
how to avoid mistakes in, 33-48
interpretation of language; 34-38
mistakes in, 37
key ingredients of, 34
legal principles behind, 33-48
rights and violations under, 42-43
state involvement in, 21
validity of, 18
writers of, 19
Contract claims, 17
Contract forms:
standard construction, 19
Contract law:
roots of, 17-24
validity of contract, 18
Contract manager:
and CPM, 30
Contracting:
changes in, 25-31
Contracting officer, in federal disputes, 91-93, 95
Contractor, 19
and responsibilities of subcontractors, 45-46
Associated General, 28
capitalizing on conflicts, 83
duty of, 22-23
liability of, 46, 80
obligations of, 44-46
rights of, 44
role in construction problems, 83-85
Contracts:
and inclusion of other documents, 34
Corps of Engineers, and dredging contracts, 33
Cost estimating, 77, 78-79
CPM, 28, 30, 73, 81
and construction disputes, 30
and MIS, 30
difference from PERT, 57
CPM consultant, 63-64, 94
CPM diagrams, 57
CPM schedules, 49, 50-56, 62-63, 64-65, 69, 74
activity and, 50-51
bar charts, 60-61
definition of, 58, 59-60
drawing, 51-52
dummy, 51
float, 55
logic, 50-51
preparation of, 50
restraints, 51

D

Davis Bacon laws, 29
Defects:
latent, 46-48
Delay claims; 38-40, 46
exculpatory clause, 41-42
prerequisite to, 38, 41
problems in, 40
Delays:
as basis for claims, 34
basic types, 40-41
combination of causes for, 41
no-fault, 41
Design, 76
Design-build contracts, 26
Design-engineer, 19
evolution of functions, 22-24
Design phase, 75-78
as critical period, 75
designer's role, 77-78
elimination of discrepancies, 76-77
owner's role, 75-77
Design professional, 18-19
Designer:
and cost estimating, 78-79
and the contract, 20
as independent consultant, 76

Designer: *(Cont.)*
as owner's agent, 26-27
liability of, 78-79, 83
interfacing, 79
responsibility of, 78
role in construction problems, 82-83
role in contract design phase, 77-79
Differing Site Conditions Claims, 42
Disposition, 112
Disputes:
federal, 87-102
non-federal, 103-113
settlement of, 22
rise of experts in, 110-111
Disputes clause, 90
Doctrine of practical or legal impossibility, 44
Drawing:
CPM schedules, 51-52
I-J, 51
precedence, PDM 51
Dredging contracts, 33
Dummy, 71-72
and CPM schedules, 51

E

Education:
as basis for change, 25-26
and AIA-type dispute, 45
Engineer:
as contract interpreter, 20-21
development of, 19-20
in federal contract disputes, 93
liability of, 45
obligations of, 45
English Law of Equity and Property, 18
Exculpatory clauses, 41-42
no-delay damage type, 41
Express warranties, 47
Experts:
use of in disputes, 110-111

F

Federal contracts:
Article 5, 45
Federal contract disputes, 87-102
denial of claims, 90
preliminary points, 88-89
presentation, 95-98
justification of damage calculation, 95-98
total cost approach, 97
processing, 89-95
claim procedure, 89-91
resources, 91-95
architects, 94
contracting officer, 91-92
inspectors, 93-94

Federal contract disputes, *(Cont.)*
lawyers, 94
resident engineers, 93
special or outside counsel, 94-95
settlement, 98-102
and litigation, 98-100
getting government attention, 101
length of time, 101-102
when to present case, 101
when to seek, 100-101
Float time, 67-69, 72, 74
ownership of, 68
Fraud, 47-48, 96

G

Gantt charts (*see* bar charts):
General Provision Article, 10, 47
Government:
and bidding, 77
and changes claim, 38-40
and claims, 88
and CPM schedules, 52-54
and latent defects, 46-48
and purchasing power, 29
and Site Inspection clause, 42
as drafter of contracts, 35
relationship with, 88
Gross mistake:
definition, of, 48
Guarantees, 46-48, 78

I

Implied duties:
concept of, 22
Inflated damages:
definition of, 96
Inspection clause, 46
Interpretation:
and prior actions, 36-37
most common rule of, 35
Interpretation of language:
basic rules, 34-38
mistakes in, 37
Interrogatories, 112-113

L

Labor:
and construction litigation, 30
and Great Society programs, 31
problems, 30
Latent defects, 46-48
Legal dilemmas:
causes of, 28

Liability:
  architect's, 45
  contractor's, 46, 80
  designer's, 78-79, 83
  engineer's, 45
  insurance for, 28, 31-32, 76
  privity as essential for, 22
  warranty type, 47
Litigation:
  and labor problems, 30
  concept of life as cause of, 32
  difference from arbitration, 107
  Great Society programs as cause of, 31
  PLI as cause of, 31-32
  settlement, 98-100

M

Management Information Systems, 28, 30
Martindale Hubbell, 104
Minority business requirement, 29, 30
Mobility:
  as cause of construction disputes, 31

N

National Society of Professional Engineers, 22
Network diagram, 51
No-damage-for-delay clause, 41, 46
Notice requirements, 44-45

O

Owner:
  basic rights, 43
  cooperation with contractor, 43-44
  liability of, 73, 83-84
  obligations of, 73
  role in construction problems, 81-82
  roll in design phase, 75-77

P

PDM (precedence diagramming), 56
PERT, 46, 57
  difference from CPM, 57
Pre-bid construction schedules, 72-73
Privity:
  as essential for liability, 22
Professional Liability Insurance, 28, 31-32, 76

R

Reconciliation of interests, 86
Records:
  review of in disputes, 111-112
Restraints:
  and CPM schedule, 51

S

Schedules (see construction schedules)
Site Inspection Clause, 42-43
Site inspection:
  necessity of, 42
Soil defects, 23
Sovereign Immunity, 21
Specifications, 78, 79
  and contractor, 44
  and owner's responsibility, 43
  and performance criteria, 47
  error in, 39
Standard contract default clause, 34
Standard form, 20, 22, 30
  American Institute of Architects Form AIA 201A, 21-22
Subcontractor:
  and contractor, 45-46, 85-86
  and schedule, 84
  rights of, 46
  role in construction problems, 85-86
Suspension of Work Clause, 88

T

Trade customs, 61
  and contract meaning, 35-36
  helpful in disputes, 35-36
Travel:
  effect on construction industry, 30-31

U

Uniform Commercial Code, 47
  Section 2, 47
United States Government *(see* government):
  as buyer of construction, 21

W

Warranties, 46-48, 76
  and Uniform Commercial Code, 47
  clause, 46-48
  express, 47
Working conditions:
  changes in, 28